Roger Lee, Gongzu Hu, and Huaikou Miao (Eds.)

Computer and Information Science 2009

T0190239

Studies in Computational Intelligence, Volume 208

Editor-in-Chief

Prof. Janusz Kacprzyk
Systems Research Institute
Polish Academy of Sciences
ul. Newelska 6
01-447 Warsaw
Poland
E-mail: kacprzyk@ibspan.waw.pl

Roger Lee, Gongzu Hu, and Huaikou Miao (Eds.)

Computer and Information Science 2009

 Springer

Roger Lee
Software Engineering & Information
Technology Institute
Central Michigan University
Mt. Pleasant, MI 48859, U.S.A.
E-mail: lee@cps.cmich.edu

Huaikou Miao
School of Computer Engineering and Science
Shanghai University
Shanghai, China
E-mail: hkmiao@shu.edu.cn

Gongzu Hu
Department of Computer Science
Central Michigan University
Mt. Pleasant, MI 48859, U.S.A.
E-mail: hu@cps.cmich.edu

ISBN   978-3-642-10174-8

e-ISBN   978-3-642-01209-9

DOI 10.1007/978-3-642-01209-9

Studies in Computational Intelligence

ISSN 1860949X

Typeset & Cover Design: Scientific Publishing Services Pvt. Ltd., Chennai, India.

Printed in acid-free paper

9 8 7 6 5 4 3 2 1

springer.com

Preface

The 8th ACIS/IEEE International Conference on Computer and Information Science, held in Shanghai, China on June 1-3 is aimed at bringing together researchers and scientist, businessmen and entrepreneurs, teachers and students to discuss the numerous fields of computer science, and to share ideas and information in a meaningful way. This publication captures just over 20 of the conference's most promising papers, and we impatiently await the important contributions that we know these authors will bring to the field.

In chapter 1, Abhijit Mustafi and P. K. Mahanti develop a contrast enhancement technique to recover an image within a given area, from a blurred and darkness specimen, and improve visual quality. The author's results are presented using developed technique on real images, which are hard to be contrasted by other conventional techniques.

In chapter 2, Shahid Mumtaz et al. use an ad-hoc behavior in opportunistic Radio, to present how the overall system performance effect in terms of interference and routing. They develop a simulation tool that addresses the goal of analysis and assessment of UMTS TDD opportunistic radio system with ad hoc behavior in coexistence with a UMTS FDD primary cellular network.

In chapter 3, Guoqing Zhang et al. propose a novel geocast routing protocol called GRUV for urban VANETs. GRUV adapts to the current network environment by dynamically switching three forwarding approaches which are used to compute forwarding zones. Their simulations show GRUV performance compared with other geocast routing protocols.

In chapter 4, Kwan Hee Han and Jun Woo Park propose an object-oriented (O-O) ladder logic development framework in which ladder code is generated automatically based on the O-O design results and can be verified by port-level simulation. To show the applicability of their proposed framework, a software tool for the design and generation of ladder code is developed.

In chapter 5, Thomas Neubauer and Mathias Kolb highlight research directions currently pursued for privacy protection in e-health and evaluates common pseudonymization approaches against legal and technical criteria. With it they support decision makers in deciding on privacy systems and researchers in identifying the gaps of current approaches for privacy protection as a basis for further research.

In chapter 6, Qitao Wu, Qimin Feng discuss learning based on the under-sampling and asymmetric bagging is proposed to classify BGP routing dynamics and detect abnormal data. They conclude the by enumerating their findings on the detection of abnormal BGP routing dynamics in relation to BGP routing products.

In chapter 7, Haiyun Bian and Raj Bhatnagar propose an algorithm that enables the mining of subspace clusters from distributed data. Their algorithm is backed by analytical and experimental validation.

In chapter 8, Juan Li proposes a novel framework for discovery Semantic Web data in large-scale distributed networks. They use simulations, which substantiate that our techniques significantly improve the search efficiency, scalability, and precision.

In chapter 9, Shiqi Li, et al. propose a method based on an extension of Kinstch's Construction-Integration (CI) model in order to extract and represent spatial semantics from linguistic expressions. Their experimental result shows an encouraging performance on two datasets.

In chapter 10, Huaikou Miao, et al. propose a declarative methodology based on F-logic for modeling OWL-S ontologies, present a formal F-logic semantics of OWL-S and specify their global properties and frame as logic formulas. Their methodology allows them to bring to bear a body of work for using first order logic based model checking to verify certain global properties of OWL-S constructed service systems.

In chapter 11, Xunli Fan, et al. present a new scheduling strategy based on Cache Offset (BCOP), which deals with the rarity and urgency of data block simultaneously. They demonstrate the effectiveness of the proposed strategy in terms of higher continuity and lower startup latency.

In chapter 12, Jichen Yang et al. propose an algorithm for story segmentation based on anchor change point from audio. Their experimental results show the effectiveness of the methodology as an accuracy of story segmentation detection.

In chapter 13, Ying Liu et al. The investigate how we can make use of Discrete Cosine Transform (DCT) and Discrete Wavelet Transform (DWT) in image sharpening to enhance image quality. Their experimental results demonstrate the effectiveness of the proposed algorithm for image sharpening purpose.

In chapter 14, Daisuke Horie et al. present a generator of ISO/IEC 15408 security target templates, named "GEST," that can automatically generate security target templates for target information systems from evaluated and certified security targets. Using GEST, designers with a little experience can easily get security target templates as the basis to create security targets of target systems.

In chapter 15, Fangqin Liu et al. develop a set of general performance metrics for streaming services and propose a modeling method that can effectively map such metrics for real measurement usages.

In chapter 16, Deah J. Kadhim et al. advance the stochastic network calculus by deriving a network service curve, which expresses the service given to a flow by the network as a whole in terms of a probabilistic bound. The authors demonstrate the benefits of the derived service curve are illustrated for the Exponentially Bounded Burstiness (EBB) traffic model.

In chapter 17, Gu-Hsin Lai et al. adopt data mining to generate spam rules and statistical test to evaluate the efficiency of them. They use only the rules found to be the most significant the other rules are purged in the name of efficiency.

In chapter 18, Jihyun Lee et al. present a business process maturity model called Value based Process Maturity Model (vPMM) that overcomes the limitations of the existing models. Overcoming means, helping an organization set priorities for improving its product production and/or service provisioning using a proven strategy and for developing the capability required to accomplish its business values.

In chapter 19, Bo Song et al. propose an event-based dependence graph models (EDGMs) to model and extract database interactions in Web applications. They develop detailed test sequences in order to instantiate and execute them.

In chapter 20, Eun Ju Park et al. seek to improve the workload efficiency and inference capability of context-aware process in real-time data processing middleware core of ubiquitous environment. The authors suggest the rule-based context-aware neural network model to solve these problems. The empirical results are then shown.

In chapter 21, Jianlong Zhou and Masahiro Takatsuka present structural relationship preservation as a technique for improving efficiency of volume rendering-based 3D data analysis. The experimental results comment on the intuitive and physical aspects of structural relationship preservation in volume rendering.

In chapter 22, Qi Zeng et al. present a quasi-synchronous frequency-hopping multiple-access communications system with OFDM-BPSK scheme. A novel frequency-hopping code, No-Hit-Zone (NHZ) code, is applied in the proposed system to acquire the numerical and simulation results for analysis.

In chapter 23, Behrouz Jedari and Mahdi Dehghan propose a novel task scheduling algorithm, called Resource-Aware Clustering (RAC) for Directed Acyclic Graphs (DAGs). They then compare simulation results of RAC algorithm with famous scheduling approaches such as MCP, MD and DSC.

In chapter 24, Dilmurat Tursun et al. study and propose the most effective solutions and ideas for linguistic ambiguity in the Uyghur, Kazak, and Kyrgyz languages. They propose a Relocated Unicode Format Method. They use this method to experimentally verify their disambiguation claims.

In chapter 25, Zhenhua Duan and Xinfeng Shu present a proof system for projection temporal logic (PTL) over finite domains. The authors give an example to illustrate how the axioms and rules work.

In chapter 26, the final chapter, J. Lillis et al. make five major contributions have been achieved: (1) Results from both wired and wireless scenarios are critically evaluated and then concluded; (2) Accurate simulations of 802.11b network in NS2 are validated with live network tests; (3) Evidence is given through both simulation and real network tests to show that current TCP standards are inefficient on wireless networks; (4) A new TCP congestion control algorithm is proposed as well as an outline for a fresh approach; (5) Industry reactions are then given to recommendations for network changes as a result of this investigation.

It is our sincere hope that this volume provides stimulation and inspiration, and that it will be used as a foundation for works yet to come.

June 2009   Roger Lee

Contents

List of Contributors

Behrouz Jedari
Islamic Azad University of Tabriz, Iran
behrouz_jedari@yahoo.com

Noor Azimah
Saitama University, Japan
azim@aise.ics.
saitama-u.ac.jp

Raj Bhatnagar
University of Cincinnati, US
raj.bhatnagar@uc.edu

Haiyun Bian
Metropolitan State College, US
hbian@mscd.edu

Chia-Mei Chen
National Sun Yat-Sen University,
Taiwan

Dao Chen
Southwest Jiaotong University,
Sichuan
cd_2005@162.com

Wu Chen
Polytechnical University, China
chenwu@nwpu.edu.cn

Jingde Cheng
Saitama University, Japan
cheng@aise.ics.saitama-
u.ac.jp

Wenqing Cheng
Huazhong University of Science and
Technology, China

Chao-Wei Chou
I-Shou University, Taiwan
choucw@isu.edu.tw

W. Dargie
Technical University of Dresden,
Germany

Mahdi Dehghan
Islamic Azad University of Tabriz, Iran

Zhenhua Duan
Xidian University, China
zhenhua duan@126.com

Xunli Fan
Loughborough University, UK

Qimin Feng
Ocean University of China, China

Marques A.Gameiro
Institute of Telecommunication,
Aveiro, Portugal

Li Gao
College of Automation, Northwestern
Polytechnical University
gaoli@nwpu.edu.cn

Yuichi Goto
Saitama University, Japan
gotoh@aise.ics.saitama-u.ac.jp

Lin Guan
Loughborough University, UK
L.Guan@lboro.ac.uk

A. Grigg
Loughborough University, UK

Askar Hamdulla
Xinjiang University, China
askar@xju.edu.cn

Kwan Hee Han
Gyeongsang National University, Jinju, Republic of Korea
hankh@gsnu.ac.kr

Qianhua He
South China University of Technology, China
eeqhhe@scut.edu.cn

Tao He
Shanghai University, China
he_tao@foxmail.com

Daisuke Horie
Saitama University, Japan
horie@aise.ics.saitama-u.ac.jp

Saba Q. Jobbar
Huazhong University of Science and Technology, China

Deah J. Kadhim
Huazhong University of Science and Technology, China
deya_naw@yahoo.com

Shiqi Li
Harbin Institute of Technology, China
sqli@mtlab.hit.edu.cn

Sungwon Kang
Information and Communications University, Korea
kangsw@icu.ac.kr

Haeng-Kon Kim
Catholic University of Daegu, Korea
hangkon@cu.ac.kr

Mathias Kolb
Secure Business Austria, Austria
kolb@securityresearch.ac.at

Gu-Hsin Lai
National Sun Yat-Sen University, Taiwan

Jihyun Lee
Information and Communications University, Korea
puduli@icu.ac.kr

Roger Y. Lee
Central Michigan University, US
leelry@cmich.edu

Danhyung Lee
Information and Communications University, Korea
danlee@icu.ac.kr

Hanjing Li
Harbin Institute of Technology, China
hjlee@mtlab.hit.edu.cn

Juan Li
North Dakota State University, US
j.li@ndsu.edu

Yanxiong Li
South China University of Technology,
China
yanxiongli@163.com

Hong Liang
College of Automation, Northwestern
Polytechnical University
hongliang@nwpu.edu.cn

Beng Keat Liew
Republic Polytechnic, Singapore
liew_beng_keat@rp.sg

J. Lillis
Loughborough University, UK

Chuang Lin
Tsinghua University, China
clin@csnet1.cs.tsinghua.edu.cn

Fangqin Liu
Tsinghua University, China
fqliu@csnet1.cs.tsinghua.
edu.cn

Wei Liu
Huazhong University of Science and
Technology, China

Ying Liu
Republic Polytechnic, Singapore
liu_ying@rp.sg

P. K. Mahanti
University of New Brunswick,
Canada
pmahanti@unb.ca

Huaikou Miao
Shanghai University, China
hkmiao@shu.edu.cn

Liping Li
Shanghai University, China
llping2000@yahoo.com.cn
Polytechnical University
mudejun@nwpu.edu.cn

Shahid Mumtaz
Institute of Telecommunication,
Aveiro, Portugal
smumtaz@av.it.pt

Abhijit Mustafi
Birla Institute of Technology, India
abhijit@bitmesra.ac.in

Thomas Neubauer
Vienna University of Technology,
Austria
neubauer@ifs.tuwien.ac.at

Tek Ming Ng
Republic Polytechnic, Singapore
ng_tek_ming@rp.sg

Ya-Hua Ou
I-Shou University, Taiwan

Eun Ju Park
Catholic University of Daegu, Korea
ejpark@cu.ac.kr

Jun Woo Park
Gyeongsang National University,
Jinju, Republic of Korea

Daiyuan Peng
Southwest Jiaotong University, Sichuan
dypeng@home.swjtu.edu.cn

Jonathan Rodriguez
Institute of Telecommunication,
Aveiro, Portugal

Dejun Mu
College of Automation, Northwestern

Xinfeng Shu
Institute of Posts and
Telecommunications, China
shuxinfeng@gmail.com

Bo Song
Shanghai University, China
songbo@shu.edu.cn

Masahiro Takatsuka
University of Sydney, Australia
masa@vislab.usyd.edu.au

Yong Ho Toh
National University of Singapore,
Singapore
u0509543@nus.edu.sg

Turdi Tohti
Xinjiang University, China

Dilmurat Tursun
Xinjiang University, China

Weining Wang
South China University of Technology,
China
weiningwang@scut.edu.cn

Wenbo Wang
Northwest University, China

X. G. Wang
Plymouth University, UK

Qitao Wu
Ocean University of China, China
whichtall@sohu.com

Yijun Xu
South China University of Technology,
China
yjx@163.com

Zhong Xu
College of Automation, Northwestern
Polytechnical University
xuzhong@mail.nwpu.edu.cn

Kenichi Yajima
Saitama University, Japan
yajima@aise.ics.saitamau.
ac.jp

Jichen Yang
South China University of
Technology, China
NisonYoung@yahoo.cn

Hao Yin
Tsinghua University, China
hyin@csnet1.cs.tsinghua.ed
u.cn

Qi Zeng
Southwest Jiaotong University,
Sichuan
zeng_qi@yahoo.com.cn

Guoqing Zhang
College of Automation,
Northwestern Polytechnical University
gniq@mail.nwpu.edu.cn

Tiejun Zhao
Harbin Institute of Technology, China
tjzhao@mtlab.hit.edu.cn

Jianlong Zhou
University of Sydney, Australia
zhou@it.usyd.edu.au

Bin Zhu
Shanghai University, China
tozhubin@163.com

An Optimal Algorithm for Contrast Enhancement of Dark Images Using Genetic Algorithms

Abhijit Mustafi and P.K. Mahanti

Abstract. This paper develops a contrast enhancement technique to recover an image within a given area, from a blurred and darkness specimen, and improve visual quality. The technique consists of two steps. Firstly determine a transform function that stretches the occupied gray scale range for the image secondly the transformation function is optimized using genetic algorithms with respect to the test image. Experimental results are presented using our developed technique on real images, which are hard to be contrasted by other conventional techniques.

Keywords: Image enhancement, Contrast Stretching, Sigmoid Function, Genetic Algorithms, Optimality.

1 Introduction

Image enhancement has always been an attractive field of research, partly because of the numerous avenues available to proceed with the task and partly because of the practical importance that image enhancement has in a wide variety of applications. In particular the enhancement of very dark and blurred images has been of particular interest as many aspects of these images are ambiguous and uncertain. Examples of these vague aspects include determining the border of a blurred object and determining which gray values of pixels are bright and which are dark [1]. Methods for image enhancement have been successfully developed in the spatial and frequency domains using various techniques. The suitability of one over the other is often a matter of subjective judgement or constrained by the input image.

In the frequency domain image enhancement has traditionally been performed by manipulating the Fourier transform of the original image with filters designed for particular purposes [2]. Filters like the Butterworth's filter have been long in existence and perform well under most conditions. Recent efforts in this field have also used the DCT [3] and wavelet based methods [4]. However it is difficult to decide on a single filter to get desired results and often masks have to be applied in a sequence to obtain results suitable for the required task. Also working with the absolute value of complex numbers may not be particularly acceptable in certain cases e.g. medical imaging, remote sensing where even minute details may be of utmost importance.

Abhijit Mustafi
Department of Computer Science, Birla Institute of Technology, India
e-mail: abhijit@bitmesra.ac.in

P.K. Mahanti
Department of CSAS, University of New Brunswick, Canada
e-mail: pmahanti@unb.ca

R. Lee, G. Hu, H. Miao (Eds.): Computer and Information Science 2009, SCI 208, pp. 1–8.
springerlink.com   © Springer-Verlag Berlin Heidelberg 2009

In the spatial domain the use of histogram equalization [5] and its variants like the local histogram equalization [6] and the adaptive histogram equalization [7][8] have dominated image enhancement methods. However ordinary histogram equalization is not a parameterized process and fails to be uniformly successful for all images. Over emphasis of slow varying gray scale planes is considered a major drawback for histogram equalization. On the other hand local histogram equalization and adaptive histogram equalization can provide better results but are computationally intensive. Recent efforts have also focussed on the use of look up tables (LUT) [9] and clustering algorithms like the k-means method [10].

This research note presents a new algorithm to optimally enhance the contrast of a dark image. The proposed algorithm is robust because it automatically tunes the various parameters that affect the enhancement process for a particular image that is to be enhanced. It also uses genetic algorithms to optimally enhance the contrast of dark blurred images with minimal over emphasis of contrast in the slow varying gray value regions of the image.

2   Problem Formulation

Contract enhancement is usually applied to input images to obtain a superior visual representation of the image by transforming original pixel values using a transform function of the form

$$g(x, y) = T[r(x, y)] \tag{1}$$

where $g(x,y)$ and $r(x,y)$ are the output and input pixel values at image position (x,y). Usually for correct enhancement it is desirable to impose certain restrictions on the transformation function T [2] in the form of (a) the nature of the function which should be monotonically increasing and (b) the output range of T which should be congruent with the input to the function. It is usual to find the function T producing an output in the range [0,1]. The first condition negates gray value reversal in the output image so that pixels having smaller values in the original image continue to be smaller in respect of higher valued pixels. The second condition ensures that the output image is scaled to the same domain as the input image. Numerous functions have been proposed to perform contrast enhancement including the exceedingly common techniques of histogram equalization, piecewise contrast stretching, specific range contrast stretching etc.

In the proposed algorithm we use a variation of the sigmoid function as the transformation function. Our choice is influenced by the mathematical consistency of the sigmoid function [11] and also the simplicity of implementing the sigmoid function in practice. Sigmoid function is a continuous nonlinear activation function. The name, sigmoid, obtained from the fact that the function is "S" shaped. Statisticians call this function the logistic function, Using f (x) for input, and with a as a gain term, the sigmoid function is:

$$f(x) = \frac{1}{1 + e^{-ax}} \tag{2}$$

The sigmoid function has the characteristics that it is a smooth continuous function, the function outputs within the range 0 to 1, mathematically the function is easy to deal with: it goes up smoothly. The interested reader can refer to [12] for a detailed treatment of the sigmoid function and its variants. Our algorithm uses a modified form of the sigmoid function given by

$$g(x, y) = \frac{1}{1 + \left[\dfrac{m}{r(x, y)}\right]^n} \tag{3}$$

Where m is some chosen threshold value in the appropriate gray scale and n is some positive constant. For 8 bit gray scale images it is common to choose m = 127. It is easy to observe that in the limiting case the transformation tends to the binarization of gray scale images. For effective contrast enhancement the output image should ideally span the entire available gray scale and also have equal distribution of pixels over all the gray scale bins. These conditions are often satisfied mathematically but cannot be essentially satisfied for discrete images. Our objective is to produce a suitable choice of m, n or both in the previous equation so as to produce an output image as close as possible to the optimal distribution without distorting the image in any appreciable manner or introducing artefacts.

Let 'd' equal the range of the gray values present in the output image after application of the transformation equation. Therefore

$$d(x, y) = \max(g(x, y)) - \min(g(x, y)) \tag{4}$$

where obviously x and y varies from 0 to N-1 and we assume the input image is square with size NxN. For non square images the images can be suitable padded before application of the algorithm. For an image with high contrast we wish to maximize 'd' ensuring that the entire available gray scale is utilized.

Let 'w' equal the ratio of the occupied gray values in the output image and 'd' where occupation of a gray value is defined as the presence of pixels in the image with a that particular gray value. Thus

$$p = \frac{number \text{ of occupied gray values}}{d} \tag{5}$$

Obviously for an even or flat histogram it is essential to maximize the quantity p. Consequently our algorithm proposes to maximize both d and p using GAs. It is interesting to note that maximizing both d and p simultaneously is mutually incoherent and maximizing p leads to a minimization of p. However for optimal contrast enhancement it is also desirable that the distribution of pixels In all occupied gray value bins is as close to equal as possible.

To simplify the implementation of the parameter 'p' discussed above an alternate expression was used in our algorithm. We define the term 't' to act as a measure of pixel distribution in the output image over gray scale bins . Mathematically 't' is given as

$$t = \sqrt{\left[\frac{1}{d}\left\{\bar{c} - c(i)\right\}\right]} \tag{6}$$

where 'I' varies over all occupied gray values in the image, 'c' is the number of pixels having gray value i, $\bar{c}$ is the average number of pixels per gray value in the output image. For our purpose we need to minimize 't'. These considerations lead us to formulate our objective function as

$$F = w_1 d - w_2 t \tag{7}$$

This equation simultaneously tries to optimize both our parameters 'd' and 't' so that the final image is as optimal as possible.

3 Overview of GA Based Optimization

Genetic algorithms are meta-heuristic optimization techniques based on natural theory and survival of the fittest [13]. The operators involved in GA tend to be

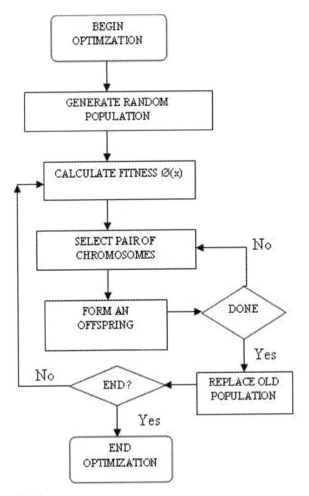

Fig. 1

heavily inspired by natural selection and consequently successive generations of the algorithm continue to propagate the best traits of the population. This leads to rapid convergence of the search. Also the introduction of the mutation operator ensures that diversity is not neglected and the search is not trapped in a local maximum. A flow chart illustrating the basics steps of GA based optimization is given below.

4   Experimental Results

To compare the performance of the proposed algorithm we chose the *pout* image from the standard MATLAB image database. The image is both dark and of low contrast and therefore ideal to demonstrate the credibility of the algorithm. The original image and it histogram is shown in Fig. 2 below.

The line across figure 2(b) shows the average gray scale bin occupancy and the compressed nature of the histogram is a consequence of the very low contrast of the image. It is of interest to note the washed appearance of the image particularly the fence in the background and the monogram on the jacket. Figure 3(a) below shows the effect of applying the proposed algorithm on the pout image using a crossover fraction of 0.8 and using uniform mutation. The algorithm usually converged quite fast (often in less than 200 iterations) because of the restriction of the gray scale range which in this case is [0,255]. Figure 3(b) shows the corresponding histogram for the output image. As a comparison the output after histogram equalization is shown in 3(c) and the corresponding histogram in 3(d).

The images produced by histogram equalization have over emphasized regions of constant pixel values while the image in fig. 3(a) which has been optimally enhanced looks better without looking over emphasized. This fact is also corroborated by the shape of the enhanced histogram in fig. 3(b) which bears a remarkable similarity to the original histogram (fig. 2b) while being spread out over the entire gray scale.

a) Original Image

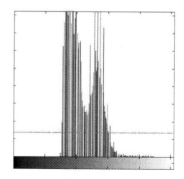

b) Histogram of (a)

Fig. 2

(a) Image after application of proposed algorithm.

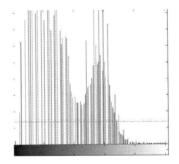

(b) Histogram of (a)

(c) Image after application of histogram equalization.

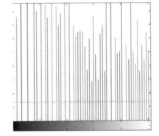

(d) Histogram of (c).

Fig. 3

To further illustrate the improvements offered by out algorithm in comparison to established methods we apply the algorithm to an image of quite good contrast and see the effects produced.

(a)Original Image

(b) Image after application of proposed algorithm

(c) Histogram Equalized Image

Fig. 4

The images in Fig. 4 show the viability of the new algorithm even for images of quite good contrast. It is of interest to note the over emphasis of the periodic noise artefacts in the image background and also the overall darkness of the final output. In comparison fig. 4(b) has much better overall brightness, and enhances the image in a more controlled manner. (Note the hand and the creases on the pant). For further evidence of the effectiveness and speed of the algorithm we produce the histogram of the final image and the convergence diagram of the GA below.

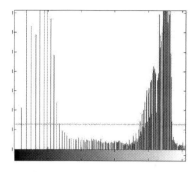

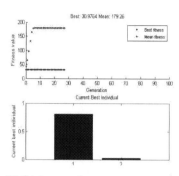

(a) Final Histogram.　　　　　　　　　　　(b) GA in execution.

Fig. 5

Figure 5 again illustrates the sensitivity of the algorithm to over emphasizing contrast by its histogram shape preserving nature and also the speedy execution and convergence because of the restrictions of the gray scale.

5 Conclusion

Dark and blurred images are very common outputs of a vast range of imaging applications e.g. medical imaging, GIS, low light photography etc. We have proposed a new algorithm for enhancing the visual quality of such images without distorting the contrast balance and introducing artefacts which is simple to implement, fast and works on a wide variety of images. The initial choice of the parameters could be restricted with some apriori knowledge.

References

[1] Haubecker, H., Tizhoosh, H.: Computer Vision and Application. Academic Press, London (2000)
[2] Gonzalez, R.C., Woods, R.E.: Digital Image Processing, 2nd edn. Pearson, London
[3] Chang, H.S., Kang, K.: A compressed domain scheme for classifying block edge patterns. IEEE Trans on Image Process 14(2), 145–151 (2005)
[4] Laine, A., Fan, J., Yang, W.: Wavelets for contrast enhacement of digital mammography. IEEE Engineering in Medicine and Biology (September/October 1995)
[5] Korpi-Anttila: Automatic color enhancement and scene change detection of digital video, Licentiate thesis, Helsinki University of Technology, Laboratory of Media Technology (2003)

[6] Pfizer, S.M., et al.: Adaptive Histogram Equalization and its Variations. Computer Vision, Graphics and Image Processing 39, 355–368 (1987)

[7] De Vries, F.P.P.: Automatic, adaptive, brightness independent contrast enhancement. Signal Processing 21, 169–182 (1990)

[8] Stark, J.A., Fitzgerald, W.J.: An Alternative Algorithm for Adaptive Histogram Equalization. Graphical Models and Image Processing 56, 180–185 (1996)

[9] Chung, K.L., Wu, S.T.: Inverse halftoning algorithm using edge-based lookup table approach. IEEE Transactions Image Processing 14(10), 1583–1589 (2005)

[10] Yang, S., Hu, Y.-H., Nguyen, T.Q., Tull, D.L.: Maximum-Likelihood Parameter Estimation for Image Ringing-Artifact Removal. IEEE Transactions on circuits and systems for video Technology 11(8), 963–974 (2001)

[11] Naglaa, Y.H., Aakamatsu, N.: Contrast Enhancement Techniques of Dark Blurred Images. In: IJCSNS, vol. 6(2A) (February 2006)

[12] Hertz, J., Plamer, R.: Introduction to the neural computation. Addison Wesley, California (1991)

[13] Paulinas, M., Usinskas, A.: A survey of Genetic Algorithms Applications for Image Enhancement and Segmentation. Information Technology and Control 36(3) (2007)

Ad-Hoc Behavior in Opportunistic Radio

Shahid Mumtaz, P. Marques A. Gameiro, and Jonathan Rodriguez

Abstract. The application of mathematical analysis to the study of wireless ad hoc networks has met with limited success due to the complexity of mobility, traffic models and the dynamic topology. A scenario based UMTS TDD opportunistic cellular system with an ad hoc behaviour that operates over UMTS FDD licensed cellular network is considered. In this paper, using an ah-hoc behavior in opportunistic Radio, we present how the overall system performance effect in terms of interference and routing.Therefore we develop a simulation tool that addresses the goal of analysis and assessment of UMTS TDD opportunistic radio system with ad hoc behaviour in coexistence with a UMTS FDD primary cellular networks.

Keywords: Opportunistic Radios, routing, ad-hoc network.

1 Introduction

Wireless communications play a very important role in military networks and networks for crisis management, which are characterised by their ad hoc heterogeneous structure. An example of a future network can be seen in Figure 1. This illustrates a range of future wireless ad hoc applications. In the heterogeneous ad hoc network, it is difficult to develop plans that will cope with every eventuality, particularly hostile threats, due to the temporary nature. Thus, dynamic management of such networks represents the ideal situation where the new emerging fields of cognitive networking and cognitive radio can play a part. Here we assume a cognitive radio 'is a radio that can change its transmitter parameters based on interaction with the environment where it operates' [1], and additionally relevant here is the radio's ability to look for, and intelligently assign spectrum 'holes' on a dynamic basis from within primarily assigned spectral allocations. The detecting of holes and the subsequent use of the unoccupied spectrum is referred to as opportunistic use of the spectrum. An Opportunistic Radio (OR) is the term used to describe a radio that is capable of such operation [2].In this paper we use the opportunistic radio system which was proposed in [3] that shares the spectrum with an UMTS cellular network. This is motivated by the fact that UMTS radio frequency spectrum has become, in a significant number of countries, a very expensive commodity, and therefore the opportunistic use of these bands could be one way for the owners of the licenses to make extra revenue.

Shahid Mumtaz, P. Marques A. Gameiro, and Jonathan Rodriguez
Institute of Telecommunication,Aveiro,Portugal
e-mail: smumtaz@av.it.pt

R. Lee, G. Hu, H. Miao (Eds.): Computer and Information Science 2009, SCI 208, pp. 9–21.
springerlink.com   © Springer-Verlag Berlin Heidelberg 2009

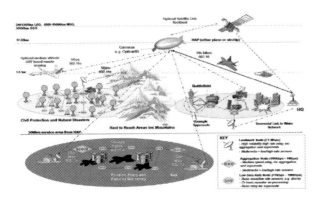

Fig. 1 Ad-hoc future network

The OR system exploits the UMTS UL bands, therefore, the victim device is the UMTS base station, likely far from the opportunistic radio, whose creates local opportunities. These potential opportunities in UMTS FDD UL bands are in line with the interference temperature metric proposed by the FCC s Spectrum Policy Task Force [4]. The interference temperature model manages interference at the receiver through the interference temperature limit, which is represented by the amount of new interference that the receiver could tolerate. As long as OR users do not exceed this limit by their transmissions, they can use this spectrum band. However, handling interference is the main challenge in CDMA networks, therefore, the interference temperature concept should be applied in UMTS licensed bands in a very careful way. In this paper we propose how an ad hoc behaviour uses in an opportunities radio and with careful selections of routing schemes, we minimize overall interference level on the victim device UMTS base station.

This paper is organized as follows: In Section 2 the scenario is defined. In Section 3 explains the opportunistic network with ad-hoc topology. In Section 4 explains coexistence analysis for a single opportunistic radio link. In Section 5 coexistence analysis for ad hoc opportunistic networks are explains and conclusions are made in Section 6.

1.1 Scenario Defenition

The UMTS is a DS-CDMA system, thus all users transmit the information spreaded over 5 MHz bandwidth at the same time and therefore users interfere with one another. Figure 2 shows a typical UMTS FDD paired frequencies. The asymmetric load creates spectrum opportunities in UL bands since the interference temperature (amount of new interference that the UMTS BS can tolerate) is not reached.

In order to fully exploit the unused radio resources in UMTS, the OR network should be able to detect the vacant channelization codes using a classification technique [5].Thus the OR network could communicate using the remaining spreading codes which are orthogonal to the used by the UMTS network. However, classify and identify CDMA's codes is a very computational intensive task for real time applications.

Fig. 2 UMTS FDD spectrum
bands with asymmetric load

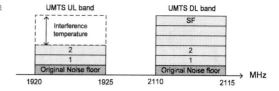

Moreover, synchronization between UMTS UL signals and the OR signals to keep the ortogonality between codes will be a difficult problem. Our approach is to fill part of the available interference temperature raising the noise level above the original noise floor. This rise is caused by the OR network activity, which aggregated signal is considered AWGN (e.g CDMA, MC-CDMA, OFDM).We consider a scenario where the regulator allows a secondary cellular system over primary cellular networks. Therefore we consider opportunistic radios entities as secondary users. The secondary opportunistic radio system can use the licensed spectrum provided they do not cause harmful interference to the owners of the licensed bands i.e.. the cellular operators.Specifically we consider as a primary cellular network an UMTS system and as secondary networks an ad hoc network with extra sensing features and able to switch its carrier frequency to UMTS FDD frequencies. Figure 3 illustrates the scenario where an opportunistic radio network operates within an UMTS cellular system. We consider an ah hoc OR network of M nodes operating overlapped to the UMTS FDD cell. The OR network acts as a secondary system that exploit opportunities in UMTS UL bands. The OR network has an opportunity management entity which computes the maximum allowable transmit power for each OR node in order to not disturb the UMTS BS.

Fig. 3 Ad hoc ORs networks operating in
a licensed UMTS UL band

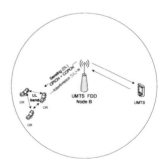

1.2 The Opportunistic Network with Ad Hoc Topology

The opportunistic network, showed in Figure 4, will interface with the link level simulator through LUTs.

The propagation models developed for the UMTS FDD network will be reused, and the entire channel losses (slow and fast fading) computed. The outputs will be

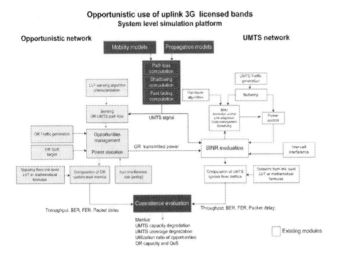

Fig. 4 Block diagram of the system level platform

the parameters that usually characterize packet transmissions: Throughput, BLER and Packet Delay. The LUT sensing algorithm characterization block contains the cyclostationary detector's performance, i.e. the output detection statistic, d, as a function of the SNR measured at the sensing antenna for different observation times [6].The sensing OR-UMTS path loss block estimates the path loss between UMTS BS and the OR location through the difference between the transmitted power and the estimated power given by cyclostationary detector (LUT sensing algorithm characterization block output). The OR traffic generation block contains real and non-real time service traffic models. OR QoS block defines the minimum data rate, the maximum bit error rate and the maximum transmission delay for each service class. The non-interference rule block compute the maximum allowable transmit power without disturbing the UMTS BS applying a simple non-interference rule (according to policy requirements).In the following, we briefly explain the opportunistic network blocks that was designed and implemented, using a C++ design methodology approach.

First of all, we assume that the OR knows a priori the UMTS carrier frequencies and bandwidths, which has been isolated and brought to the baseband. In order to get the maximum allowable power for OR communications the OR nodes need to estimate the path loss from its location to the UMTS BS. i.e.. the victim device. The opportunistic user is interested in predefined services which should be available every time. This motivates the proposal of defining a set of usable radio front end parameters in order to support the demanded services classes under different channel conditions. Basically, at the beginning of each time step the opportunistic radio requires certain QoS guarantees including certain rate, delay and minimum interference to the primary user (non interference rule policy).

The opportunistic network has an opportunity management entity which computes the maximum allowable transmit power for each opportunistic node in order the aggregated interference do not disturb the UMTS BS. The aggregated transmit power allowed to the opportunistic network can be computed using a simple non-interference rule

$$10\log\left(\sum_{k=1}^{K}10^{\frac{P_{OR}(k)+G_{OR}+G_{BS}-\hat{L}p(k)}{10}}\right) \leq 10\log\left(10^{\frac{Nth+\mu}{10}}-10^{\frac{Nth}{10}}\right) - \Gamma$$

Where G_{OR} is the OR antenna gain, G_{BS} is the UMTS BS antenna gain, Lp is the estimated path loss between the OR node and the UMTS BS, K is the Number of ORs, performed by a sensing algorithm, and Nth is the thermal noise floor. μ is a margin of tolerable extra interference that, by a policy decision, the UMTS BS can bear. Finally, Γ is a safety factor to compensate shadow fading and sensing s impairments. Notice if the margin of tolerable interference $\mu=0$ the OR must be silent. Γ is a safety factor margin (e.g. 6-10 dB) to compensate the mismatch between the downlink and uplink shadow fading and others sensing's impairments. The margin of tolerable interference is defined according to policy requirements.

Employing scheduling algorithms, we can provide a good tradeoff between maximizing capacity, satisfying delay constraint, achieving fairness and mitigating interference to the primary user. In order to satisfy the individual QoS constraints of the opportunistic radios, scheduling algorithms that allow the best user to access the channel based on the individual priorities of the opportunistic radios, including interference mitigation, have to be considered. The objective of the scheduling rules is to achieve the following goals:

- Maximize the capacity;
- Satisfy the time delay guarantees;
- Achieve fairness;
- Minimize the interference caused by the opportunistic radios to the primary user.

A power control solution is required to maximize the energy efficiency of the opportunistic radio network, which operates simultaneously in the same frequency band with an UMTS UL system. Power control is only applied to address the non-intrusion to the services of the primary users, but not the QoS of the opportunistic users.

A distributed power control implementation which only uses local information to make a control decision is of our particular interest. Note that each opportunistic user only needs to know its own received SINR at its designated receiver to update its transmission power. The fundamental concept of the interference temperature model is to avoid raising the average interference power for some frequency range over some limit. However, if either the current interference environment or the transmitted underlay signal is particularly non uniform, the maximum interference power could be particularly high.

Following we are going to explain why we consider Ad-hoc topology for the opportunistic radio system in cellular scenario. Mobile ad-hoc network is an autonomous system of mobile nodes connected by wireless links; each node operates as an end system and a router for all other nodes in the network. Mobile ad-hoc network fits for opportunistic radio because the following features:

Infrastructure

MANET can operate in the absence of any fixed infrastructure. They offer quick and easy network deployment in situations where it is not possible. Nodes in mobile ad-hoc network are free to move and organize themselves in an arbitrary fashion. This scenario is fit in the Opportunities in UMTS bands which are local and may change with OR nodes movement and UMTS terminals activity.

Dynamic Topologies

Ad hoc networks have a limited wireless transmission range. The network topology which is typically multi-hop may change randomly and rapidly at unpredictable times, and may consist of both bidirectional and unidirectional links which fits the typical short range opportunities which operate on different links in UMTS UL bands.

Energy-constrained operation

Some or all of the nodes in a MANET may rely on batteries or other exhaustible means for their energy. For these nodes, the most important system design criteria for optimization of energy conservation. This power control mechanisms for energy conversion (power battery) also helps to avoid harmful interference with the UMTS BS.

Reconfiguration

Mobile ad-hoc networks can turn the dream of getting connected "anywhere and at any time" into reality. Typical application examples include a disaster recovery or a military operation. As an example, we can imagine a group of peoples with laptops, in a business meeting at a place where no network services is present. They can easily network their machines by forming an ad-hoc network. In our scenario OR network reconfigure itself, as the interference coming from licensed users (PUs) causes some links being dropped. Ad hoc multi hop transmission allows decreases the amount of the OR's transmitted power and simultaneously decreases the interference with the UMTS BS.

Bandwidth-constrained, variable capacity links

Wireless links will continue to have significantly lower capacity. In addition, the realized throughput of wireless communications after accounting for the effects of multiple access, fading, noise, and interference conditions, etc. is often much less than a radio's maximum transmission rate. This constrained also fit in our scenario where maximum transmission rate of ORs is less than the UMTS base station after the effects of multiple access, fading, noise and interference conditions.

Security

Mobile wireless networks are generally more prone to physical security threats than are fixed cable nets. The increased possibility of eavesdropping, spoofing, and denial-of-service attacks should be carefully considered. Existing link security techniques are often applied within wireless networks to reduce security threats. As a benefit, the decentralized nature of network control in MANETs provides additional robustness against the single points of failure of more centralized approaches. By using this property of MANETs, we avoid single point failure in ORs.

2 Coexistence Analysis For a Single Opportunistic Radio Link

In this section we consider the simplest case where a single OR link operates within a UMTS FDD cell. Simulations were carried out to compute the coexistence analysis between the OR link and the UMTS network. The main parameters used for the simulations are summarized in Table 1. We consider an omnidirectional cell with a radius of 2000 meters. Each available frequency, in a maximum

Table 1 Main parameters used for the simulations

Parameter Name	Value
UMTS system	
Time transmission interval (T_{tl})	2 ms
Cell type	Omni
Cell radius	2000 m
Radio Resource Management	
Nominal bandwidth (W)	5 MHz
Maximum number of available frequencies (N_{fmax})	12
Data rate (R_b)	12.2 kbps
E_b/N_o target	9 dB
SIR target (γ)	-16 dB
Spreading factor	16
Spectral noise density (N_o)	-174 dBm/Hz
Step size PC	Perf. power ctrl
Channel Model	Urban
Carrier frequency	2 GHz
Shadowing standard deviation (σ)	8 dB
Decorrelation length (D)	50 m
Channel model	ITU vehicular A
Mobile terminals velocity	30 km/h
Primary User (PU)	
Number of primary user(s) terminals per cell/frequency (K)	64
Sensibility/Power received	-117 dBm
UMTS BS antenna gain	16 dBi
Noise figure	9 dB
Orthogonally factor	0
Opportunistic Radio (OR)	
Number of opportunistic radio(s) in the cell coverage area	2
Maximum/Minimum power transmitted ($P_{o[max/min]}$)	10/-44 dBm
Antenna gain	0 dBi
Duration call	90 s

of 12, contains 64 primary user terminals. Each of these primary users receives the same power from the UMTS base station (perfect power control). We assume the primary users data rate equal to 12.2 kbps (voice call); the E_b/N_o target for 12.2 kbps is 9 dB. Thus, and since the UMTS receiver bandwidth is 3840 kHz, the signal to interference ratio required for the primary users is sensibly -16 dB. There is (minimum one) opportunistic radio in the cell coverage area, which has a transmitted power range from -44 to 10 dBm. The opportunistic radio duration call is equal to 90 seconds. We furthermore consider load characteristics identical in every UMTS cellular system and the frequencies are close enough so that the same statistical models apply.

2.1 Simulation Results for a Single UMTS Frequency

In order to calculate Cumulative Distribution Function (CDF) for the interference at UMTS BS we consider 64 UMTS licensed UMTS terminals in each cell (with radius equal to R= 2000 m), as shown in the following Figure 5. The OR receiver gets interference from the PUs located in the central UMTS cell and in 6 adjacent cells. The ORs are within an ad-hoc network service area (with radius equal to R= 100 m); the OR receiver is 10 m away from the OR transmitter. The OR transmitter is constrained by the non-interference rule.

Fig. 5 Ad-hoc Single Link scenario

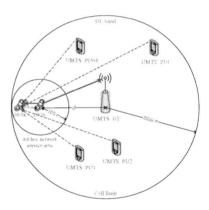

Based on the capacity's Shannon formula, the OR's link capacity that can be achieved between two OR nodes is given by:

$$C_{Mbps} = B \log_2 \left(1 + \frac{L_2 P_{OR\_Tx}}{Nth + I_{UMTS}} \right) \qquad \begin{array}{l} B = 5 \, MHz \\ Nth = -107 \, dBm \end{array}$$

Where B=5 MHz, L_2 is the path loss between the OR_Tx and the OR_Rx. Nth is the average thermal noise power and I_{UMTS} is the amount of interference that the UMTS terminals cause on the OR_Rx. On the other hand, the total interference at the UMTS BS caused by the OR activity can not be higher than the UMTS BS interference limit, -116 dBm [13].

The following Figure 6 shows the CDF of the interference computed at the UMTS BS due the OR network activity. The results show that an 8 Mbps OR's link capacity is guaranteed for approximately 98% of the time without exceeding the UMTS BS interference limit (-116 dBm). However, this percentage decreases to 60% when an OR link with 32 Mbps is established.

Fig. 6 Interference at UMTS BS

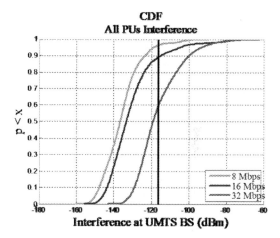

**CDF
All PUs Interference**

Interference at UMTS BS (dBm)

3 Coexistence Analysis for Ad Hoc Opportunistic Networks

In the previous section we have considered a single link between two nodes ORs, now we extend the coexistence analysis to the case of an ad hoc OR network with several nodes, multiple hop communication and routing mechanisms.

Routing by definition is a process of selecting paths. Conventional routing algorithms normally find the route with shortest path to improve efficiency [8]. However, in a wireless ad hoc network, shortest path routing is not necessarily the best solution. In such a wireless scenario, other criteria, such as interference, capacity, etc., should be considered while making the routing decisions. We assume a cognitive network 'is a network with a cognitive process that can perceive current network conditions, and then plan, decide, and act on those conditions [9]. For the routing mechanism to be 'cognitive', it must have three elements of processes: observing, reasoning and acting [10,11]. The observing process refers to how necessary information is gathered for implementation. The reasoning process is where the cognitive entity, e.g. a node (ORs), considers (orient, plan, decide, and learn) the way to behave for its various goals based on the information gathered from observing. The acting process is about how to implement the decision made by reasoning. We focus on the reasoning process mainly, assuming necessary information for each node is available through observing and different adjustment can be carried out in terms of acting. In order to serve a goal of the network level (e.g. finding the shortest path), there must be a certain mechanism to comprehensively link the nodes in the system and enable them to function in a collective way. For a routing problem, the route discovery algorithm is required to find the ideal route in the system. The classic Dijkstra's algorithm [12] is chosen for this work for the route discovery problem. It is originally derived for solving the single-source shortest path problem for a graph with non-negative path costs. The functionality of Dijkstra's original algorithm can be extended for various purposes. For example, OSPF (open shortest path first) protocol is an implementation of Dijkstra's algorithm for Internet routing [12]. Dijkstra's algorithm can effectively choose the route with lowest accumulative cost. In order to achieve higher level goals in the system, we can exploit the optimization function of Dijkstra's algorithm by redefining the cost. That is to say, the definition of cost is manipulated in order to serve a different purpose other than finding the 'shortest' path. In the following sections, we demonstrate how this algorithm can be used for OR system operating over the licensed UMTS band and routing strategies base on hop count and capacity information are investigate and compare.

3.1 Hop Based Routing

To find the shortest path for a routing mission in terms of hop count, the cost can be defined as:

$$W_{OR\_Tx, OR\_Rx} = 1 \quad , \quad \forall \; OR\_Tx, OR\_Rx \in N \tag{1}$$

Where W_{OR_Tx, OR_Rx} denotes the cost to transmit from link OR_Tx to OR_Rx. N is the total number of ah hoc nodes in the OR system This routing scheme is often known as the 'hop count' or 'shortest path routing', and is the most commonly used routing metric in existing routing protocols.

3.2 Capacity-Based Routing

With an ad hoc wireless network in our UMTS system, the scalability of the network is a major constraint due to the relaying burden each OR node has to carry. Heterogeneous OR nodes in the system can potentially improve the scalability of the network if the higher capacity OR nodes are placed in the right positions where more traffic has to be relayed. Intuitively, it is desirable to divert relaying traffic away from the OR nodes which have limited capacity to ones which are more capable to relay. For this purpose, we can define the cost as follows:

$$W_{OR\_Tx, OR\_Rx} = \frac{1}{C_{OR\_Rx}} , \forall \; OR\_Tx, OR\_Rx \in N \tag{2}$$

C_{OR_Rx} denotes the virtual capacity of the receiving node OR_Rx a node with higher capacity will be more likely to be chosen for the object route because the cost is less compared with a lower capacity node.

We consider that the transmission power of each OR node is identical, then C_{OR_Rx} is determined by two factors:

- The interference suffered by the UMTS base station.
- The spectrum resource that can be utilized by the OR.

In a cognitive radio context, having more available spectrum for the OR node implies that the OR node is capable of finding a larger spectrum hole (or more spectrum holes) and exploiting it for transmission. If we only consider a system without bandwidth constraint at each node as a special case, interference is then the only concern. It is a worst-case scenario, in which the interference problem is dealt with in the most conservative way and has the most severe impact on capacity. Assuming power levels and available bandwidth at each OR node, we let both of them equal to 1,

$$W_{OR\_Tx.OR\_Rx} = \frac{1}{\frac{1}{I_{OR\_Rx}+1}} < -116\,dbm\,.\,\forall\,OR\_Tx\,.OR\_Rx \in N$$

$$I_{OR\_RX} = 10\log\left(\sum_{k=1}^{M}10^{\frac{P_{OR}(k)+G_{OR}+G_{BS}-Lp(k)}{10}}\right)$$

I_{OR_Rx} represents aggregated interference from the UMTS base station which node OR_Rx suffers.

3.3 Routing Performance Analysis

We show the effectiveness of different routing mechanisms proposed in section V. A uniform geographic distribution of traffic is assumed to be generated by the ORs nodes simultaneously in the UMTS system. This is a worst scenario, where the system is running at its maximum overall capacity and most severe interference level is experienced from the UMTS base station. We define the capacity required to serve one route end-to-end without the impairment caused by interference and sharing with other routes as one *Erlang*. By using *Erlang* as the capacity unit, we can simply focus on the theoretic capacity performance without loosing the generality.

First, we look at the scenario without capacity constraint in the system. In other words, there is no shortage of bandwidth for each transmission in the system and all the originating traffic of each node will be delivered without constraint. We look at a case where the sink (landmark node) of the system is placed in the corner as shown in the Figure 7 using hop-based routing. Equation (1) is used in the route discovery process. We can see that the closer a node is to the sink, the more traffic going through the node and most severe interference level is experienced with victim device, i.e. the UMTS base station, as indicated by higher requested capacity of the node. It shows the requested capacity of each node is dependent on the topological location of the OR node in respect of the topological location of the sink. Instead of using conventional hop-based routing mechanism, we can

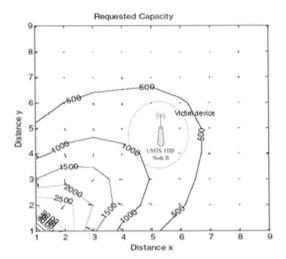

Fig. 7 Contour plot with gradient arrows to show the requested capacity performance using hop-based routing strategy in a system

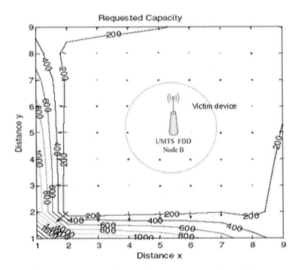

Fig. 8 Contour plot with gradient arrows to show the requested capacity performance using capacity-based routing strategy in a system

implement the capacity-based routing in the previous scenario. In a system without capacity constraints, equation (2) is utilized in route discovery. Figure 8 shows the requested capacity performance for the case in which the sink is in the corner. We can see that this time the routing pattern has been dramatically changed compared with the one with hop-based routing strategy. The route discovery process in this scenario tends to choose the routes along the edge of the system, in order to reduce severe interference from UMTS base station. It shows that by adjusting the cost function based on interference, the capacity-based routing strategy can effectively reduce the overall interference level in the system by shifting traffic to the edge of the network.

4 Conclusion

In this paper, we have considered an ah hoc behavior in the opportunists radio (ORs) and suggested that by implementing ah hoc features in the ORs will improve the overall performance of system. We implemented routing feature of an ad hoc network in ORs, which dramatically reduce the server interference from the Victim device UMTS BS. Routing strategies based on hop count and capacity information are investigated and compared. The capacity-based routing strategy can reduce the overall interference level in the system by shifting traffic to the edge of the network. OR networks with fully implemented an ad Hoc features are the areas of future research.

Acknowledgments. The work presented in this paper was supported by the European project IST-ORACLE and Portuguese Foundation for Science and Technology (FCT) through project AGILE.

References

[1]   FCC, ET Docket No 03-222 Notice of proposed rule making and order (December 2003)

[2]   Mitola, J.: Cognitive Radio for Flexible Multimedia Communications. In: MoMuC 1999, pp. 3–10 (1999)

[3]   Marques, P.: Opportunistic use of 3G uplink Licensed Bands Communications, 2008. In: ICC IEEE International Conference (May 2008)

[4]   FCC, ET Docket No 03-237 Notice of inquiry and notice of proposed Rulemaking, ET Docket No. 03- 237 (2003)

[5]   Huang, C., Polydoros, A.: Likelihood methods for MPSK modulation classification. IEEE Transaction on Communications 43 (1995)

[6]   ORCALE WP2, http://www.ist-oracle.org

[7]   Pereira, A.: Opportunistic Radio and Licensed Users Coexistence in Cellular Networks. In: Wireless Pervasive Computing, 2007. ISWPS (2007)

[8]   Royer, E.M., Toh, C.-K.: A Review of Current Routing Protocolsfor Ad Hoc Mobile Wireless Networks. IEEE Personal Communications 6, 46–55 (1999)

[9]   Thomas, R.W., Friend, D.H., DaSilva, L.A., MacKenzie, A.B.: Cognitive Networks Adaptation and Learning to Achieve End-to-End Performance Objectives. IEEE Commun. Mag. 44, 51–57 (2006)

[10] Haykin, S.: Cognitive Radio: Brain-Empowered Wireless Communications. IEEE Journal on Selected Areas in Communications 23(2), 201–220 (2005)

[11] Mitola, J., Maguire Jr., G.Q.: Cognitive radio: making software radios more personal. IEEE Personal Communications 6, 13–18 (1999)

[12] Pioro, M., Medhi, D.: Routing Flow and Capacity Design in Communication and Computer Networks. Morgan Kaufmann, San Francisco (2004)

[13] Marques, P., et al.: Procedures and performance results for interference computation and sensing. IST-ORACLE project report D2.5 (May 2008)

Geocast Routing in Urban Vehicular Ad Hoc Networks

Guoqing Zhang, Wu Chen, Zhong Xu, Hong Liang, Dejun Mu, and Li Gao

Abstract. Vehicular Ad hoc Networks (VANETs) using wireless multi-hop technology have recently received considerable attention. Because of frequent link disconnection caused by high mobility and signal attenuation due to obstacles (e.g. tall buildings) in urban areas, it becomes quite challenging to disseminate packets to destinations locating in a specified area. We proposed a novel geocast routing protocol called GRUV for urban VANETs. GRUV adapts to the current network environment by dynamically switching three forwarding approaches which are used to compute forwarding zones. According to the locations, vehicles are categorized into crossroads nodes and in-road nodes, and different next-hop node selection algorithms are proposed respectively. Simulations with relatively realistic vehicle movement model and probability-based obstacle model show that GRUV has a higher performance compared with other geocast routing protocol.

1 Introduction

Recently there has been increasing interest in exploring computation and communication capabilities in transportation systems [1]. Many automobile manufactures started to equip GPS, digital map and the wireless device with new vehicles. Existing cars can also be easily upgraded. With the rapid development of the wireless technology, it is easy to support low-cost inter-vehicle communications. Without need of fixed infrastructure, Ad hoc communication architecture was introduced to construct a new variation of MANETs (Mobile Ad hoc Networks) called VANETs (Vehicular Ad hoc Networks). One typical example service VANET provides is to disseminate messages to drivers in a certain area (geocast region). These messages may help the drivers to choose the fast route to their destinations by making a detour of jamming roads or to find near commercial places (i.e., gas station, supermarket, and restaurant) as advertisements.

In VANETs, the movement of each vehicle is constrained by the roads, so the trajectory of the vehicle usually can be predicted to a certain extent based on the road topology and speed limits. As vehicles move on the road at the high speeds the topology of the network changes rapidly which means frequent link disconnection may happen. This makes it challenging to set up a robust path between

Guoqing Zhang, Wu Chen, Zhong Xu Hong Liang, Dejun Mu, and Li Gao
College of Automation, Northwestern Polytechnical University, Xi'an, 710072, China
e-mail: {gniq,xuzhong}@mail.nwpu.edu.cn,
{chenwu,hongliang,mudejun,gaoli}@nwpu.edu.cn

R. Lee, G. Hu, H. Miao (Eds.): Computer and Information Science 2009, SCI 208, pp. 23–31.
springerlink.com   © Springer-Verlag Berlin Heidelberg 2009

the source and the destination. In the city environments, there are a lot of different types of obstacles along the roads which may block and attenuate radio signals. Especially in an urban area crowed tall buildings around intersections, links are not stable between two vehicles locating at different road segments shadowed by these buildings.

We focused on the delivery of data packet from the source to a group of destinations locating in the same area (geocast region) in vehicular networks in city environments. This paper proposed a novel geocast routing for urban VANETs called GRUV based on the consideration of a network and scenario characteristics. GRUV uses a *mesh* to build multiple and redundant paths between the source vehicle and geocast region in a source routing manner. Block-extended forwarding zones and a dynamical switch strategy of three forwarding approaches decrease the routing overhead. According to the locations, vehicles are categorized into *crossroads nodes* and *in-road nodes*, and the different next-hop node selection algorithms were proposed respectively. Simulation results show that GRUV outperforms than the compared geocast routing protocol [2].

The reminder of this paper is organized as follows. In Section 2 we investigate the related work. Section 3 describes the detailed GRUV. Section 4 shows the performance evaluation results. Finally, Section 5 concludes this paper.

2   Related Work

In recent years various ad hoc routing protocols have been proposed and can classified into two categories: reactive (on-demand) protocols and proactive protocols. Because of less control overhead reactive protocols like AODV [3] and DSR [4] have gained much more attention and popularity. Many location-based protocols in ad hoc networks are also proposed to make use of geographic location information to help making routing decisions. Location-Aided Routing (LAR [5]) protocol and the Distance Routing Effect Algorithm for Mobility (DREAM [6]) are typical examples which use a forwarding zone, instead of the whole network, to flood unicast route request packets, thus the routing overhead of them decrease due to the limit of forwarding areas.

Recently, a few of unicast protocols applied to VANET have been proposed. Most of them assume that node can obtain their positions via GPS, even can access digital maps, e.g. Anchor-Base Street and Traffic Aware Routing (A-STAR [7]) and Geographic Source Routing (GSR [8]). There has been recent interest in developing multicast protocols in MANET. Examples of multicast include On-Demand Multicast Routing Protocol (ODMRP [9]) and Core-Assisted Mesh Protocol (CAMP [9]), both of which use mesh to create multiple and redundant paths to route multicast packets. Geocast is a variation of multicast with all nodes in the destination group locating in the same area called geocast region. Geocast Adaptive Mesh Environment for Routing (GAMER [2]) is a typical geocast routing protocol in ad hoc networks. GAMER also build multiple paths between the source and geocast region. Forwarding zones including CONE, CORRIDOR and FLOOD are defined to decrease the overhead. The similar idea is also involved in GRUV.

3 The GRUV Protocol

3.1 Overview

GRUV uses vehicle position and digital map to help making decisions when setting up a route, so we assume the vehicle equips GPS receiver and digital map device. This assumption is reasonable due to the rapid development of automotive industry. GRUV uses the following three basic concepts: source routing [5], forwarding zones and meshes. In source routing, each packet carries a full route (a sequenced list of nodes) that should be able to traverse in its header. GRUV can also easily be modified to use local routing state instead of using source routing.

A forwarding zone, which was usually a rectangle and used to route unicast packets in LAR [6], is used to route geocast packets in GRUV. Route request messages are flooded only in the forwarding zone instead of the whole network. Three forwarding approaches (FAs) -BOX, Extended-Box (E-Box) and FLOOD are proposed in GRUV to define different types of forwarding zones, as well as a dynamical FA switching strategy according to the feedback of route response packets from geocast region.

A mesh is a subset of network topology that provides multiple paths between sender and multiple receivers. It is used in many multicast routing protocols [9, 9] to maintain network connectivity effectively. We use the mesh to establish multiple paths between source node and geocast region in forwarding zones. Since geocast is a variation of multicast that all destination nodes are located in the same area, it is less costly to provide redundant paths from a source to destination nodes.

Considering characteristics of VANET and city environments we categorized vehicles into crossroads nodes and in-road nodes according to their locations and proposed different next-hop selection algorithms respectively. Once a node locating in geocast region receives the geocast data packet, it broadcasts locally to confirm the delivery of that packet to other destinations in the same area.

3.2 Block-Extended Forwarding Zone

Routing Request (RREQ) packets are only flooded in the forwarding zone instead of the whole network. To reduce routing overhead, GRUV provides three forwarding approaches (FAs) to define different forwarding zones. They are BOX, Extended-BOX (E-BOX) and FLOOD. We assume each vehicle knows its current position and can access the digital map. In the following, we use S and P_S to denote the source node and its position, and C_{GR} to denote the center of geocast region.

Fig.1 shows an example of a mesh created by GRUV when using BOX FA. The forwarding zone computed by S is the minimum rectangle covering both its position and geocast region. If the boundary of the zone locates in road, the rectangle enlarges to cover the road.

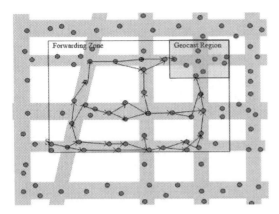

Fig. 1 Forwarding zone defined by BOX FA

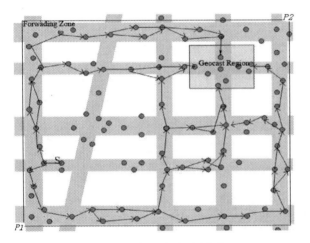

Fig. 2 Forwarding zone defined by E-BOX FA

Fig.2 provides a mesh example generated by E-BOX FA. The rectangular forwarding zone enlarges based on that defined by BOX. Firstly, the forwarding area is increased one block along the direction $C_{GR}S$ until contains a new junction. Similarly, the forwarding zone enlarges one block towards the direction SC_{GR} until encounters a junction. In Fig.2, there are two vertical streets and two horizontal streets more than in Fig.1 created by BOX FA.

FLOOD FA extends the forwarding zone further in the same way than E-BOX based on BOX, while the enlargement scale is two blocks at both endpoints of the segment C_{GR} S. We think the forwarding zone is large enough to find best route in most cases, so once S cannot receive the response packet before wait time expires, it is proper to drop the packets or buffer them in a moment. In fact in our implementation we simplified this case by flooding routing request packet in the whole simulating network scenarios.

3.3 Forwarding Approaches Switching

While a source node in GRUV has geocast packet to transmit, RREQ packet is periodically sent to the geocast region. GRUV uses FA to define forwarding zone whose size can infect the number of forwarding nodes generally, so an adaptive FAs switching algorithm is necessary to keep the routing overhead as low as possible.

GRUV dynamically switches the FAs from BOX, E-BOX and FLOOD according to the current network environment. The source initializes a timer (FA_TIMER) for each RREQ packet before it sends and destroys it when expires. BOX FA is selected at the beginning. If the BOX FA fails to create a mesh (i.e., no RREP returns to the source node) within FA_TIMER, then the next RREQ packet is sent via E-BOX FA. If E-BOX FA fails to setup a route before FA_TIMER expires, then the next RREQ packet is sent via FLOOD. IF FLOOD also fails, the following RREQ is transmitted via FLOOD continually. Once at least one path between the source node and the geocast region is found, the source node tries to send the next RREQ via a smaller forwarding zone. Specifically, if the BOX FA is successful, this FA is used continually. If E-BOX FA is successful, the next RREQ is transmitted using the BOX FA. If the FLOOD FA succeeds, the next RREQ packet is sent via the E-BOX FA.

3.4 Next-Hop Node Selection Algorithms

Since obstacles such as buildings in city environment may block radio signals especially when two communicating vehicles are located in different road segments around the same junction, data packets should always be forwarded to a node on a junction rather than be forwarded across a junction. Besides, vehicle at the crossroads always speeds down; therefore route lifetime may increase if the vehicle is selected in the route. For the nodes not near a junction, speed is the key factor to influencing the quality of communication. GRUV categorizes vehicles into *crossroads* nodes and *in-road* nodes according to their locations and applies different next-hop node selection algorithms respectively.

In source routing source node always prefers to select the route recorded in the earlier response packet, while the path may not be the stable one. By varying the delay time T_d which intermediate nodes should delay to forward the RREQ packet, stable path can be collected firstly by the source node. For an intermediate node, the smaller T_d is, the more possibility it is selected in the final path. Here, $T_d \in [0, C]$ and C is the maximum delay time.

3.4.1 Crossroads Node Selection Algorithm

A crossroads area is defined as a circle which covers the whole junction. The circle center is the center of the junction and its radius is a constant which is much more than the road width and is set 50m in our experiment. For an intermediate node which is finally selected in the path in source routing, *round trip time* (RTT) is the interval from the time it firstly receives the RREQ to the time it begins to forward the data packet. It should be avoided that the existed route is broken for signal blocking at the crossroads at the stage of data transmission. Therefore, GRUV tries to find nodes which will stay in the circle longest after RTT by predicting the vehicle's position according to the position of crossroads center and current speed.

Given vehicle X position $P(x, y)$ at the time t_0 and its velocity $\vec{v}$, then computes X predicted position $P'(x', y')$ at the time $t_0 + T_{RTT}$ and the estimated distance d which the vehicle needs to run out of the circle. The formulations are:

$$P'(x', y') = P(x, y) + \vec{v} * T_{RTT} \tag{1}$$

$$d = \begin{cases} R + |P' - P_o| & \text{if } X \text{ runs to the circle center} \\ R - |P' - P_o| & \text{else} \end{cases} \tag{2}$$

where

P_o is the position of the circle center;

R is the radius of the circle;

T_{RTT} is the value of *RoundTripTime* field recorded in the RREQ header. Both RREQ and RREP packets have a *SendTime* field to record the RREQ sent time at the source node. The source obtains the RTT which indicates the latest network condition and then updates the *RoundTripTime* field with this value before the next RREQ transmission.

According to the current velocity $\vec{v}$, X can compute $t' = d/v$ and the delay time T_d for forwarding the RREQ packet.

$$T_d = \begin{cases} C & t' < T_a \\ C(t' - T_b)/(T_a - T_b) & T_a \leq t' \leq T_b \\ 0 & t' > T_b \end{cases} \tag{3}$$

where $T_a = 5C$ and $T_b = 12C$ are configurable parameters.

3.4.2 In-Road Node Selection Algorithm

In linear roads data dissemination path between vehicles is much more stable when involved nodes move towards the same direction due to the small relative speed. In GRUV, a node running to the same direction as its last hop takes priority of that running to the opposite direction. Vehicles moving in roads are grouped in two groups (G0 and G1) based on their velocity vectors. $\overrightarrow{SC_{GR}}$ is the unit vector from source node S to the center point of geocast region, and its perpendicular divides the Cartesian coordinates into G0 and G1.

Let velocity vector $\overrightarrow{V'}$ belongs to G0 and $\overrightarrow{V''}$ belongs to G1. In our implementation, when an intermediate node X firstly receives RREQ, it calculates $\overrightarrow{SC_{GR}}$ according to *SourcePosition* field and *GeocastRegion* field recorded in the packet header, and then computes the dot product (*dir*) of its velocity $\overrightarrow{V}$ and $\overrightarrow{SC_{GR}}$. If *dir* >0, let G=0; else let G=1. Here, G=0 means X belongs to G0 and G=1 indicates X belongs to G1. Next, X compares G with the field of *LDG* (*Last Direction Group*) in RREQ packet header which indicates which group the last hop node belongs to. If they are equal, X has the same direction with its last hop vehicle and let $T_d = 0$; else let $T_d = C$. Before forwarding the request packet to the next-hop node, X updates the *LDG* field with G in the packet header.

4 Performance Evaluation

4.1 Simulation Environment

We use ns2 simulator to evaluate our GRUV protocol. The detailed network and scenario parameters are shown in Table 1. Manhattan Model [11] with grid road topology is used to describe vehicles movement pattern. It supports multiple lanes, probabilistic approaches, and interface to access simple digital maps. In our previous work, a probability-based obstacle model (POM) was designed to evaluate the impact on data transmission when two communicating nodes across the streets in presence of obstacles (i.e., buildings) between them. We integrated our POM into Manhattan, so the mobility model is relatively realistic for urban vehicular environment. To avoid that there's no destination nodes (no vehicle running in the geocast region), we place one static node at the center of geocast region in our simulation. At the same time, we let the source node stay at the furthest junction (300,300) from the geocast region. We also evaluated GAMER in the same scenario and compared with GRUV in four aspects: packet delivery ratio (PDR), average hop count (AHC), average end-to-end delay (AED) and overhead.

4.2 Simulation Results and Analysis

The PDR measures the accuracy of transmissions. Fig.3 (a) illustrates the ratio as a function of vehicle speed. As shown, the PDR of both GRUV and GAMER decrease when vehicles maximum speeds increase, while GRUV provides higher delivery ratio at all speeds than GAMER, even more than 19% at 20m/s (72km/h).

Table 1 The scenario and network parameters

Scenario parameter	Value	Network parameter	Value
Map size (m^2)	1200×1200	Simulation time(s)	600
vehicles	150	Transmission range(m)	250
Geocast region (m^2)	250×250	MAC protocol	802.11b
Horizontal streets	3	Bandwidth (Mb)	2
Vertical streets	3	Propagation model	TwoRayGround
Junctions	9	Mobility model	Manhattan with POM
Lanes	2	Source node number	1
Max speed(m/s)	1,5,10,15,20	CBR load(bytes)	64
Max acceleration(m/s^2)	0.1,0.5,1,1.5,2.0	Packet rate (packet/s)	4

There are three reasons: a) The forwarding zones (CONE and CORRIDOR) defined by GAMER are not realistic due to the road topology in city; b) Intermediate nodes in the path are selected to form stable route; c) GRUV reduces the impact of signal attenuation between communicating vehicles near crossroads.

Since in our experiment the distance between the source and geocast region is fixed, the average hop count measures the ability to find routes to a certain extent. As shown in Fig. 3 (b), GRUV has a little more hops than GAMER because our protocol prefers to select nodes at the crossroads instead of across a junction. While average end-to-end delay caused by hop increase is very little as shown in Fig. 3 (c), AED of GRUV is acceptable after long distance (>6 hops).

Overhead is the ratio of all control packets over the total amount of packets. Fig. 3 (d) illustrates the overhead comparison with GRUV and GAMER. Due to the frequent link broken caused by vehicles high speeds, both protocols creates more control packets to build and maintain the route, so their overhead increases with vehicle speeds increase.

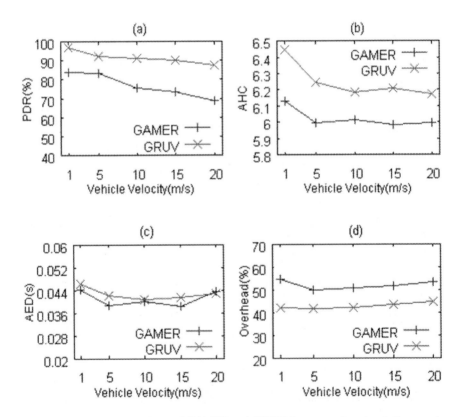

Fig. 3 Performance comparison of GAMER and GRUV in terms of packet delivery ratio (PDR), average hop count (AHC), average end-to-end delay (AED) and overhead

5 Conclusion

In this paper, we proposed a novel geocast routing protocol called GRUV for the urban vehicular ad hoc networks based on the consideration of a network and scenario characteristics. Block-extended forwarding zones and a dynamical switch strategy of three forwarding approaches decrease the routing overhead. To find a robust path, vehicles are classified into crossroads nodes and in-road nodes according to their positions, and different next-hop node selection algorithms are proposed respectively. Simulation results show that GRUV performs better compared with other geocast routing protocol.

Acknowledgments. Our work was supported by National Nature Science Foundation of China (No. 60803158) and Science Foundation of Aeronautics of China (NO. 05F53029). Thanks for the great help.

References

1. Enkelmann, W.: FleetNet- applications for inter-vehicle communication. In: Proceedings of Intelligent Vehicles Symposium, pp. 162–167 (2003)
2. Camp, T., Liu, Y.: An adaptive mesh-based protocol for geocast routing. Journal of Parallel and Distributed Computing Routing in Mobile and Wireless Ad Hoc Networks 63, 196–213 (2003)
3. Perkins, C.E., Royer, E.M.: The Ad Hoc on-demand distance-vector protocol. Ad hoc networking (2001)
4. Johnson, D.B., Maltz, D.A., Broch, J.: DSR: the dynamic source routing protocol for multihop wireless ad hoc networks. In: Ad hoc networking. Addison-Wesley Longman Publishing Co., Inc., New Jersey (2001)
5. Ko, Y.B., Vaidya, N.H.: Location-aided routing (LAR) in mobile ad hoc networks. Wireless Networks 6, 307–321 (2000)
6. Basagni, S., Chlamtac, I., Syrotiuk, V.R., et al.: A distance routing effect algorithm for mobility (DREAM). In: Proceedings of 1998 Mobile computing and networking, pp. 76–84 (1998)
7. Seet, B.C., Liu, G., Lee, B.S., et al.: A-STAR: A Mobile Ad Hoc Routing Strategy for Metropolis Vehicular Communications. In: Mobile and Wireless Communications. Springer, Heidelberg (2004)
8. Lochert, C., Hartenstein, H., Tian, J., et al.: A routing strategy for vehicular ad hoc networks in city environments. In: Proceedings of 2003 Intelligent Vehicles Symposium, pp. 156–161 (2003)
9. Chiang, C.C., Gerla, M.: On-demand multicast in mobile wireless networks. In: Proceedings of the 1998 International Conference on Network Protocols, pp. 262–270 (1998)
10. Madruga, E.L., Garcia-Luna-Aceves, J.J.: Multicasting along meshes in ad-hoc networks. In: Proceedings of International Conference Communications, pp. 314–318 (1999)
11. Bai, F., Sadagopan, N., Helmy, A.: The IMPORTANT framework for analyzing the Impact of Mobility on Performance Of RouTing protocols for Adhoc NeTworks. Ad Hoc Networks 1, 383–403 (2003)

Object-Oriented Ladder Logic Development Framework Based on the Unified Modeling Language

Kwan Hee Han and Jun Woo Park

Abstract. In order to improve current PLC programming practices, this paper proposes an object-oriented (O-O) ladder logic development framework in which ladder code is generated automatically based on the O-O design results and can be verified by port-level simulation. Proposed ladder logic development framework consists of two phases: First is the design phase. Second is the generation and verification phase. During the first phase, O-O design model is built, which consists of three models: functional model, structure model and interaction model. Two steps are conducted during the second phase. Firstly, ladder code is generated automatically using the interaction model of design phase. Secondly, generated ladder code is verified by input/output port simulation. In order to show the applicability of proposed framework, a software tool for the design and generation of ladder code is also developed.

1 Introduction

Most enterprises are struggling to change their existing business processes into agile, product- and customer-oriented structures to survive in the competitive and global business environment. Among their endeavor to overcome the obstacles, one of the frequently prescribed remedies for the problem of decreased productivity and declining quality is the automation of factories [1]. As the level of automation increases, material flows and process control methods of the shop floor become more complicated. Currently, programmable logic controllers (PLC) are mostly adopted as controllers of automated manufacturing systems, and the control logic of PLC is usually programmed using a ladder diagram (LD). More recently, manufacturing trends such as flexible manufacturing facilities and shorter product life cycles have led to a heightened demand for reconfigurable control systems. Therefore, logic control code must be easily modifiable to accommodate changes in manufacturing plant configuration with minimal down time [2].

Kwan Hee Han and Jun Woo Park
Department of Industrial & Systems Engineering, Engineering Research Institute,
Gyeongsang National University, Jinju, Republic of Korea
e-mail: hankh@gsnu.ac.kr

R. Lee, G. Hu, H. Miao (Eds.): Computer and Information Science 2009, SCI 208, pp. 33–45.
springerlink.com   © Springer-Verlag Berlin Heidelberg 2009

To cope with these challenges, a new effective and intuitive method for logic code generation is needed. However, currently there are no widely adopted systematic logic code development methodologies to deal with PLC based control systems in the shop floor. So, the control logic design phase is usually omitted in the current PLC programming development life cycle though it is essential to reduce logic errors in an earlier stage of automation projects before the implementation of control logic.

Moreover, PLC ladder logic gives only microscopic view of the system processes, and lacks semantic and conceptual integrity. As a result, it is difficult for factory automation (FA) engineers to have overall perspectives about the interaction of system components intuitively. To deal with frequent configuration changes of modern manufacturing systems, it is required that logic code can be generated automatically from the design results without considering complicated control behavior. Therefore, current PLC ladder programming practices require a more integrated way to design and generate the control logic.

In order to improve current PLC programming practices, significant efforts have been made in researches on O-O technologies in manufacturing systems. O-O modeling has been mainly used as a method for the analysis and design of software system. Recently, it was presented that O-O modeling is also appropriate for the real-time system design like an AMS (Automated Manufacturing System) as well as the business process modeling. The most typical features of O-O modeling techniques include the interaction of objects, hierarchical composition of objects, and the reuse of objects [3].

The main objective of this paper is to propose an object-oriented ladder logic development framework in which ladder code is generated automatically based on the UML (Unified Modeling Language) design results and can be verified by port-level simulation. And a software tool for the design and generation of ladder code is also developed.

The rest of the paper is organized as follows. Section 2 reviews related works. Section 3 describes a proposed O-O ladder logic development framework. Finally, the last section summarizes results and suggests directions for future research.

2 Related Works

Several researches were made regarding the O-O modeling methods for the manufacturing system: Calvo et al (2002) proposed an O-O method for the design of automation system, but they only showed the static structure comprised of a class diagram and a use case diagram [4]. Young et al (2001) proposed UML modeling of AMS and its transformation into PLC code, but they did not presented the method of PLC code generation [5]. Bruccleri and Diega (2003) presented UML modeling of flexible manufacturing system and its simulation implementation, but they restricted the control level to the supervisory control level [6].

Among researches about design and validation tools for the PLC control logic, Spath and Osmers (1996) proposed a simulation method integrating plant layout sub-model and control sub-model, and also a PLC code generation from simulation result, but they omitted details of generation procedure [7]. Baresi et al (2000) presented the procedure of control logic design using IEC function block diagram (FBD), its transformation into Petri net, and the validation of control logic using SIMULINK simulation system, and C code generation [8]. But, they confined their modeling scope to simple control logic which can be represented by FBD. Authors of this paper developed O-O design tool based on the extension of UML and showed usefulness of O-O approach to ladder logic design [9].

In the area of automatic ladder logic generation method, there exist mainly three approaches as follows: First approach is Petri net-based ([10], [11], [12], [13]). Second approach is finite state machine-based ([14], [15], [16], [17]). Last approach is flow chart-like-based ([14], [18]). Among three approaches, first and second approaches have a state explosion problem when complexity of control logic increases.

Among three approaches, third approach is relatively easy to use by its sequential and intuitive nature to control logic programmers. However, the result of ladder code generated by the third approach proposed by Jack (2007) is different from the code directly written by FA engineers due to its automatic generation features [14]. Therefore, it is not natural and revealed difficulties to understand the generated ladder code. Hajarnavis and Young (2005) explained functionalities of Enterprise Controls commercial package of Rockwell Automation. FA engineer designs the ladder logic in the form of flow chart using Enterprise Controls, and ladder code is generated automatically [18]. However, they did not show how the ladder code is generated. Proposed generation method in this paper belongs to the third category, in which ladder logic code is generated from the UML activity diagram which is a kind of flow chart.

3   Object-Oriented Ladder Logic Development Framework

Proposed ladder logic development framework consists of two phases: First is the design phase for ladder logic. And second is the generation and verification phase for ladder logic. In the design phase, O-O design tool developed by authors is used, which is based on the extension of UML. Main characteristics of UML are its graphical notation, which is readily understood, and a rich set of semantics for capturing key features of object-oriented systems.

3.1   Phase I: Design Phase for Ladder Logic

O-O design for ladder logic is conducted based on system specifications such as drawings and problem descriptions. During the design phase, FA engineers develop three models for describing the various perspectives of manufacturing systems: 1) a functional model for representing functional system requirements of AMS, 2) a structure model for representing the static structure and relationships of system components, and 3) an interaction model for representing the dynamic behavior of system components.

A functional model is constructed using an UML use case diagram in which each functional requirement is described as a use case. A PLC as a plant controller is represented by a 'system' element, and input or output part of PLC such as a sensor, actuator, and operator is represented by an 'actor' element of a use case diagram. Since the UML stick man icon of 'actor' is not appropriate for representing the resource of AMS, new icons are introduced in a functional model using UML stereotype property. Therefore, in the extended UML use case diagram, four types of actor (i.e., operator, actuator, sensor and MMI) are newly used instead of standard stickman symbol. PLC input part such as sensor and operator are located at the left side of 'system' symbol, and PLC output part such as actuator and MMI (Man Machine Interface) are located at the right side of 'system' symbol.

The details of each use case are described in a use case description list. The use case description list includes the pre-/post-condition of a use case and interactions of a PLC with its actors such as sensors and actuators. For realizing a use case, related domain classes accomplish an allocated responsibility through the interactions among them. These related classes are identified in the structure model. And the system-level interactions in a use case description list are described in more detail at the interaction model.

Fig. 1 and 2 shows a functional model example in the form of use case diagram and use case description list for the example system. This example application prototype is a kind of conveyor-based material handling system which identifies defective products according to their height, extracts defective products, and sorts good products according to their material property. It has 6 use cases for describing major functions from power control to product counting as depicted in Fig. 1. Fig. 2 shows the use case description of use case 4 (defects extraction) in Fig. 1. It describes the high-level interactions between system (PLC) and its actors such as photo sensors and cylinder.

A generic AMS is comprised of 4 parts: there is a 'plant' for manufacturing products. A plant is controlled by a 'controller' (PLC) which is managed by an 'operator' who monitors plant through MMI. A 'work piece' flows through a plant. A plant is further decomposed into standard resource groups hierarchically.

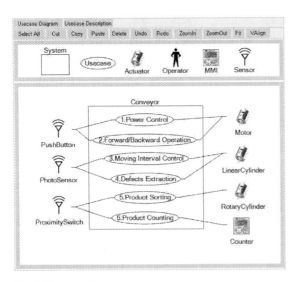

Fig. 1 Use case diagram

Usecase Description	
System : ConveyorSystem Name : Defects extration	
Scenario : Identifies defective products using 2 photo sensors. If defect product is identified, controller actuators extraction cylinder for the removal of product.	
Pre-condition : identification of defective products	
Post-condition : extraction of defective products	
Typical course or Events	
Actor	System
1. Product arrives 　1.1 high&low level sensors are all ON 　1.2 high level sensor is OFF and low level sensor is ON 　1.3 high&low level sensor are all OFF 3. Extraction point sensor senses product 5. Extract defective product by forward stroke 6. Proximity switch senses good product	2. Identifies state of product 　2.1 identifies good product 　2.2/2.3 identifies defective product 4. Controller control cylinder according to the state of product 　4.1 in case of defective products, sends forward stroke signal 　4.2 initialize product status memory 7. Initialize product status memory

Fig. 2 Use case description list

Any standard resources can be classified using 3-level hierarchy of resource group-device group-standard device: A plant is composed of 'resource group' such as mechanical parts, sensor, actuator, and MMI. A resource group consists of 'device group'. For example, actuator resource group is composed of solenoid, relay, stepping motor, AC servo motor, and cylinder device group and so on. Sensor resource group is composed of photo sensor, proximity switch, rotary encoder, limit switch, ultrasonic sensor, counter, timer, and push button device group and so on. Finally, Device group consists of 'standard devices' which can be acquired at the market.

To facilitate the modular design concept of modern AMS, the structure of AMS is modeled using an UML class diagram based on the generic AMS structure. By referencing this generic AMS structure, FA engineers can derive the structure model of specific AMS reflecting special customer requirements easily. Fig. 3 represents a static structure model of an example application prototype. Various kinds of device group class such as proximity switch and counter are inherited from resource group class such as sensor.

Since the real FA system is operated by the signal sending and receipt among manufacturing equipments such as PLC, sensors, and actuators, it is essential to describe the interactions of FA system components in detail for the robust design of device level control. This detail description of interactions is represented in the interaction model. UML provides the activity diagram, state diagram, sequence diagram, and communication diagram as a modeling tool for dynamic system behaviors. Among these diagrams, the activity diagram is most suitable for the

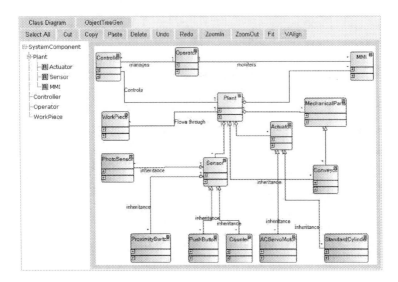

Fig. 3 Class diagram

control logic flow modeling because of following features: 1) it can describe the dynamic behaviors of plant with regard to device-level input/output events in sequential manner. 2) It can easily represent typical control logic flow routing types such as sequential, join, split, and iteration routing. The participating objects in the activity diagram are identified at the structure model.

In order to design and generate ladder logic, modification and extension of standard UML elements are required to reflect the specific features of ladder logic. First of all, it should be tested whether UML activity diagram is suitable for the description of control logic flow, especially for the ladder logic flow. The basic control flow at the ladder logic is sequence, split and join. Especially, three types of split and join control flow must be provided for ladder logic: OR-join, AND-join, AND-split. UML activity diagram can model basic control flows of ladder logic well.

Basically, ladder diagram is a combination of input contact, output coil and AND/OR/NOT logic. Since 'NOT' (normally closed) logic flow in the ladder logic cannot be represented directly in standard UML activity diagram, new two transition symbols for representing normally closed contact and negated coil are added as normal arcs with left-side vertical bar (called NOT-IN Transition) or with right-side vertical bar (called NOT-OUT transition) as depicted in Fig. 4. In the extended UML activity diagram, logic and time sequence flow from the top to the bottom of diagram.

Fig. 5 represents the interaction model for the identification and extraction of defective parts according to the height of products at the example application prototype. (refer the use case number 4 in Fig. 1 and use case description in Fig. 2)

base	extension		symbol
Transition	Normal Transition		⟶
	Not Transition	IN	⊢⟶
		OUT	⟶⊢▸

Fig. 4 Extensions of transitions in AD

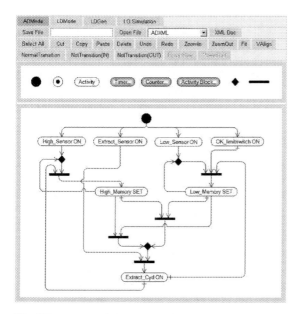

Fig. 5 Extended activity diagram for use case 4 in Fig.1

The control logic of Fig. 5 is as follows: 1) High_Memory:= (High_Sensor + High_Memory)*Extract_Cyl, 2) Low_Memory:= (Low_Sensor + Low_Memory) * !Extract_Cyl * !OK_LimitSwitch, 3) Extract_Cyl:= {(High_Memory * Low_Memory) + (!High_Memory * !Low_Memory)} * Extract_Sensor where "!" means negation (NOT), '*' means conjunction (AND), and '+' means disjunction (OR).

3.2 Phase II: Generation and Verification Phase for Ladder Logic

The following two steps are conducted during phase II: Firstly, ladder code is generated automatically using the result of interaction model at the design phase.

Secondly, generated ladder code is verified by input/output port-level simulation. In the second phase, a software tool developed by authors is also used.

For the automatic generation of ladder logic, the mapping scheme of an UML activity diagram to a ladder diagram is established. IEC61131-3 standard ladder diagram have 5 major elements: contact, coil, power flow, power rail and function block (FB). Contact is further classified to normally open and normally closed contact. Coil is further classified to normal and negated coil. Power flow is further classified to vertical and horizontal power flow. Power rail is further classified to left and right power rail.

Elements of an activity diagram are classified to two types: an activity type and a transition type. Activity type is decomposed into start/stop activity, normal activity, special activity such as counter and timer, and block activity (refer Fig. 5). Transition type is decomposed into normal transition, NOT-IN transition for normally closed contact, NOT-OUT transition for negated coil, and logic flow transition. Logic flow transition is further decomposed into OR-join, AND-join and AND-split.

Fig. 6 shows mapping scheme from an activity diagram to a ladder diagram. In order to store graphical activity diagrams and ladder diagrams in computer readable form, XML schema called AD-XML and LD-XML is devised for each diagram. In particular, LD-XML is an extension of PLCopen XML format [19].

After the activity diagram for specific control logic is stored in the form of AD-XML, AD-to-LD transformation procedure is conducted. Since basic ladder lung is a combination of input contact and output coil, an activity diagram is needed to be decomposed into several transformation units which having input(s) and output(s) corresponding to each ladder lung. This basic transformation unit is called

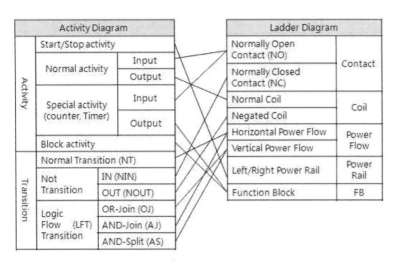

Fig. 6 Mapping scheme from AD to LD

IOU (Input Output Unit) which is a 1:1 exchangeable unit to ladder lung except start/stop activity. For example, the activity diagram depicted in Fig. 7, which describes of power control logic (use case number 1 in Fig. 1), has three IOUs. The control logic of Fig. 7 is as follows: Conveyor Motor: = (Power ON Button + Conveyor Motor) * !Power OFF Button.

The transformation procedure is as follows: 1) After the creation of an activity diagram graphically, store it in the form of AD-XML. 2) Decompose an activity diagram into several input/output units called IOUs, and store it in the form of two-dimensional table called IOU-Table. IOU-table has four columns named input activity, transition, output activity and IOU pattern type. Each row of IOU-Table becomes a part of ladder lung after the transformation process. 3) Determine the pattern type for each identified IOU. There are five IOU pattern types of activity diagram from the start/stop IOU type to the concatenation of logic flow transition IOU type. Generated IOU table for Fig. 7 is shown at Table 1. 4) Finally, generate ladder lungs using IOU table and node connection information of AD- XML.

Fig. 8 shows five IOU types and their corresponding LD patterns. IOU pattern type is classified to two types. One is simple type that is transformed to several

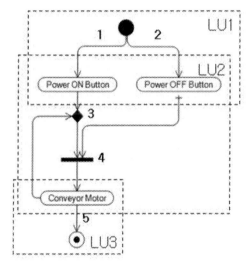

Fig. 7 IOU (Input/Output Unit) decomposition

Table 1 IOU table for Fig. 7 (use case 1-power control in Fig.1)

No.	Input Activity	Transition	Output Activity	Pattern
1	Start	1 : NT, 2 : NT	Power ON Button Power OFF Button	Type 1
2	Power ON Button Power OFF Button Conveyor Motor	3 : OJ, 4 : AJ	Conveyor Motor	Type 5
3	Conveyor Motor	5 : NT	Stop	Type 1

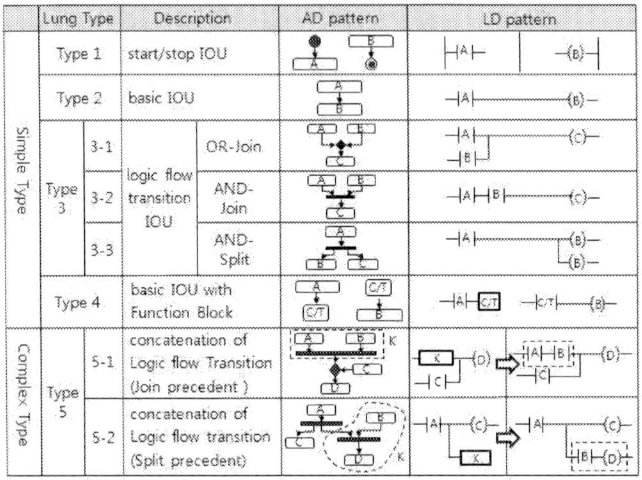

Fig. 8 Five IOU types

basic ladder elements. The other is complex type that is a combination of simple types. Simple type is further classified to four types according to their corresponding lung structure: Type-1 (start/stop IOU), Type-2 (basic IOU), Type-3 (logic flow transition IOU: OR-join, AND-join, AND-split), and Type-4 (basic IOU with function block).

Since complex type is combination of several consecutive logic flow transitions, it has most sophisticated structure among 5 IOU types. Complex type is further classified to two types: Type 5-1 (join precedent) and Type 5-2 (split-precedent). Classification criteria is whether 'join' logic flow transition is precedent to other logic flow transitions or 'split' transition is precedent.

In order to transform the type-5 IOU to ladder pattern, hierarchical multi-step procedure is needed. The type-5 IOU is grouped hierarchically into several macro blocks for simplifying the consecutive control logic. A macro block is considered as a kind of block activity. Later, one macro block is transformed to one of five LD patterns. In other words, in order to simplify inputs for succeeding logic flow transition, firstly a macro block is built including precedent or succeeding logic flow transition. Later, a macro block is substituted by one of 5 ladder lung pattern. Fig. 9 shows the example of transformation procedure for the join-precedent type 5-1.

Ladder code is automatically generated based on the IOU table and node connection information of AD-XML. The generated ladder code is stored in the form of LD-XML, and is graphically displayed by reading LD-XML file as depicted in Fig. 10.

After ladder code is generated, it is necessary to verify the generated code. The simulation for code verification is conducted by input/output port level. The ladder diagram in Fig. 10 is generated from the activity diagram of Fig. 5. As depicted in Fig. 10, one can simulate the logic flow by closing or opening an input contact of specific lung, and monitoring the result of output coils and input contacts of other lungs.

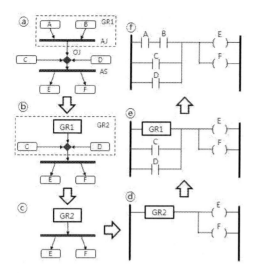

Fig. 9 Transformation procedure of join-precedent type 5-1

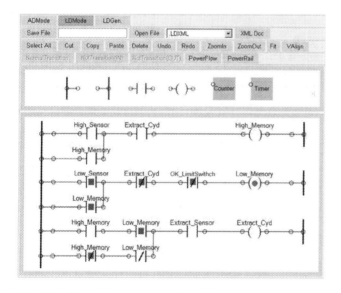

Fig. 10 Ladder code generation and port-level simulation

4 Conclusions

Currently, most enterprises do not adopt systematic development methodologies for ladder logic programming. As a result, ladder programs are error-prone and require time-consuming tasks to debug logic errors. In order to improve current PLC programming practices, this paper proposes an object-oriented ladder logic development framework in which ladder code is generated automatically and verified based on the o-o design results.

Proposed ladder logic development framework consists of two phases: First is the design phase. Second is the generation and verification phase. During the phase I, object-oriented design model is built, which consists of three sub-models: functional sub-model, structure sub-model and interaction sub-model. Two steps are conducted during the phase II. Firstly, ladder code is generated automatically using the interaction model of design phase. Secondly, generated ladder code is verified by input/output port simulation.

A framework in this paper facilitates the generation and modification of ladder code easily within a short time without considering complicated control behavior to deal with current trend of reconfigurable manufacturing systems. In addition, this framework serves as a helpful guide for systematic ladder code development life cycle.

As a future research, reverse transformation method from a ladder diagram to an activity diagram is needed for the accumulation of ladder logic design documents since design documents of control logic are not well prepared and stored in the shop floor.

Acknowledgments. This research was conducted as a part of project for the development of digital based real-time adaptive manufacturing system platform (Development of generation & verification technology for manufacturing control program) sponsored by Ministry of Knowledge Economy and VMS Solutions co., ltd.

References

[1] Zhou, M.C., Venkatesh, K.: Modeling, simulation and control of flexible manufacturing systems. World Scientific, Singapore (1999)

[2] Lee, S., Ang, M.A., Lee, J.: Automatic generation of logic control. Ford Motor Company (2006)

[3] Maffezzoni, C., Ferrarini, L., Carpanzano, E.: Object-oriented models for advanced automation engineering. Control Engineering Practice 7(8), 957–968 (1999)

[4] Calvo, I., Marcos, M., Orive, D., Sarachaga, I.: Using object-oriented technologies in factory automation. In: Proceedings of 2002 IECON Conference, Sevilla, Spain, pp. 2892–2897 (2002)

[5] Young, K.W., Piggin, R., Rachitrangsan, P.: An object-oriented approach to an agile manufacturing control system design. Int. J. of Advanced Manufacturing Technology 17(11), 850–859 (2001)

[6] Bruccoleri, M., Diega, S.N.: An object-oriented approach for flexible manufacturing control systems analysis and design using the unified modeling language. Int. J. of Flexible Manufacturing System 15(3), 195–216 (2003)

[7] Spath, D., Osmers, U.: Virtual reality- an approach to improve the generation of fault free software for programmable logic controllers. In: Proc. of IEEE International Conference on Engineering of Complex Computer Systems, Montreal, Canada, pp. 43–46 (1996)

[8] Baresi, L., Mauri, M., Monti, A., Pezze, M.: PLCTools: design, formal validation, and code generation for programmable controllers. In: Proc. of 2000 IEEE Conference on Systems, Man and Cybernetics, Nashville, USA (2000)

[9] Han, K.H., Park, J.W.: Development of object-oriented modeling tool for the design of Industrial control logic. In: Proc. of the 5th International Conference on Software Engineering Research, Management and Applications (SERA 2007), Busan, Korea, pp. 353–358 (2007)

[10] Peng, S.S., Zhou, M.C.: Ladder diagram and petri net based discrete event control design methods. IEEE trans. on Systems, Man and Cybernetics-Part C 34(4), 523–531 (2004)

[11] Lee, G.B., Zandong, H., Lee, J.S.: Automatic generation of ladder diagram with control Petri net. J. of Intelligent Manufacturing 15(2), 245–252 (2004)

[12] Frey, G., Minas, M.: Internet-based development of logic controllers using signal interpreted petri nets and IEC 61131. In: Proc. of the SCI 2001, Orlando, FL, USA, vol. 3, pp. 297–302 (2001)

[13] Taholakian, A., Hales, W.M.M.: PN <-> PLC: a methodology for designing, simulating and coding PLC based control systems using Petri nets. Int. J. of Production Research 35(6), 1743–1762 (1997)

[14] Jack, H.: Automating manufacturing systems with PLCs (2007).
     http://clay-more.engineer.gvsu.edu/~jackh/books.html

[15] Manesis, S., Akantziotis, K.: Automated synthesis of ladder automation circuits based on state diagrams. Advances in Engineering Software 36(4), 225–233 (2005)

[16] Sacha, K.: Automatic code generation for PLC controllers. In: Winther, R., Gran, B.A., Dahll, G. (eds.) SAFECOMP 2005. LNCS, vol. 3688, pp. 303–316. Springer, Heidelberg (2005)

[17] Liu, J., Darabi, H.: Ladder logic implementation of Ramadge-Wonham supervisory controller. In: Proc. of Sixth International Workshop on Discrete Event Systems, pp. 383–389 (2002)

[18] Hajarnavis, V., Young, K.: A comparison of sequential function charts and object modeling with PLC programming. In: Proc. of American Control Conference, pp. 2034–2039 (2005)

[19] PLC Open, XML Formats for IEC 61131-3 (2005). http://www.plcopen.org

An Evaluation of Technologies for the Pseudonymization of Medical Data

Thomas Neubauer and Mathias Kolb

Abstract. Privacy is one of the fundamental issues in health care today. Although, it is a fundamental right of every individual to demand privacy and a variety of laws were enacted that demand the protection of patients' privacy, approaches for protecting privacy often do not comply with legal requirements or basic security requirements. This paper highlights research directions currently pursued for privacy protection in e-health and evaluates common pseudonymization approaches against legal and technical criteria. Thereby, it supports decision makers in deciding on privacy systems and researchers in identifying the gaps of current approaches for privacy protection as a basis for further research.

1   Introduction

Privacy is not only a fundamental issues in health care today but a trade-off between the patient's demands for privacy as well as the society's need for improving efficiency and reducing costs of the health care system. Electronic health records (EHR) improve communication between health care providers and access to data and documentation, leading to better clinical and service quality. The EHR promises massive savings by digitizing diagnostic tests and images (cf. [4]). With informative and interconnected systems comes highly sensitive and personal information that is often available over the Internet and — what is more concerning — hardly protected. It is a fundamental right of every individual to demand privacy because the disclosure of sensitive data may cause serious problems. Insurance companies

Thomas Neubauer
Vienna University of Technology
e-mail: neubauer@ifs.tuwien.ac.at

Mathias Kolb
Secure Business Austria
e-mail: kolb@securityresearch.ac.at

R. Lee, G. Hu, H. Miao (Eds.): Computer and Information Science 2009, SCI 208, pp. 47–60.
springerlink.com   © Springer-Verlag Berlin Heidelberg 2009

or employers could use this information to deny health coverage or employment. Therefore, a variety of laws were enacted that demand the protection of privacy: In 2006 the United States Department of Health & Human Service Health issued the Health Insurance Portability and Accountability Act (HIPAA) which demands the protection of patients data that is shared from its original source of collection (cf. [22]). In the European Union the Directive 95/46/EC [5], Article 29 Working Party [6], and Article 8 [3] of the European Convention for the Protection of Human Rights and Fundamental Freedoms demand the protection of health data. In order to protect patients' privacy when using, transferring and storing medical records, a variety of privacy enhancing techniques (cf. [7]) are proposed. However, only a few of these approaches comply (i) with the current legal requirements and (ii) basic security requirements. Researchers agree that more needs to be done to protect consumers' privacy against the onslaught of rapidly advancing technologies that track, store, and share sensitive data. Bruce Schneier states it this way: "If we ignore the problem and leave it to the "market" we'll all find that we have almost no privacy left". This development has profound implications for our society and our freedoms; it influences the way we think and live. In this discussion privacy is often not the main concern, but surveillance, and the effects it has - both positive and negative - on human values, relationships, and daily practice.

This paper presents an evaluation of current privacy enhancing technologies that specifically aim at protecting medical data and, thus, can be used as a basis for EHR systems. In the scope of this paper we regard evaluation as the "systematic assessment of the operation and/or the outcomes of a program or policy, compared to a set of explicit or implicit standards, as a means of contributing to the improvement of the program or policy" (cf. [25]). We use a combination of a testing programs approach and objectives-based approach (cf. [11]). The objectives used for the evaluation are taken from the legal acts HIPAA and the EU Directive as well as from literature. This evaluation provides management decision makers such as chief privacy officers and chief security officers with a funded decision-making basis for the selection of privacy-enhancing technologies in the e-health area. As literature does not provide evaluations focusing on the comparison of PETs in e-health so far, this paper provides a major contribution to the research area of privacy.

2   Description of Selected Pseudonymization Approaches

Legal requirements and the explosion in privacy invasive technologies have encouraged the investment of substantial effort in the development of privacy enhancing technologies: *Anonymization* is the removal of the identifier from the medical data. It is realized by deleting the patient's identification data and leaving the anamnesis data for secondary use. Although this approach (and improvements) are often used in research projects due to its simplicity, it has the major drawback that patients cannot profit from the results made in the research project. *Depersonalization* comprises the removal of as much information as needed to conceal the patient's identity. *Encryption* assures patient's privacy by encrypting the anamnesis data with the patient's private key. Encrypted data cannot be used for research projects (secondary use) without explicit allowance by the patient who has to decrypt the data and, thus, unconceals his identity. *Pseudonymization* allows an association with a patient only under specified and controlled circumstances. Pseudonymization is a technique where identification data is replaced by a specifier (pseudonym) that can only be associated with the identification data by knowing a certain secret. This chapter describes major pseudonymization approaches in detail. Thereby, we differentiate between pseudonymization approaches that store patient's data (i) encrypted and (ii) unencrypted.

2.1 Peterson Approach

Peterson [13] claims to provide a system for making available personal medical information records to an individual without jeopardizing privacy. The main ideas behind the approach are (i) the encryption of patient's data, (ii) the universal access to medical records by any (also unauthorized) person while (iii) the patient is responsible for granting privacy. The user registers at the provider's website, receives a unique Global Key (GK) and server side key ($SSID$) generated by the provider and has to provide a unique Personal Encryption Key (PEK) as well as a password. GK, PEK and password are stored in the "Data Table". The user is demanded to enter a PEK until he provides a unique one. After registration the user may print the GK on an ID Card (on paper).

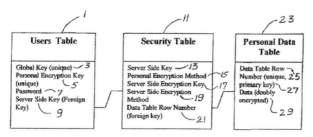

Fig. 1 Peterson Approach

This approach consists of three database tables: a "Security Table" that links the data in the "Data Table" (using attribute SSID) to the appropriate entries in the "User Table" (using attribute data table row number). Data is stored double encrypted in the database. If the user wants to retrieve data from the database, the user enters the GK or PEK which are sent to the server through a firewall and checked if the match any entry in the database. The user enters an arbitrarily key and gets immediate access to the records without authentication. The server looks up the $SSID$ and all corresponding data table row numbers needed for retrieving the (medical) data entries from the database. The records are decrypted and delivered to the user. At this time, internal staff or malware on the health care provider's computer could monitor the patients' key. If a person knows the global key GK or PEK or both, but does not have a password, she is able to view medical data sets. To be able to add, modify or delete medical datasets, the person has to provide an additional password. Peterson arguments, that this access levels protect the privacy of a patient, because the data does not contain any identifying information. So, for an attacker, it would be uninteresting to receive anonymous data. The approach of Peterson [13] provides a fall-back mechanism, if the patient has lost or used her GK. Therefore the patient has to login to the system with her PEK and password. Afterwards she requests a new GK, which could be printed on a new card. The new GK assures, that her medical data is protected against unauthorized access using the old key.

2.2 Pseudonymization of Information for Privacy in e-Health

Pseudonymization of Information for Privacy in e-Health (PIPE) [15, 16, 12] introduces a new architecture that provides the following contributions compared to other methodologies: (i) authorization of health care providers or relatives to access defined medical data on encryption level, (ii) secure fall-back mechanism, in case the security token is lost or worn out, (iii) data storage without the possibility of data profiling, and (iv) secondary use without establishing a link between the data and the owner. The client is a service, which provides an interface to legacy applications, manages requests to local smart card readers and creates a secure connection to the server. The server, also called Logic (L), handles requests from clients to the storage. The data in the storage is divided into two parts, the personal data and the pseudonymized medical data. As shown in figure 2, the link between personal data and pseudonymized medical data is protected through a hull-architecture. The hull-architecture contains a minimum of three security-layers: the authentication layer (outer hull), the user permission layer (inner hull) and the concealed data layer. To reach the next hull, there are one or more secrets, for example, symmetric or asymmetric keys or hidden relations, in every hull-layer. PIPE defines users with different roles comprising patient A, relative B, health care provider C or operator O. The patient is the owner of her data and has full control of her datasets. She is able to view her medical data, add and revoke health care providers and she may define relatives, who have the same rights as herself. Health care providers can be authorized to see and create subsets of anamnesis data by the patient. The operators are the administrators of the system.

- The authentication layer contains an asymmetric key pair, for example the patient outer public key K_A and outer private key K_A^{-1}. These keys are stored on a smart card and are protected with a pin code. The outer private key is used to decrypt the keys of the permission hull-layer.
- The permission layer contains an asymmetric key pair and a symmetric key, for example the patient inner public key $\hat{K}_A$, inner private key $\hat{K}_A^{-1}$ and symmetric key $\overline{K}_A$. The symmetric key is encrypted with the inner private key and is used to en-/decrypt pseudonyms in the concealed data layer. If a patient associates

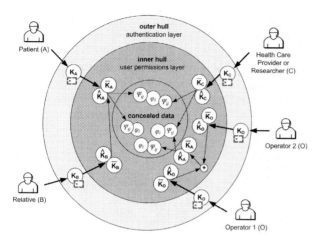

Fig. 2 PIPE: Layered model representing the authorization mechanism

a relative, her inner private key $\hat{K}_A^{-1}$ will be encrypted with the relative's inner public key $\hat{K}_B$. So, the relative is able to decrypt the patient's symmetric key $\overline{K}_A$ with her inner private key $\hat{K}_B^{-1}$, until the patient's inner private key $\hat{K}_A^{-1}$ is changed.

* The concealed data layer contains hidden relations, which are called pseudonyms. Each medical data set is associated with one or more pseudonyms ψ_{i_j}. As the patient is the owner of her medical data and the person with security clearance, she owns the so called root-pseudonym ψ_{i_0}. These pseudonyms are calculated with an algorithm, which on the user's symmetric key. Only instances, who are able to decrypt one of these pseudonyms ψ_{i_j}, can rebuild the link between the patient and her medical data.

To find the pseudonyms to rebuild the link to the medical data, the authors introduced keywords. Keywords are selected on creation time of the medical data or when another user is authorized. They are encrypted with the symmetric key of the root user and the authorized users. After the keywords are stored in the database, the user can select any of these keywords to find the pseudonym. As all data would be lost if a patient loses her smart card or the smart card is worn-out, PIPE implements a fall-back mechanism to replace the smart card. Therefore, "operators" O have been introduced, who share the patient's inner private key $\hat{K}_A^{-1}$. Therefore the patient's inner private key $\hat{K}_A^{-1}$ is divided into shared secrets by the use of Shamir's threshold scheme [18]. This scheme allows sharing keys between several operators. The key is shared with N_A ($N_A \subset N$) randomly assigned operators and to recover the key, N_k ($N_k \subseteq N_A$) operators are needed. The operators have no knowledge which keys they hold.

2.3   Electronic Health Card (eGK)

The electronic health card [8, 2] is an approach of the Fraunhofer Institute supported by the Federal Ministry of Health in Germany. EGK is designed as a service-oriented architecture (SOA) with some restrictions. One of these restrictions is, that the health card can only be accessed locally on the client side. Another restriction is, that services should use remote procedure calls for communication due to performance and availability issues. Therefore, the system architecture is divided into five layers.

1. The *presentation* layer defines interfaces to communicate with the user.
2. The *business logic* layer combines different services, which are processed automatically.
3. The *service* layer provides special functional uncoupled services.
4. The *application* layer primarily realizes the user right and data management.
5. The *infrastructure* layer contains all physical hardware and software management, for example, data storage, system management, virtual private networks, etc.

With this layered architecture, the system provides several service applications such as emergency data, electronic prescription, electronic medical report or a electronic health record system. The system includes a ticketing concept to realize some uncoupled action in combination with security mechanisms, to comply with the privacy policy: All data, which will be stored in the virtual filesystem is encrypted with a one-time symmetric key, called session key. This session key is encrypted with the public key of the patient. To decrypt the data, the patient has to decrypt the session key with his private key and finally the data is decrypted with this session key. A user is authenticated by using a challenge-response approach. Therefore the system generates a random number that is encrypted with the user's public key. Only the user is allowed to decrypt this random number with his private key, which is stored on her health card and can send it back to the eGK system. Furthermore, the ticketing concept manages the access rights to the system. A file or directory in this virtual filesystem has a default ticket-toolkit and any amount of private ticket-toolkits, called t-node). The user defines a private ticket-toolkit for every other user in the system. This private ticket-toolkit can have stronger or looser access policies than the default ticket-toolkit. The ticket-toolkit contains a ticket-building tool, a ticket-verifier, the access policy list and a encrypted link to the directory or file. Every user holds a root directory in the virtual filesystem. Any directory contains unencrypted links to the ticket-toolkits of their child nodes. This technique enables the system to perform a fast selection of sub nodes (select * from t-nodes where parentID = directoryID). To be able to find the root node of a specific user, the query service maps a unique identifier and returns a ticket-toolkit containing an encrypted link to the root node. If there is no private ticket-toolkit available for the user, who performed the request, the system returns a default ticket-toolkit, which is based on a challenge. If the user is able to solve this challenge, she will get the access rights, which have been defined in the default access policy. Both, the hybrid encryption

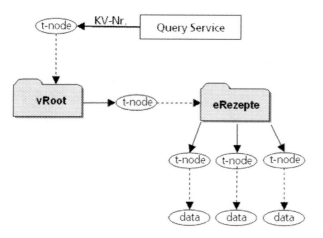

Fig. 3 eGK: Virtual filesystem [8]

and the challenge response technique are based on the asymmetric key pair, which is stored on the patients' health card. Neither the operating company nor any public administration organization could recover the data, if the patient lost the smart card or the card is worn out. To overcome this problem, the eGK architecture optionally allows to store a second private ticket-toolkit for every entry. This private ticket-toolkit uses an asymmetric key pair, which is stored on an emergency card. The architecture does not specify this emergency card, but recommends to use the card of a family member or a notary. In case the card has been lost, the patient requests a new health card. Therefore, the emergency card is used to decrypt the session keys of the second ticket-toolkit and finally the session keys are encrypted with the keys of the new health card. After this process, the system does not accept the old health and emergency cards anymore.

2.4 Thielscher Approach

Thielscher [21] proposed a electronic health record system, which uses decentralized keys stored on smart cards. The medical data is split into identification data and the anamnesis data and stored in two different databases. The key stored on the patient's smart card is used to link the patient identity to her datasets. Therefore, this key generates a unique data identification code (DIC), which is also stored in the database. Such a DIC does not contain any information to identify an individual. Data identification codes are shared between the patient and health care providers to authorize them to access the medical data set. For more security the authorization is limited to a certain time period. After this period any access attempt is invalid. The keys to calculate the data identification code (DIC) are stored on smart cards. In case these smart cards are lost, a fall-back mechanism is provided by Thielscher.

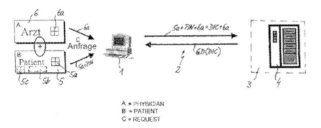

A = PHYSICIAN
B = PATIENT
C = REQUEST

Fig. 4 Thielscher: Architecture [21]

Every pseudonym hold by a patient is stored in a list, which is stored at an off-line computer. In case the smart card is lost or destroyed, this list could be used to re-link the data to the patient.

2.5 *Approach of Slamanig and Stingl*

The approach of Slamanig and Stingl [20, 19] stores the data in a centralized database and uses smart cards for authentication. The system keeps the pseudonyms of a user secret. Each pseudonym realizes a sub-identity of the user and is encrypted with a public key. In case the user wants to view the datasets of one of his sub-identities, she has to login into the system with her general pin code and she has to enter the pin code of the sub-identity to activate the private key on the smart card. Furthermore a public pseudonym of each user is available, which is used for authorization purposes. The system is divided into two repositories, the user repository and document repository. The link between these repositories is done by holding a 5-tuple dataset $(U_S, U_R, U_C, U_P, D_i)$, which contains the sender U_S, the receiver U_R, the creator U_C, the concerning user U_P (e.g., the patient) and the document D_i. To ensure, that on creation time no linkage between the concerned user is possible, all elements in the tuple, are encrypted with the public key of the receiver, except of the receiver element U_R. Until the receiver has not logged into the system, the receiver element U_R will be the public pseudonym. On the next login of the receiver, the system will replace the receiver element U_R with a secret pseudonym of the user and re-encrypts the other elements of the tuple. As shown in figure 5, this tuple dataset can also be used for exchanging documents between users. There are six possible variations for exchanging the message:

1. $(\_,\_,U_C,U_P,D_i)$: Creator and concerning user are known
2. $(\_,\_,U_C,U_P*,D_i)$: Creator is known and concerning user is pseudonymized
3. $(\_,\_,U_C,\_,D_i)$: Creator is known and concerning user is anonymized
4. $(\_,\_,\_,U_P,D_i)$: Concerning user is known
5. $(\_,\_,\_,U_P*,D_i)$: Concerning user is pseudonymized
6. $(\_,\_,\_,\_,D_i)$: fully anonymized

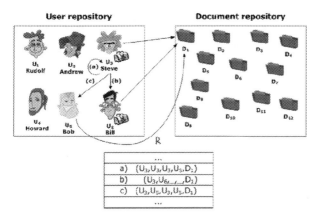

Fig. 5 Slamanig and Stingl: Repositories and Shares [19]

As fall-back mechanism, the authors mention, that a distributed key backup to N users using a (t, N)-threshold secret sharing scheme could be implemented.

2.6 Pommerening Approaches

Pommerening [14] proposes different approaches for secondary use of medical data. The first approach is based on data from overlapping sources for one-time secondary use. To connect the data a unique identifier (PID) is introduced. A pseudonymization service encrypts the PID with a hash algorithm and the medical data is encrypted with the public key of the secondary user (cf. Figure 6). The secondary user can decrypt the medical data and merge the data of a person, but cannot identify it. The second approach is also based on one-time secondary use, but with the possibility to re-identify the patient. Therefore, Pommerening extends the first approach with a PID service, which stores a reference list containing the patient's identity and the associated PIDs. In case the patient should be notified, the pseudonymization service decrypts the pseudonym and sends the request to the PID service, which allows to notify the data source owner. The third approach fits the need of a research network with many secondary users and it also supports long-term observation of patients. The research results can be send to the patient or her responsible health care

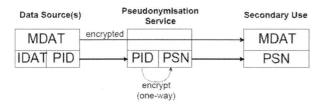

Fig. 6 Pommerening: Data Flow for One-Time Secondary Use [14]

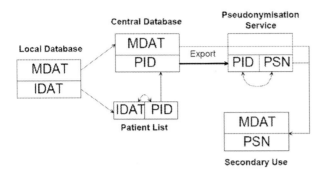

Fig. 7 Pommerening: Data Flow for multiple Secondary Uses [14]

provider. Therefore a physician export his local database to the central researcher database (cf. Figure 7). The identification data is replaced with a PID using the PID service. For each secondary use the data is exported through the pseudonymization service. The PID is encrypted by the pseudonymization service with a project specific key to ensure that different projects get different pseudonyms.

3 Evaluation

Pseudonymization approaches (e.g., used for securing electronic health record systems) have to adhere certain requirements to accord with privacy laws in the European Union and United States. The following set of requirements has been extracted from legal acts (cf. [5, 9, 10, 24, 23]).

- *User authentication*: The system has to provide adequate mechanisms for user authentication. This could be done, for example with smart cards or finger print.
- *Data ownership*: The owner of the medical data has to be the patient. The patient should be able to define who is authorized to access and create her medical records.
- *Limited access*: The system must ensure that medical data is only provided to authenticated and authorized persons.
- *Protection against unauthorized and authorized access*: The medical records of an individual have to be protected against unauthorized access. This includes system administrators who should not be able to access these medical records, for example, through compromising the database.
- *Notice about uses of patients data*: The patient should be informed about any access to her medical records.
- *Access and copy own data*: The system has to provide mechanisms to access and copy the patients own data.

Additionally, pseudonymization approaches have to implement certain technical requirements in order to provide an appropriate level of security (cf. [16, 1]).

- *Fallback mechanism*: The system should provide mechanisms to backup and restore the security token used for pseudonymization. Therefore, the system has to guarantee, that the security token could only be restored under the compliance of a four-eye-principle.
- *Unobservability*: means, that pseudonymized medical data can not be observed and linked to a specific individual in the system.
- *Secondary use*: The system should provide a mechanism to (i) export pseu-do-nymi-zed data for secondary use and (ii) notify the owner of the exported data, if new medicaments or treatment methods are available.
- *Emergency access*: In case of an emergency, the rescue service or emergency physician should have access to an emergency dataset, in which important information is saved. For example, blood group, informations about medication, allergic reactions to specific medicaments, etc.
- *Insider abuse*: Medical personal may abuse their access rights for own purposes. For example they want to know how their family members or celebrities are being treated. Insider do not only abuse their privileges for own purposes. They may release informations to outsiders for spite, revenge or profit [17].
- *Physical compromise of the database*: The system should grant that an attacker who gets physical access to the systems database cannot identify the owners of the stored data records.
- *Modification of the database*: If an attacker breaks into the system, the system must detect modifications and inform the system administrator about this attack.

Table 1 applies the legal and technical criteria defined above to the selected pseudo-nymi-zation approaches. Characteristics that are accurate with the law or fully implemented are denoted with *x*, whereas characteristics that are not accurate with the law or not implemented are denoted with — and *o* indicates properties that are partially implemented.

Table 1 Evaluation of pseudonymization approaches

Legal Requirements	DPA	HIPAA	PIPE	eGK	Po	Pe	Th	St
User authentication	x	x	x	x	-	o	x	x
Data ownership	x	x	x	x	-	-	x	x
Limited access	x	x	x	x	o	-	x	x
Protection against unauthorized and authorized access	x	x	x	x	o	-	o	x
Notice about uses of patients data	x	x	x	x	-	-	-	-
Access and copy own data	x	x	x	x	o	x	x	x
Technical Requirements	DPA	HIPAA	PIPE	eGK	Po	Pe	Th	St
Fallback mechanism	-	-	x	x	-	o	x	x
Unobservability	x	x	x	x	x	-	x	x
Secondary use	-	x	x	o	x	-	-	x
Emergency access	-	-	x	x	-	x	-	-
Insider abuse	-	-	x	x	x	x	x	-
Database modifications	-	-	x	x	x	x	x	x

The approaches of Pommerening and Peterson only pseudonymize data on export. The approaches of Pommerening have the drawback that the generated pseudonyms from the PID service are stored in a reference patient list, to be able to re-build the link to the patient. To enhance the security, this list will be stored at a third party institution, but this measure does not prevent an abuse of the list through an insider of the third party institution. An attacker could bribe an insider of the third party institution to get access to the patient list or the identifying data of some pseudonyms. The Peterson approach has some major security issues. As all keys needed for decrypting the medical data are stored in the database, an attacker getting access to the database can decrypt all information and the attacker may change data stored in the database unnoticedly, as the password and the keys are stored in the database. The PEK is selected by the user but must be unique in the system. This behavior does not only open a security leak because the user trying to chose a key is informed about the keys that already existing in the system. An attacker could use the keys reported as existing for immediate access to the medical data associated with this key. Moreover, this behavior is impractical and inefficient in practice as the user might have to select dozens of keys before he enters a valid one. Although the data is doubly encrypted an attacker getting access to the database gets access to all data stored on the server because the keys needed for decrypting the data are (i) also stored in the same database and (ii) the relation between the tables (thus between the identification data and the medical data) are stored in clear text.

Thielscher's approach comes with the shortcoming, that a centralized pseudonym is centrally stored in the patient mapping list for recovery purposes. To prevent attacks to this list, Thielscher keeps this list off-line, but this mechanism cannot prevent insider abuse or social engineering attacks. Furthermore, it does not provide protection if the attacker gets physical access to the computer. Another drawback of the system is the emergency call center, that . This call center can abuse their access privileges to get access to medical data of any patient. The drawback of the approach of Slamanig and Stingl is that an attacker may authorize other users, send faked medical documents or disclose medical data. This attack is possible, because the authors use a weak mechanism for authorization and disclosure. For example, the requirements to send a faked medical document are, (i) access to the database, (ii) the public pseudonym U_P of the user, which the attacker wants to harm. (iii) any public pseudonym to fake the sender U_S and creator U_C, (iv) the public pseudonym and the public key K_R of the receiver U_R, for example the employer, and (v) a harmful document D_i. After the attacker has all the required information, she inserts a new tuple into the authorization table. After the next login of the receiver, the system replaces the public pseudonym of the user with a private pseudonym of the receiver. In contrast, PIPE and eGK store the data pseudonymized in the database. Attackers who get access to the database or system administrators cannot link the data to individuals. Both approaches provide a high level of security because even if the attacker breaks into the database, she would not be able to link and read the stored data. The only way to link the data to an individual is to fake an official photo ID in order to get a new smart card that allows to access the system. Another method to link data to an individual is by doing a data mining or data profiling attack. This attack could be done by identified keywords, which are not encrypted in the eGK approach or by identifiable words in the anamnesis data in the PIPE approach.

4 Conclusions

Health care requires the sharing of patient related data in order to provide efficient patients' treatment. As highly sensitive and personal information is stored and shared within interconnected systems, there is increasing political, legal and social pressure to guarantee patients' privacy. Although, legislation demands the protection of patients' privacy, most approaches that lay claim to protect patient's privacy fail in fulfilling legal and technical requirements. This paper gave an overview of research directions that are currently pursued for privacy protection in e-health and identified pseudonymization as the most promising approach. By evaluating common pseudonymization approaches against legal and technical criteria taken from legal acts and literature, this paper answered the question which approaches fulfill the current legal requirements regarding the protection of medical data. From the six candidates that were evaluated, only two can be seriously considered for use in practice. The result show that more contemporary approaches fulfill more of the legal requirements of the European Union and the United States. Whereas the eGK approach encrypts patients' data, PIPE leaves the decision of encrypting patients' data up to the user. Therefore, PIPE turns out to be the more appropriate option if secondary use is demanded. The results of the evaluation can support decision makers (such as chief security officers) especially in health care in their decision process when it comes to the selection of a system for protecting patients' data according to legal requirements posed by HIPAA or the EU Directives. Furthermore, the results may assist researchers in identifying the gaps of current approaches for privacy protection as a basis for further research.

Acknowledgements. This work was supported by grants of the Austrian Government's FIT-IT Research Initiative on Trust in IT Systems under the contract 816158 and was performed at the Research Center Secure Business Austria funded by the Federal Ministry of Economics and Labor of the Republic of Austria (BMWA) and the City of Vienna.

References

1. Barrows, R.C., Clayton, P.D.: Privacy, confidentiality, and electronic medical records. Journal of the American Medical Informatics Association 13, 139–148 (1996)
2. Caumanns, J.: Der Patient bleibt Herr seiner Daten. Informatik-Spektrum, pp. 321–331 (2006)
3. Council of Europe: European Convention on Human Rights. Martinus Nijhoff Publishers (1987)
4. Ernst, F.R., Grizzle, A.J.: Drug-related morbidity and mortality: Updating the cost-of-illness model. Tech. rep., University of Arizona (2001)

5. European Union: Directive 95/46/EC of the European Parliament and of the council of 24 October 1995 on the protection of individuals with regard to the processing of personal data and on the free movement of such data. Official Journal of the European Communities L 281, 31–50 (1995)
6. European Union, Article 29 Working Party: Working document on the processing of personal data relating to health in electronic health records (EHR) (February 2007)
7. Fischer-Hübner, S.: IT-Security and Privacy: Design and Use of Privacy-Enhancing Security. Springer, Heidelberg (2001)
8. Fraunhofer Institut: Spezifikation der Lösungsarchitektur zur Umsetzung der Anwendungen der elektronischen Gesundheitskarte (2005)
9. Hinde, S.: Privacy legislation: A comparison of the US and european approaches. Computers and Security 22(5), 378–387 (2003)
10. Hornung, G., Götz, C.F.J., Goldschmidt, A.J.W.: Die küenftige Telematik-Rahmenarchitektur im Gesundheitswesen. Wirtschaftsinformatik 47, 171–179 (2005)
11. House, E.R.: Assumptions underlying evaluation models. Educational Researcher 7(3), 4–12 (1978)
12. Neubauer, T., Riedl, B.: Improving patients privacy with pseudonymization. In: Proceedings of the International Congress of the European Federation for Medical Informatics (2008)
13. Peterson, R.L.: Encryption system for allowing immediate universal access to medical records while maintaining complete patient control over privacy. US Patent Application Publication, No.: US 2003/0074564 A1 (2003)
14. Pommerening, K., Reng, M.: Secondary use of the Electronic Health Record via pseudonymisation. In: Medical And Care Compunetics 1, pp. 441–446. IOS Press, Amsterdam (2004)
15. Riedl, B., Neubauer, T., Boehm, O.: Patent: Datenverarbeitungssystem zur Verarbeitung von Objektdaten. Austrian-Patent, No. A 503 291 B1, 2007 (2006)
16. Riedl, B., Neubauer, T., Goluch, G., Boehm, O., Reinauer, G., Krumboeck, A.: A secure architecture for the pseudonymization of medical data. In: Proceedings of the Second International Conference on Availability, Reliability and Security, pp. 318–324 (2007)
17. Rindfleisch, T.C.: Privacy, information technology, and health care. Commun. ACM 40(8), 92–100 (1997)
18. Shamir, A.: How to share a secret. Commun. ACM 22(11), 612–613 (1979)
19. Slamanig, D., Stingl, C.: Privacy aspects of e-health. In: Proceedings of the Third International Conference on Availability, Reliability and Security, pp. 1226–1233 (2008)
20. Stingl, C., Slamanig, D.: Berechtigungskonzept für ein e-health-portal. In: Schreier, G., Hayn, D., Ammenwerth, E. (eds.) eHealth 2007 - Medical Informatics meets eHealth, vol. 227, pp. 135–140. Österreichische Computer Gesellschaft (2007)
21. Thielscher, C., Gottfried, M., Umbreit, S., Boegner, F., Haack, J., Schroeders, N.: Patent: Data processing system for patient data. Int. Patent, WO 03/034294 A2 (2005)
22. United States Department of Health & Human Service: HIPAA administrative simplification: Enforcement; final rule. Federal Register / Rules and Regulations 71(32) (2006)
23. U.S. Congress: Health Insurance Portability and Accountability Act of 1996. In: 104th Congress (1996)
24. U.S. Department of Health & Human Services Office for Civil Rights: Summary of the HIPAA Privacy Rule (2003)
25. Weiss, C.H.: Evaluation: Methods for studying programs and policies, 2nd edn. Prentice-Hall, Englewood Cliffs (1998)

Abnormal BGP Routing Dynamics Detection by Active Learning Using Bagging on Neural Networks

Qitao Wu and Qimin Feng

Abstract. Because of BGP's critical importance as the de-facto Internet inter-domain routing protocol, accurate and quick detection of abnormal BGP routing dynamics is of fundamental importance to internet security where the classes are imbalanced. Alougth there exist many active learning methods, few of them were extended to solve BGP problems. In this paper, avtive learning based on the under-sampling and asymmetric bagging is proposed to classify BGP routing dynamics and detect abnormal data. Under-sampling is used in training neural networks and asymmetric bagging is used to improve the accuracy of the algorithm. Our BGP data is the RIPE archive, which is a huge archive of BGP updates and routing tables that are continuously collected by RIPE monitors around the world. The experimental results suggest that the accuracy of the detection of abnormal BGP routing dynamics is satisfying and applicable to BGP products. We emphasize that this is a promising direction to improve security, availability, reliability and performance of internet security by detecting and preventing abnormal BGP routing dynamics traffic.

Keywords: Oversmaple, Undersample, Neural Networks, Abnormal BGP Events.

1 Introduction

The Border Gateway Protocol (BGP) is the de-facto standard inter-domain routing protocol on the Internet. As the size, complexity, and connectivity of the Internet increase, the analysis of operational BGP dynamics becomes more and more important. In particular, abnormal BGP routing dynamics, including attacks [1], misconfigurations [2], and large-scale power failures [3], have always been a major

Qitao Wu
College of Environmental Science and Engineering ,Ocean University Of China

International Business School, Qingdao University , No. 308 Ningxia Road,
Qingdao, China, 266071
e-mail: whichtall@sohu.com

Qimin Feng
College of Environmental Science and Engineering , Ocean University Of China,
Qingdao, China 266100

R. Lee, G. Hu, H. Miao (Eds.): Computer and Information Science 2009, SCI 208, pp. 61–72.
springerlink.com   © Springer-Verlag Berlin Heidelberg 2009

concern of the Internet engineering community. And the reason is that they can not only cause high bandwidth and processing overhead on routers, but may also lead to packet forwarding failures, including packet delay, jitter, drop, reordering, duplication, or other difficulties in reaching destinations. Therefore at the time when the Internet is indispensable to modern communications and the economy, it is critical to detect abnormal BGP routing dynamics. Concretely, given a large set of BGP update messages, we can accurately categorize them as "normal" or "abnormal". And with this categorization, we can detect and avoid abnormal ones.

Two criteria jointly define an anomaly. One criterion is related to BGP performance. For example, a router's slow convergence to reach a stable view of the Internet's available routes [4]–[6] as it announces many invalid routes to downstream BGP routers, belongs to this type. The other criterion refers to a statistical anomaly (also called "relative anomaly")—significant deviations of current routing behavior from expected routing behavior.

In the past few years, several well-known BGP events have been reported. For example, in April 2001, a misconfiguration caused AS 3561 to propagate more than 5,000 invalid route announcements from one of its customers, causing connectivity problems throughout the entire Internet [7]. In January 2003, the Slammer worm caused a surge of BGP updates [8]. In August 2003, the East Coast electricity blackout affected 3,175 networks and many BGP routers were shut down [9]. Also, smaller scale anomalies, although probably unnoticeable, can happen even more frequently, further raising concerns for such events on a daily basis.

Because of BGP's critical importance, accurate and quick detection of abnormal BGP routing dynamics is of fundamental importance to internet security. In this paper, we study empirically the effect of active learning using bagging on neural networks to classify BGP routing dynamics and detect abnormal ones. Because BGP classes are imbalanced, under-sampling [10] is used in training neural networks. Compared with over-sampling, under-sampling method reduces the time complexity since it samples a small number of patterns from the majority class. We propose a bagging approach [11] [15]–[17] to overcome the disadvantage of the under-sampling method. We compared the results with other 4 methods in our experiment, and the new method achieves a more satisfying accuracy.

We consider that this paper illustrates an application of a powerful technique to a field, which is different from those it has been previously applied to. At the meantime, we are anticipating further work exploiting the techniques we describe as well as encouraging more research to bring new techniques to abnormal BGP routing dynamics detection.

The rest of this paper is organized as follows. Section 2 introduces BGP on the Internet. Section 3 presents the learning methods studied in this paper. Section 4 reports on the empirical study. Section 5 concludes.

2 BGP on the Internet

Routing information provides details about the path along which data must be sent from a source to a destination. This information is exchanged between the networks that form the Internet, as well as internally within these smaller networks.

The smaller networks are able to manage their own internal routing using a variety of routing protocols, and thus it is necessary to use a common protocol between them. And the Border Gateway Protocol (BGP) becomes the standard protocol for exchanging routing information on the Internet [1]. BGP routing transmits data between Autonomous Systems (AS), which are made up of a network or group of networks that have a common administration and routing policies [11]. At a broader level, BGP is able to transmit routing information between many AS, and provide a high level of control and flexibility for inter-domain routing while it can still successfully manage policy and performance criteria. The use of AS allows BGP devices to store only the routes between neighboring AS, rather than the entire set of paths between all hosts on the Internet.

BGP routers exchange lists of advertised routing paths, informing neighboring routers about routes that are currently used by a router for transmitting data. Each path exchanged between the routers includes the full list of hops between the routers, which eliminates the possibility of loops. BGP routers exchange their full set of routing paths on initiation of a session, after which incremental updates are sent to inform of any changes. The protocol does not rely on any scheduled updates.

The BGP protocol runs on top of the TCP protocol [16], which already provides fragmentation, retransmission, acknowledgement and sequencing. TCP is already supported by BGP routers since it is the most used transmission protocol for delivering data on the Internet.

There are four types of messages supported by the BGP-4 protocol [12] [13]:

1) The Open message is the first command sent between BGP devices and is used to establish a BGP session. Once the session has been established, the devices are able to send other BGP messages.
2) The Update message provides the latest information for the advertised routes used by the BGP device. It is used to simultaneously advertise new routes (known as announcements), and to inform of routes that have become unavailable (known as withdrawals). Advertised routes are those routes that the device is actively using for transmitting data, and does not include routing paths that it is aware of but not using.
3) Notification messages are sent when an error forces the closure of the BGP session.
4) The Keep-Alive message is transmitted between BGP devices to provide notification that the devices are still active. This is sent at timed intervals to ensure that the BGP session doesn't expire.

Two specific features of BGP are of specific interest to this project – the Minimum Route Advertisement Interval (MRAI) property and BGP route flap damping. MRAI is part of the original BGP protocol, and is implemented on the sender side to limit the time between successive BGP Update messages that provide information about a specific AS [12]. This is designed to limit route exploration [14], and deal with route instability in a time scale of tens of seconds [15]).

BGP route flap damping was introduced in the mid 1990s to reduce the effect of localized edge instability propagating outwards through the Internet [15]. It caters for instability on a longer time scale than MRAI, by reducing the effect of route flapping, which occurs when route availability constantly changes. These changes cause a router to continuously send frequent updates to neighboring routers, which potentially indicates localized instability. The research performed by Lad and his colleagues indicates that BGP route flap damping has not been fully deployed across the Internet, however if it is fully deployed the convergence time to recover from instability would be extended by route flap damping [8].

3 Learning Methods

3.1 Under-Sampling

Under-sampling method balances the ratio of the classes by sampling a small number of patterns from the majority classes. It randomly removes data of the non-target class until it is about the same size of the target class data. To ensure that non-target class data retain their ratio in the original data, stratified sampling may be performed for this purpose [16].

Compared with over-sampling, under-sampling method reduces the time complexity since it samples a small number of patterns from the majority class. However under-sampling will throw away many useful data. If we employ multiple training sets, majority class patterns would have better chances to be included in the training sets. In this paper, we use bagging to overcome the disadvantage of under-sampling [17] [18].

3.2 Asymmetric Bagging

Bagging is one of the traditional ensemble methods, which uses bootstrap samples to produce the diversity of individuals and uses major voting to obtain the final decision results for classification problems [19]. It works well for "unstable" learning algorithms.

Input: Training data set $S_r\left(x^1, x^2, \cdots, x^D, C\right)$;

    Number of individuals T

Procedure:

        For k=1: T

           1.   Generate a training subset S_{rk}^- from negative training set S_r^- by using Bootstrap sampling algorithm, the size of S_{rk}^- is the same with that of S_r^+.

           2.   Train the individual model N_k the training subset $S_{rk}^- \cup S_r^+$ by using neural networks.

Fig. 1 The architecture of the asymmetric bagging

Bagging helps improve the stableness of a single learning machine, however imbalance also reduces its generalization performance; therefore, we propose to employ asymmetric bagging [20] to handle the imbalanced problem, which only executes the bootstrapping on the negative examples since there are far more negative examples than positive ones. Tao et al. [21] applied asymmetric bagging to another unbalanced problem of relevance feedback in image retrieval and obtained satisfactory results. The pseudo code of the asymmetric bagging is shown in Fig. 1.

3.3 Active Learning Using Bagging on Neural Networks

Active learning is very effective in some real applications. In this paper, an active learning using bagging on neural networks approach is proposed. The flow is shown in Fig. 2. Given a pool $U = \{x_1, \ldots, x_n\}$ of unlabeled instances where each x_i is a vector in some space X. Instances are assumed to be distributed according to some unknown fixed distribution $P(x)$. Each instance x_i has a label $y_i \in Y$ (where in our case $Y = \{\pm 1\}$) distributed according to some conditional distribution $P(y \mid x)$. At the beginning, a few of the labels is known to the learner. At each stage of the active learning, let L be the set of labeled instances already known to the learner. An active learner consists of a classifier learning algorithm A (in Fig.2 is neural network), and a querying function Q (here we use bagging neural networks), which is a mapping $Q : L \times U \rightarrow U$. The querying function determines one (or some) unlabeled instance in U to be labeled by the teacher. On each trial $t = 1, 2, \ldots$, the active learner first applies Q to choose one (or some) unlabeled instance x from U. The label y of x is then revealed, the pair (x, y) is added to L and x is removed from U. Then the learner applies A to induce a new classifier C_t using L as a training set and a new trial begins, etc. Thus, the active learner generates a sequence of classifiers $C_1, C_2, \ldots$ [22].

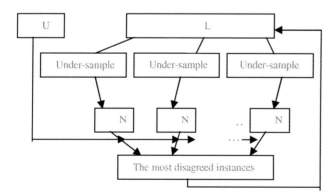

Fig. 2 The flow of the combined approach

Fig. 3 shows the pseudo code of the combined approach. In the combined method, under-sampling is used to the imbalanced training set to train N neural network classifiers i.e. the requested bagging number N (statement 3, 4). Under-sampling is very effective in handling imbalanced problems [17] [18]. The ensemble of the neural networks predicts the labels for the unlabeled instances (statement 5).Each neural network classifier predicts the labels for the instances in U respectively. The most uncertain instances are picked out using majority voting and then labeled—this is the querying stage of the active learning(statement 6). The voted instances are labeled by the teacher (statement 7). After they are labeled, they will be moved from the unlabeled set to the training set (statement 8). The new training set is used in the next phase to detect abnormal BGP data. The process repeats until all the instances are classified [22] [23].

Input: Unlabeled data set U, Querying number M, Bagging number N
Output: Labeled training set L
Procedure:
   1、 Select an initial subset L as the training set.
   2、 For each trial phase
   3、 Under-sample the training set N times to get N training subsets.
   4、 Generate N neural network classification models using the N different training subsets.
   5、 The ensemble of the N classifications predicts the labels for the instances in the unlabeled data set U.
   6、 The ensemble of the classifications voting for M instances by the majority voting rule.
   7、 Label the M instances by the expert.
   8、 Move the M instances from the unlabeled set U to the training set L.
   9、 Repeat until all instances in U are labeled.

Fig. 3 Shows the pseudo code of the combined approach

The important parameters that can be varied in the method in Fig. 3 are the number of trial phases, the number of instances to be selected for labeling in each phase (querying number M) and the number of bagging of each phase(classifiers number N). The bigger the number of neural network classifiers N is, the better the experimental results are, however the more time is needed to run the algorithm.

In this method, under-sampling balances the training sets for neural networks. Active learning using bagging improves the accuracy effectively. The previous method combined perfectly.

4 Empirical Study

4.1 Configuration

Back propagation (BP) neural network [24] was used in the empirical study. Each network has one hidden layer containing ten units, and is trained to 200 epochs. Note that since the relative instead of absolute performance of the investigated methods are concerned, the architecture and training process of the neural networks have not been finely tuned. After under-sampling the training data, neural networks were achieved. Here we set 60% of the data as the training set, 10% as the testing set, and the left as the unlabeled set. Each trial we trained 5 classifiers and were repeated 10 times. At last, we got the average results.

4.2 Data Set

The BGP data archive that we used to prepare database tables was the RIPE archive. It is a huge archive of BGP updates and routing tables that are continuously collected by RIPE monitors around the world. We used the BGP update data from six randomly selected peers. (We did not use the Oregon Route Views archive, as it does not contain BGP updates for the CodeRed and Nimda worm periods that we would like to study).

BGP data was cleaned and processed according to [27]. To obtain worm events that can affect BGP, we collected BGP data from an eight-hour period immediately after each worm started to propagate. We also prepared data from ten randomly chosen "normal" days (dispersed within a two-year period from July 2001 to August 2003), in which no major events were known to have happened.

To test the rules obtained from the training, we further prepared data from the half day when the Slammer worm was active (January 25, 2003). To provide a basis of comparison for testing the rules, data from another set of four randomly chosen "normal" days were also collected. We chose 2001.12.10 2001.12.12 20011003 2002.8.10 as our BGP normal data set.

To eliminate the effect of boundary, we deleted the first row and the last row of data set. In total, the data used contained 5733 rows of normal data and 504 rows of (CodeRed and Nimda) worm data. And AS 4777, AS 513 and AS 13129 were used as the analysis examples in the following section. Each worm and normal period was further divided into 1-minute bins, with each bin represented by exactly one data row. As a result, for each bin, a new row was added to the corresponding database table used for training or testing. When used for training, a new row was also labeled as either "abnormal" or "normal." As we did not calculate the information gain of parameters for attribute relevance analysis, we used all 35 parameters to classify.

Numbers of BGP updates (parameters 1–3) are certainly indicators of the Internet routing dynamics. As one BGP update message could speak for multiple prefixes, we also considered the number of updated prefixes (parameters 4–6). Labovitz et al defined a method to classify BGP updates into different types; with our further refinement, this led to nine parameters (parameters 7–15), each counting a different type of BGP updates. Furthermore, parameters that may capture

Table 1 PARAMETER LIST

ID	Parameter	Definition
1	Announce	# of BGP announcements
2	Withdrawal	# of BGP withdrawals
3	Update	# of BGP updates (=Announce + Withdrawal)
4	AnnouPrefix	# of announced prefixes
5	WithdwPrefix	# of withdrawn prefixes
6	UpdatedPrefix	# of updated prefixes (=AnnouPrefix + WithdwPrefix)
7	WWDup	# of duplicate withdrawals
8	AADupType1	# of duplicate announcements (all fields are the same)
9	AADupType2	# of duplicate announcements (only AS-PATH and NEXT-HOP fields are the same)
10	AADiff	# of new-path announcements (thus implicit withdrawals)
11	WADupType1	# of re-announcements after withdrawing the same path (all fields are the same)
12	WADupType2	# of re-announcements after withdrawing the same path (only AS-PATH and NEXT-HOP fields are the same)
13	WADup	WADupType1 + WADupType2
14	WADiff	# of new path announced after withdrawing an old path
15	AW	# of withdrawals after announcing the same path
16 ... 35		the mean and the standard deviation of ten different types of inter-arrival time

temporal characteristics of BGP updates are also important. Corresponding to every parameter from parameters 6–15, the inter-arrival time of a particular BGP update type can be studied. For example, related to parameter 10, AADiff, we can have the inter-arrival time of two announcements that declare two different paths for reaching a specific prefix. We introduced the mean and the standard deviation of every such inter-arrival time as two new parameters, leading to another twenty parameters (parameters 16–35).

Using training data as the input, we applied the procedure described in Section 3 to classify data as associated with a worm event or normal. Note that we did not differentiate data from different types of worms. On the contrary, we sought to obtain rules for just two classes——normal and abnormal.

4.3 Algorithm Effectiveness

The BGP data sets are imbalanced. The simple accuracy as an objective function used in most classification tasks is inadequate for the task with data imbalance. For example, let us consider a classification problem in which there are two classes, 1% of the patterns belonging to the minority class and 99% of the patterns belonging to the majority class. If a classier made a decision that all patterns should be classified into the majority class, it would achieve 99% of accuracy. This can be considered as a good performance in terms of simple accuracy; however this is of no use since the classier does not catch any important information on the patterns of the minority class [26].

We introduced a performance measure appropriate for imbalanced data sets. Suppose that positive patterns are the patterns belonging to the minority class and that negative patterns are the patterns belonging to the majority class. Usual classification tasks use simple accuracy computed by $\dfrac{(TP+TN)}{(TP+FN+FP+TN)}$ where TP, TN, FP, FN represent true positive, true negative, false positive, and false negative respectively. However, simple accuracy heavily relies on TN (True Negative) rather than TP (True Positive) when data are imbalanced. Thus, the classifier tends to classify most patterns as negative to achieve a high simple accuracy. In order to prevent this, some other performance measures are being considered. In this paper, we adopt Geometric Mean, which considers both the accuracies of the minority class and the majority class equally. A+, the accuracy of the minority class, is computed by $\dfrac{(TP)}{(TP+FN)}$. A-, the accuracy of the majority class, is computed by $\dfrac{(TN)}{(FP+TN)}$. Then, geometric mean is computed by $\sqrt{(A+)\times(A-)}$.

4.4 Experiment Results in BGP Data Set

Here, we evaluate the performance of our proposed combined approach—active learning using bagging on neural networks approach. We compare the geometric mean of the combined approach (g-mean-under-al) with those of four other methods: 1) asymmetric bagging NN with random sampling (g-mean-asb-rand); 2) NN with under-sampling (g-mean-under-rand); 3) active learning using asymmetric bagging based on NN with random sampling (g-mean-asb-al); 4) bagging NN based on features (g-mean-bfs-al).

Table 2 PARAMETER LIST

-	g-mean-asb-rand	g-mean-under-rand	g-mean-asb-al	g-mean-under-al	g-mean-bfs-al
Que_10	0.416316	0.421992	**0.49153**	**0.58628**	0.084017
Que_30	0.424774	0.485471	**0.83387**	**0.80712**	0.033806
Que_50	0.480174	0.556326	**0.8526**	**0.87451**	0.487616
Que_80	0.621623	0.742209	**0.93862**	**0.9102**	0.868081
Que_100	0.573138	0.680179	**0.95379**	0.89756	**0.93859**
Que_200	0.767319	0.812733	**0.96534**	**0.95187**	0.865263

Table 3 PARAMETER LIST

~	g-mean-asb-rand	g-mean-under-rand	g-mean-asb-al	g-mean-under-al	g-mean-bfs-al
Que_10	0.477463	**0.59768**	0.332525	**0.57695**	0.070203
Que_30	0.606281	**0.66582**	0.47253	**0.78415**	0.305902
Que_50	0.499657	**0.68857**	0.662964	**0.83636**	0.508289
Que_80	0.583958	0.706005	**0.76292**	**0.88696**	0.760444
Que_100	0.640663	0.735681	**0.88539**	**0.90321**	0.743352
Que_200	0.856709	0.76157	0.897698	**0.93858**	**0.942**

Our experimental results of AS 513, AS 4777, AS 13129 are shown in the 3 tables below. The first column of the tables is the numbers of each query. The bold numbers are the top two g-means respect each query.

Results in terms of g-mean of AS 513 are shown in Table 2.

From the Table 1 we can see that the combined approach (g-mean-under-al) and the active learning using asymmetric bagging based on NN with random sampling (g-mean-asb-al) reach better performances than the other three methods. Especially, when query number is chosen to 30, both the two approaches have great improvement in geometric mean which are respectively 0.80712 and 0.83387.Meanwhile, the geometric mean using g-mean-bfs-al is only up to 0.033806. On the whole, the two approaches perform well in this data set and seem to have equal good performance, but it also suggests that this data set is not representative of the overall set of Internet flows.

Results in terms of g-mean of AS 4777 are shown in Table 3.

Table 3 shows that the combined approach (g-mean-under-al) always achieves at the first or second top of the algorithms in geometric mean. When query numbers equal to 30, 50, 80, 100, the combined approach performs the best of all. Especially, when query number is chosen to 50, the geometric mean of g-mean-under-al is 0.83636, which is higher by 0.15 than that of the g-mean-under-rand method with geometric mean 0.66852.

Results in terms of g-mean of AS 13129 are shown in Table 4.

The results of Table 4 always suggest that on the average, the combined approach (g-mean-under-al) performs better in geometric mean than the other approaches. Especially, when query numbers equal to 30, 50, 80, 200, the combined approach performs the best.

Table 4 PARAMETER LIST

~	g-mean-asb-rand	g-mean-under-rand	g-mean-asb-al	g-mean-under-al	g-mean-bfs-al
Que_10	0.495095	0.718854	**0.78737**	**0.75163**	0.409419
Que_30	0.694932	0.731935	**0.85913**	**0.88233**	0.207707
Que_50	0.65479	0.870417	**0.87999**	**0.95757**	0.370532
Que_80	0.829052	0.804768	**0.89782**	**0.97085**	0.470179
Que_100	0.829962	0.886298	**0.92819**	**0.92527**	0.72801
Que_200	0.806101	**0.9273**	0.901809	**0.97625**	0.783481

Overall, our experiment shows that the combined approach is more effective in detecting abnormal BGP Routing Dynamics. It can reach a stable and good result and this is applicable to BGP product.

5 Conclusion

In this paper, the active learning using bagging on neural networks is proposed and used to classify BGP routing dynamics and detect the abnormal data. Bagging improves the accuracy of the algorithm. Under-sampling is also used in training neural networks. The experimental results suggest that the approach is effective on two-class imbalanced BGP data sets.

There are many important challenges and current activities on BGP security. Due to many threats, challenges, and the huge amount of work going on, we were only able to verify the availability of the active learning method and to give a perspective on one important aspect of BGP security — abnormal BGP routing dynamics detection from the view of the traffic statistics. Even we only apply the classification methods to identify the abnormal BGP routing dynamics, there are the stabilities of some other methods such as clustering, active learning and semi-supervised learning to be verified. Many problems for BGP security have not yet been solved satisfactorily. Thus, BGP security will remain an active and interesting research area in the near future. And the abnormal traffic detection and prevention should also be a promising direction to supply BGP security.

Acknowledgment. Thanks to Innovation Team's support in QingDao R&D Center, Alcatel-Lucent Technologies. The authors would like to thank Prof. Jun Li and Yibo Wang for their discussion about the calculating the parameters of BGP.

References

1. Wang, L., Zhao, X., Pei, D., Bush, R., Massey, D., Mankin, A., Wu, S., Zhang, L.: Observation and analysis of BGP behavior under stress. In: Proceedings of Internet Measurement Workshop (November 2002)
2. Mahajan, R., Wetherall, D., Anderson, T.: Understanding BGP misconfiguration. In: Proceedings of ACM SIGCOMM (August 2002)
3. Wu, Z., Purpus, E.S., Li, J.: BGP behavior analysis during the August 2003 blackout. In: International Symposium on Integrated Network Management, Extended abstract (2005)
4. Labovitz, C., Ahuja, A., Bose, A., Jahanian, F.: Delayed Internet Routing Convergence. In: Proceedings of ACM Sigcomm (August 2000)
5. Labovitz, C., Wattenhofer, R., Venkatachary, S., Ahuja, A.: The Impact of Internet Policy and Topology on Delayed Routing Convergence. In: Proceedings of the IEEE INFOCOM (April 2001)
6. Griffin, T., Premore, B.: An Experimental Analysis of BGP Convergence Time. In: Proceedings of ICNP (November 2001)
7. Misel, S.: Wow, AS7007!, http://www.merit.edu/mail.archives/nanog/1997-04/msg00340.html

8. Lad, M., Zhao, X., Zhang, B., Massey, D., Zhang, L.: An analysis of BGP update surge during Slammer attack. In: Proceedings of the International Workshop on Distributed Computing (IWDC) (2003)
9. Cowie, J., Ogielski, A., Premore, B., Smith, E., Underwood, T.: Impact of the 2003 blackouts on Internet communications. Technical report, Renesys (November 2003)
10. Zhou, Z.-H., Liu, X.-Y.: Training Cost-Sensitive Neural Networks with Methods Addressing the Class Imbalance Problem. IEEE Transactions On Knowledge And Data Engineering 18(1), 63–77 (2006)
11. Rekhter, Y., Li, Y.: A border gateway protocol 4, BGP-4 (1995)
12. Cowie, J., Ogielski, A., et al.: Global Routing Instabilities during Code Red II and Nimda Worm Propagation (2001)
13. Rekhter, Y., Li, Y.: A border gateway protocol 4 (BGP-4). RFC-1771 (1995)
14. Labovitz, C., Ahuja, A., et al.: Delayed Internet routing convergence. ACM SIGCOMM, Stockholm, Sweden (2000)
15. Chen, E.: Route Refresh Capability for BGP-4. RFC-2918 (2000)
16. Japkowicz, N.: The class imbalance problem: significance and strategies. In: Proceedings of the 2000 International Conference on Artificial intelligence: Special Track on Inductive Learning, Las Vegas, Nevada, pp.111–117 (2000)
17. Kang, P., Cho, S.: EUS SVMs: Ensemble of Under-Sampled SVMs for Data Imbalance Problem. In: King, I., Wang, J., Chan, L.-W., Wang, D. (eds.) ICONIP 2006. LNCS, vol. 4232, pp. 837–846. Springer, Heidelberg (2006)
18. Li, C.: Classifying Imbalanced Data Using a Bagging Ensemble Variation. In: Proseedings of the 45th annual southeast regional conference, pp. 203–208 (2007)
19. Schapire, R.: The strength of weak learn ability. Machine learning 5(2), 197–227 (1990)
20. Li, G.-Z., Meng, H.-H., Yang, M.Q., Yang, J.Y.: Asymmetric Bagging and Feature Selection for Activities Prediction of Drug Molecules. In: Proceedings of Second International Multi-symposium on Computer and Computational Sciences (IMSCCS 2007), August 2007, pp. 108–114. Iowa (2007)
21. Tao, D., Tang, X., Li, X., Wu, X.: Asymmetric bagging and random subspace for support vector machines based relevance feedback in image retrieval. IEEE Transactions on Pattern Analysis and Machine Intelligence 28(7), 1088–1099 (2006)
22. Baram, Y., El-Yaniv, R., Luz, K.: Online Choice of Active Learning Algorithms. Journal of Machine Learning Research 5, 255–291 (2004)
23. Iyengar, V.S.: Chidanand Apte, and Tong Zhang: Active Learning using Adaptive Resampling
24. Rumelhart, D.E., Hinton, G.E., Williams, R.J.: Learning internal representations by error propagation. In: Rumelhart, D.E., McClelland, J.L. (eds.) Parallel Distributed Processing: Explorations in The Microstructure of Cognition, vol. 1, pp. 318–362. MIT Press, Cambridge (1986)
25. Li, J., Dou, D., Wu, Z., Kim, S., Agarwal, V.: An Internet Routing Forensics Framework for Discovering Rules of Abnormal BGP Events. ACM SIGCOMM Computer Communication Review 35(5) (October 2005)
26. Kang, P., Cho, S.: EUS sVMs: Ensemble of under-sampled sVMs for data imbalance problems. In: King, I., Wang, J., Chan, L.-W., Wang, D. (eds.) ICONIP 2006. LNCS, vol. 4232, pp. 837–846. Springer, Heidelberg (2006)

Mining Subspace Clusters from Distributed Data

Haiyun Bian and Raj Bhatnagar

1 Introduction

Many real world applications have datasets consisting of high dimensional feature spaces. For example, the gene expression data record the expression levels of a set of thousands of genes under hundreds of experimental conditions. Traditional clustering algorithms fail to efficiently find clusters of genes that demonstrate similar expression levels in all conditions due to such a high dimensional feature space. Subspace clustering addresses this problem by looking for patterns in subspaces [1] instead of in the full dimensional space. A lot of work has been done in developing efficient subspace clustering algorithms for datasets of various characteristics [1, 6].

However, to the best of our knowledge, all existing subspace clustering algorithms assume that all data reside at a single site. While in fact, data in many applications are distributed at multiple sites in the emerging cyber infrastructure. The simplest method to find subspace clusters from such distributed datasets is to merge all data at a central location, and then apply the conventional subspace clustering algorithms to the centralized database. However, this simple scheme is not viable for several reasons. First, sending the raw datasets over networks is not only infeasible and costly but also insecure. Even when the merge is affordable, different parties participating in the joint mining task may not want to share the raw data with others. Therefore, an alternate solution is to allow each party to keep its own raw data, while only exchange the minimum required amount of information between sites to achieve the global mining objective. Many distributed mining algorithms for various data mining tasks have been proposed over the past decade [3, 5]. Most

Haiyun Bian
Department of Math & CS, Metropolitan State College of Denver, Denver, CO, 80217
e-mail: hbian@mscd.edu

Raj Bhatnagar
Department of CS, University of Cincinnati, Cincinnati, OH, 45221
e-mail: raj.bhatnagar@uc.edu

R. Lee, G. Hu, H. Miao (Eds.): Computer and Information Science 2009, SCI 208, pp. 73–82.
springerlink.com   © Springer-Verlag Berlin Heidelberg 2009

of them adopt a divide-and-conquer approach, where each participant is assigned a sub-division of the whole hypothesis space. The division is designed such that there is minimum coupling and communication between the parties.

In this paper we focus on mining one type of subspace clusters, *the closed patterns*, from horizontally partitioned data. A subspace cluster (pattern) is a closed pattern if it cannot be inferred from other patterns. The objective is to find the exact same set of closed patterns from the distributed data as would be possible from the implicit global dataset available at a single site. The global dataset is implicit in the sense that it is not constructed explicitly. The rest of this paper is organized as follows: Section 2 presents a formal description of the problem. Section 3 gives the algorithms for finding closed subspace clusters in horizontally partitioned datasets. Section 4 presents the experimental results.

2   Closed Patterns and Distributed Data

Let $\mathcal{DS}$ be a horizontally partitioned dataset, which contains a set of rows (objects) $\mathcal{O}$, and a set of attributes (columns) $\mathcal{A}$. d_{ij} denotes the value at row i and column j. Each horizontal partition $\mathcal{D}_i$ contains all the attributes in $\mathcal{A}$ and a subset of rows from $\mathcal{O}$, denoted as $\mathcal{D}_i = <\mathcal{O}_i, \mathcal{A}>$. We use k to denote the total number of partitions. We assume that there are no shared objects between partitions, that is, $\mathcal{O}_i \cap \mathcal{O}_j = \phi$ for $i \neq j$, and $\mathcal{O} = \mathcal{O}_1 \cup \mathcal{O}_2 \cup \ldots \cup \mathcal{O}_k$. We use *site* and *partition* in an interchangeable manner throughout this paper. An example data with two partitions, $\mathcal{D}_1$ and $\mathcal{D}_2$, is shown in Tables 1 and 2 respectively. The global dataset $\mathcal{DS}$ is implicit, which equals $\mathcal{D}_1 \cup \mathcal{D}_2$.

We focus on mining one type of subspace clusters, the *closed patterns*.[1] Examples of closed patterns include frequent closed itemsets [7], formal concepts [4], maximal bi-cliques [8]. All discussion in the following assumes that the datasets are binary, that is, $d_{ij} = 0, 1$.

For binary-valued data, given an arbitrary subset of attributes A, $\psi^{\mathcal{D}}(A)$ returns the set of all those objects that have an entry of '1' for each of the attributes in A in dataset $\mathcal{D}$. $\varphi^{\mathcal{D}}(O)$ returns the set of all those attributes that are shared by all objects (have a value '1') in O in dataset $\mathcal{D}$. A closed pattern is defined as a pair of an attribute set A and an object set O that satisfies the two conditions such that $\psi^{\mathcal{D}}(A) = O$ and $\varphi^{\mathcal{D}}(O) = A$. A is the *attribute set* of closed pattern C in $\mathcal{D}$, and O is the *object set* of closed pattern C in $\mathcal{D}$. $\varphi^{\mathcal{D}} \circ \psi^{\mathcal{D}}$ is called a closure operator [4].

As we can see, the decision on whether a pattern is closed depends on the dataset $\mathcal{D}$. Thus, the set of closed patterns in $\mathcal{DS}$ may be different from those that are closed in each partition $\mathcal{D}_i$. Since the global dataset $\mathcal{DS}$ is not available as a whole at any one site for computation, the inference of the global closed patterns can only be computed from all the explicit partitions, that is, the $\mathcal{D}_i$s. In the following section, we present our algorithm using $k = 2$ (two horizontal partitions) as an example for clarification purpose.

---

[1] We use *pattern* and *cluster* in an interchangeable manner.

Table 1 $\mathscr{D}_1$

	a	b	c	d
1	0	0	1	1
2	1	0	1	1
3	1	1	1	0

Table 2 $\mathscr{D}_2$

	a	b	c	d
4	0	0	1	1
5	1	1	0	0

Table 3 List of closed patterns

$\mathscr{C}(\mathscr{DS})$	$\mathscr{C}_1(\mathscr{D}_1)$	$\mathscr{C}_2(\mathscr{D}_2)$
$< \{1234\}, \{c\} >$	$< \{123\}, \{c\} >$	$< \{4\}, \{cd\} >$
$< \{124\}, \{cd\} >$	$< \{12\}, \{cd\} >$	$< \{5\}, \{ab\} >$
$< \{235\}, \{a\} >$	$< \{23\}, \{ac\} >$	
$< \{23\}, \{ac\} >$	$< \{2\}, \{acd\} >$	
$< \{2\}, \{acd\} >$	$< \{3\}, \{abc\} >$	
$< \{35\}, \{ab\} >$		
$< \{3\}, \{abc\} >$		

3   Case of $k = 2$ Horizontal Partitions

Let $\mathscr{C}_i$ be the set of locally closed patterns (subspace clusters) in partition $\mathscr{D}_i$, and $\mathscr{C}$ be the set of globally closed patterns in $\mathscr{DS}$. For the example shown above, all closed patterns in each partition as well as all those in the implicit global data space are shown in Table 3. $\mathscr{C}_i$ can be found using any closed pattern mining algorithms [9], and our objective is to find $\mathscr{C}$ given $\mathscr{C}_1$ and $\mathscr{C}_2$.

3.1   Mining Globally Closed Attribute Sets

For an arbitrary attribute set $A \subseteq \mathscr{A}$, $\psi^{\mathscr{DS}}(A)$ returns the set of objects that have an entry of '1' for each of the attributes in A in dataset $\mathscr{DS}$. Since $\mathscr{O}_1 \cap \mathscr{O}_2 = \phi$ and $\mathscr{O}_1 \cup \mathscr{O}_2 = \mathscr{O}$, $\psi^{\mathscr{DS}}(A)$ equals $\psi^{\mathscr{D}_1}(A) \cup \psi^{\mathscr{D}_2}(A)$. Furthermore, $\psi^{\mathscr{D}_i}(A) = \psi^{\mathscr{DS}}(A) \cap \mathscr{O}_i$ holds.

Since the global dataset $\mathscr{DS}$ is implicit, it is not available for directly evaluating $\varphi^{\mathscr{DS}} \circ \psi^{\mathscr{DS}}(A)$. However, we can observe from Table 3 that any closed attribute set in either $\mathscr{D}_1$ or $\mathscr{D}_2$ is also closed in the global data $\mathscr{DS}$. Lemma 1 below proves this observation formally.

Lemma 1. *If* $C = < O, A >$ *is a closed pattern in horizontal partition* $\mathscr{D}_1$, *then* $\overline{C} = < \overline{O}, A >$ *is closed in* $\mathscr{DS}$, *where* $\overline{O} = O \cup \psi^{\mathscr{D}_2}(A)$.

Rationale of proof: We prove this lemma by contradiction. Suppose $\overline{C} = < \overline{O}, A >$ is not a closed pattern in $\mathscr{DS}$, then there must exist at least one attribute $a_i \notin A$, such that $\psi^{\mathscr{DS}}(a_i) \supset \psi^{\mathscr{DS}}(A)$. Since $\mathscr{D}_1$ contains a subset of the objects in $\mathscr{DS}$, $\psi^{\mathscr{D}_1}(a_i) = \psi^{\mathscr{DS}}(a_i) \cap \mathscr{O}_1$ and $\psi^{\mathscr{D}_1}(A) = \psi^{\mathscr{DS}}(A) \cap \mathscr{O}_1$. This is equivalent to saying that $\psi^{\mathscr{D}_1}(a_i) \supset \psi^{\mathscr{D}_1}(A)$, which contradicts the assumption that $C = < O, A >$ is closed in $\mathscr{D}_1$.   **The proof is complete.**

Lemma 1 states that any locally closed pattern in any partition has its corresponding globally closed pattern. Therefore, by finding closed patterns from each partition independently, we can find closed patterns in the implicit global data space. Consider the example shown in Table 3. It can be seen that all closed attribute sets in $\mathscr{D}_1$ and $\mathscr{D}_2$ are also closed in $\mathscr{DS}$, so we can compute the union of the set of closed attribute sets from each of the two partitions, which gives us 6 closed attribute sets in the global data space as follows: $\{\{c\}, \{cd\}, \{ac\}, \{acd\}, \{abc\}, \{ab\}\}$.

However, as we can see from Table 3, $\{a\}$ is a closed attribute set in the implicit global data space, but it is not closed in either $\mathcal{D}_1$ or $\mathcal{D}_2$. This suggests that some globally closed attribute sets may not be locally closed in any single partition, thus taking the union of all closed attribute sets from local partitions is not sufficient to find all globally closed attribute sets. But $\{a\}$ can be seen as the intersection result of $\{ac\}$ from $\mathcal{D}_1$ and $\{ab\}$ from $\mathcal{D}_2$. Lemma 2 below proves that this is true for all globally closed but not locally closed attribute sets, that is, they can all be represented as the intersection of two closed attribute sets from the two partitions.

Lemma 2. *Suppose an attribute set A_x is not closed in horizontal partitions $\mathcal{D}_1$ and $\mathcal{D}_2$. Let $\varphi^{\mathcal{D}_1} \circ \psi^{\mathcal{D}_1}(A_x) = A_i$ and $\varphi^{\mathcal{D}_2} \circ \psi^{\mathcal{D}_2}(A_x) = A_j$, then A_i and A_j must be the attribute sets of two closed patterns in $\mathcal{D}_1$ and $\mathcal{D}_2$. Then A_x is a closed attribute set in $\mathcal{D}\mathcal{S}$ if and only if $A_x = A_i \cap A_j$.*

The proof of Lemma 2 is omitted due to limited space. Based on Lemma 2, an attribute set that is closed in the global dataset but is not closed in either one of the partitions, it must equal the intersection of its two closures, each from one of the partitions. At the same time, intersections of any pair of closed attribute sets, each of which comes from one of the partitions, must be a closed attribute set in the global dataset.

Lemma 1 and Lemma 2 together prove that each closed attribute set in the implicit global dataset $\mathcal{D}\mathcal{S}$ is either closed in at least one of the partitions, or can be computed as intersection of two closed sets taking one from each partition. Furthermore, those attribute sets that are closed in one partition can also be seen as the intersection of two closed attribute sets from different partitions when we perform the following augmentation: if there is no object in $\mathcal{D}_i$ for which all the attributes in $\mathcal{A}$ have value 1, add one null pattern that contains all the attributes and zero objects, that is, add $< \{\}, \{abcd\} >$ to the $\mathcal{D}_1$ column and add $< \{\}, \{abcd\} >$ to the $\mathcal{D}_2$ column in Table 3. Then any attribute set that is closed in one of the partition can be seen as the intersection of itself and the null pattern from the other partition. In the following, we assume that all $\mathcal{C}_i$s are augmented as stated above.

3.2   Computing the Object Set

Next we compute the closed object sets. If a globally closed attribute set is closed in both of the partitions, its object set can be easily computed as the union of the two object sets, each coming from one of the partitions. For the example in Table 3, the global object set of $\{cd\}$, $\{1,2,4\}$, equals the union of $\{1,2\}$ (in $\mathcal{D}_1$) and $\{4\}$ (in $\mathcal{D}_2$).

The second case is when the globally closed attribute set can only be derived as the intersection of two locally closed attribute sets. Notice that the intersections of multiple pairs of closed attribute sets may generate the same attribute set. For example, $\{a\}$ can be seen as the intersection of $\{ac\}$ from $\mathcal{D}_1$ and $\{ab\}$ from $\mathcal{D}_2$, and it can also be seen as the intersection as $\{acd\}$ from $\mathcal{D}_1$ and $\{ab\}$ from $\mathcal{D}_2$.

Therefore, we need to keep track of the largest object set union for each closed attribute set.

In many applications, we do not need the explicit list of all the objects contained in each of the patterns, but only the number of objects contained in it. For this case, instead of keeping track of the largest global object set, we only need to record the largest count result. We summarize the computation steps defined above into the following $\otimes$ operator:

Definition 1. $\otimes(\mathcal{C}_1, \mathcal{C}_2, \ldots \mathcal{C}_k) := \{(X, A_x) | A_x = A_1 \cap A_2 \cap \ldots \cap A_k, X = |O_1| + |O_2| + \ldots + |O_k|, where (A_1, O_1) \in \mathcal{C}_1, (A_2, O_2) \in \mathcal{C}_2, \ldots, (A_k, O_k) \in \mathcal{C}_k, A_x \neq \phi$ and $X > 0\}$.

Given the set of all locally closed patterns in each of the two partitions, $\otimes(\mathcal{C}_1, \mathcal{C}_2)$ returns all globally closed attribute sets and the sizes of their object sets. Since it is possible to have multiple entries in $\otimes(\mathcal{C}_1, \mathcal{C}_2, \ldots \mathcal{C}_k)$ that have the same attribute set A_x, we then apply the following $\bigwedge$ operator to remove these duplicated entries by keeping only one entry for each A_x, with its object set size set to the largest object set size among all entries that have A_x as their attribute set.

Definition 2. $\bigwedge(\otimes(\mathcal{C}_1, \mathcal{C}_2, \ldots \mathcal{C}_k)) := \{(X, A_x) | \forall (X_i, A_x) \in \otimes(\mathcal{C}_1, \mathcal{C}_2, \ldots \mathcal{C}_k), X = max(X_1, X_2, \ldots, X_l)\}$ $(1 \leq i \leq l)$.

3.3 Density Constraint

Finding all closed patterns is very time consuming. A common strategy is to focus on mining large clusters, guided by the density property [1]. Density of a pattern is defined to be the ratio between the number of objects/rows contained in a pattern divided by the total number of objects in the dataset. Given a minimum density threshold δ which is a real number in the range of $[0, 1]$, we need to find all globally closed patterns that contain at least $|\mathcal{O}| \times \delta$ objects.

The difficulty lies in the fact that locally dense patterns do not provide sufficient information about globally dense patterns. However, any globally dense cluster must be dense in at least one of the partitions. This idea is summarized in the following lemma:

Lemma 3. *For any closed pattern $C = < O, A >$ in the implicit global dataset $\mathcal{DS}$ for which $dens(C) \geq \delta$ holds, there must exist at least one horizontal partition $\mathcal{D}_i$ in which $\frac{|\psi^{\mathcal{D}_i}(A)|}{|O_i|} \geq \delta$.*

Lemma 3 states that there is no need to intersect two closed patterns from two partitions if neither of them has enough density. Therefore, each partition only needs to send its locally dense and closed patterns to the other partition in order to find all global patterns. However, when one partition receives the list of dense patterns from the other site, it needs to intersect all its locally closed patterns (both dense and non-dense ones) with the received list, since a globally dense and closed pattern may be generated by intersecting one locally non-dense pattern with one received

Algorithm 1. Distributed Dense Closed Pattern Miner (*DDCPM*)

Partition 1	Partition 2
1. $\mathscr{D}_1$ finds all closed patterns $\mathscr{C}_1$	1. $\mathscr{D}_2$ finds all closed patterns $\mathscr{C}_2$
2. $\mathscr{D}_1$ sends $\mathscr{F}_1$ to $\mathscr{D}_2$	2. $\mathscr{D}_2$ sends $\mathscr{F}_2$ to $\mathscr{D}_1$
3.	3.
$\quad$ 3.1 $\mathscr{R}_1 = \bigwedge(\bigotimes(\mathscr{F}_2, \mathscr{C}_1))$	$\quad$ 3.1 $\mathscr{R}_2 = \bigwedge(\bigotimes(\mathscr{F}_1, \mathscr{E}_2))$
$\quad$ 3.2 $\mathscr{R}_1 = \{C \mid C \in \mathscr{R}_1 \text{ and } den_C > \delta\}$	$\quad$ 3.2 $\mathscr{R}_2 = \{C \mid C \in \mathscr{R}_2 \text{ and } den_C > \delta\}$
4. $\mathscr{D}_1$ sends $\mathscr{R}_1$ to $\mathscr{D}_2$	4. $\mathscr{D}_2$ sends $\mathscr{R}_2$ to $\mathscr{D}_1$
5. $\mathscr{R} = \bigwedge(\mathscr{R}_1 \cup \mathscr{R}_2)$	5. $\mathscr{R} = \bigwedge(\mathscr{R}_2 \cup \mathscr{R}_1)$

dense pattern. Algorithm *DDCPM* shown in Algorithm 1 summarizes this idea for the case of $k = 2$. This algorithms requires sites to share only their locally dense and closed patterns ($\mathscr{F}_i$'s) instead of sharing all locally closed patterns ($\mathscr{E}_i$'s). Since usually $|\mathscr{E}_i|$ is much larger than $|\mathscr{F}_i|$, our algorithm saves a great amount of communication cost. In step 1, each site computes its local patterns independently. Then they share with each other their locally dense and closed patterns in step 2. Notice that only locally dense patterns need to be sent between sites. If the density threshold is relatively high, the total number of patterns having enough density in both partitions is expected to be much smaller than the total number of patterns, which saves a lot of communication cost. Furthermore, this also helps cut down the number of intersections performed in step 3. In step 3.1, site 1 intersects $\mathscr{F}_2$ with all patterns in $\mathscr{C}_1$, while site 2 intersects $\mathscr{F}_1$ only with infrequent patterns in site 2. This is because both sites now have $\mathscr{F}_1$ and $\mathscr{F}_2$ (locally dense and closed patterns), and the intersection between $\mathscr{F}_1$ and $\mathscr{F}_2$ needs only to be performed once at one of the two sites. Step 3.2 removes those patterns that do not meet the minimum density threshold. Steps 4 and 5 are the final merge to obtain the full list of global patterns. $\bigwedge(\mathscr{R}_1 \cup \mathscr{R}_2)$ returns the union of all the closed attribute sets appearing in at least one of the $\mathscr{R}_i$s. For each closed attribute set, its object set size is set to be the largest object set size of all its appearances in any of $\mathscr{R}_i$. For example, if an attribute set occurs in one of $\mathscr{R}_1$ and $\mathscr{R}_2$, add the attribute set and its object size to $\mathscr{R}$; if an attribute set occurs in both $\mathscr{R}_1$ and $\mathscr{R}_2$, add the attribute set and the larger one of $|O_1|$ and $|O_2|$ to $\mathscr{R}$. The reason why the final merge step is needed is explained in Section 3.4.

3.4 The Final Merge Step

We first analyze the effect of density pruning on the computation of the size of the object set. Notice that each site only has its full list of locally closed patterns and the frequent patterns sent from the other site. Consider a globally closed attribute set A_x that is not closed in either one of the partitions. Suppose $\varphi^{\mathscr{D}_1} \circ (\psi^{\mathscr{D}_1}(A_x)) = A_i$ and $\varphi^{\mathscr{D}_2} \circ (\psi^{\mathscr{D}_2}(A_x)) = A_j$, then the object set size of A_x is $|\psi^{\mathscr{D}}(A_x)| = |\psi^{\mathscr{D}_1}(A_i)| + |\psi^{\mathscr{D}_2}(A_j)|$. If $\frac{|\psi^{\mathscr{D}_1}(A_i)|}{|O_1|} \geq \delta$ and $\frac{|\psi^{\mathscr{D}_2}(A_j)|}{|O_2|} \geq \delta$, it means that

Table 4 List of closed patterns ($\delta = \frac{3}{5}$)

	$\mathscr{DS}$	$\mathscr{D}_1$	$\mathscr{L}_2$
$\mathscr{F}_i s$ ($\delta \geq \frac{3}{5}$)	$< 4, \{c\} >$ $< 3, \{cd\} >$ $< 3, \{a\} >$	$< 3, \{c\} >$ $< 2, \{cd\} >$ $< 2, \{ac\} >$	
$\mathscr{E}_i s$ ($\delta < \frac{3}{5}$)	$< 2, \{ac\} >$ $< 2, \{ab\} >$ $< 1, \{acd\} >$ $< 1, \{abc\} >$	$< 1, \{acd\} >$ $< 1, \{abc\} >$	$< 1, \{cd\} >$ $< 1, \{ab\} >$

Table 5 $\mathscr{R}_i$ ($\delta = \frac{3}{5}$)

$\mathscr{R}_1$	$\mathscr{R}_2$
$< 3, \{c\} >$	$< 4, \{c\} >$
$< 2, \{cd\} >$	$< 3, \{cd\} >$
$< 2, \{ac\} >$	$< 3, \{a\} >$

both of them are locally dense, thus they belong to $\mathscr{F}_1$ and $\mathscr{F}_2$ respectively. In order to avoid duplicate computations, $\mathscr{F}_1 \otimes \mathscr{F}_2$ is only computed at one of the sites, thus the correct size of the $\psi^{\mathscr{DS}}(A_x)$ can only be correctly computed at one of the sites where $\otimes(\mathscr{F}_1, \mathscr{F}_2)$ is computed. Similarly, if $\frac{|\psi_1(A_j)|}{|O_1|} \geq \delta$ and $\frac{|\psi_2(A_j)|}{|O_2|} < \delta$, then they belong to $\mathscr{F}_1$ and $\mathscr{E}_2$ respectively. Thus the correct object set for A_x can only be computed at site $\mathscr{D}_2$ where $\otimes(\mathscr{F}_1, \mathscr{E}_2)$ is computed. Consequently, the local intersection results from both sites must be combined in order to get the correct object sets for the global patterns.

For example, suppose we set the minimum density threshold to be $\frac{3}{5}$ for the data shown in Table 1 and Table 2. Table 4 divides all the patterns into dense ones and non-dense ones. As we can see, $\mathscr{DS}$ contains 3 dense patterns, as their object sets have at least $\frac{3}{5} \times 5 = 3$ objects. $\mathscr{D}_1$ contains 3 patterns whose densities are at least $\frac{3}{5} \times 3 = 1.8$, and $\mathscr{D}_2$ does not have any pattern whose density is at least $\frac{3}{5} \times 2 = 1.2$. So in order to find the 3 closed patterns for $\mathscr{DS}$, we need to send the three dense patterns in $\mathscr{D}_1$ to $\mathscr{D}_2$. Since $\mathscr{D}_2$ does not have any dense patterns, there is no need for it to send any patterns to $\mathscr{D}_1$.

Table 5 lists the results from both sites after the intersection step (step 3 in Algorithm 1). Since $\mathscr{D}_2$ sends no pattern to $\mathscr{D}_1$, $\mathscr{R}_1 = \mathscr{D}_1$. But $\mathscr{R}_2$ is different from $\mathscr{D}_2$ because of the intersections. For example, $< 4, \{c\} >$ is found by taking the intersection of $< 3, \{c\} >$ from $\mathscr{D}_1$ and $< 1, \{cd\} >$ from $\mathscr{D}_2$. As we can see, although $\mathscr{R}_1$ has $\{c\}$ as a closed attribute set, its object set size is wrong. So we need a final step to merge $\mathscr{R}_1$ and $\mathscr{R}_2$ to find the correct set of global patterns. When we merge the two lists of patterns from the two sites, and whenever there is an attribute set that is contained in both lists, such as $\{c\}$, we pick the largest object set size and set that as the global object set size.

3.5 Generalization to $k >= 2$

Our discussion so far assumes that there are only two horizontal partitions. However, all lemmas can be generalized to handle multiple partitions. Due to limited space, we will not present the details here.

4   Experiments

All experiments were performed on Intel 2.0GHz CPU computers with 1GB of
RAM in Windows XP. We tested the scalability of Algorithm 1 with different minimum density threshold values. Figure 1 shows the running time for different minimum density threshold values for the Mushroom dataset [2], and Figure 2 shows the
total amount of messages (measured in Kilo-Bytes) being exchanged between sites.
The running time being reported here is the average local computation time spent
by each site, which does not include the network communication time. Since we
could not find any existing distributed closed and dense pattern mining algorithm,
we compared our algorithm with the base method mentioned in Figure 2, which
refers to the case where all patterns found in each local site are sent all other sites,
and then each site mine the global patterns independently. As we can see, our algorithm has much lower communication cost compared with this base method. Since
the running time and the size of the messages being exchanged depend on how the
dataset is partitioned, we used the mean value over 30 runs, where for each run the
dataset was randomly partitioned into two parts.

The performance of the algorithm also depends on the relative sizes between
different partitions. If the partitions have very different sizes, the local computation
time is larger than the case where all partitions are of the same size. For the case of
two partitions, we define p to be the ratio between the number of rows in partition
1 and the total number of rows in the implicit global dataset, that is, $p = \frac{|c_1|}{|c|}$. Then
$p = 0.5$ is the case when the two partitions have the same size, and with p becoming
closer to 0 or 1, the imbalance between the sizes of the partitions becomes more and
more significant. Figure 4 shows that the running time increases when p becomes
more distant from 0.5. However, Figure 3 shows that the total size of messages is
about the same for all p values.

We also tested the scalability of the algorithm with various other parameters of
the problem. Figure 5 shows that the running time increases linearly with the number
of patterns being found, and so does the size of messages being exchanged between
the two sites as shown in Figure 6. Figure 7 shows that the running time increases

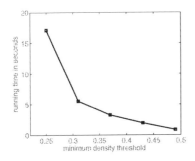

Fig. 1 Run time for different δ

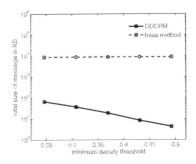

Fig. 2 Message exchange for different δ

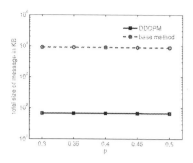

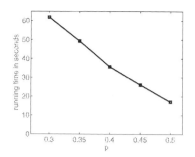

Fig. 3 Message exchange for different p values **Fig. 4** Running time for different p values

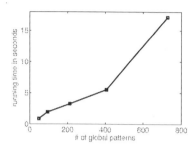

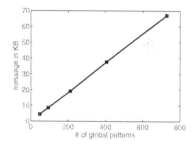

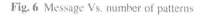

Fig. 5 Running time Vs. number of patterns   **Fig. 6** Message Vs. number of patterns

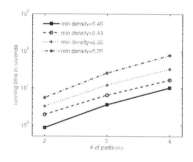

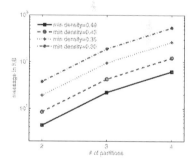

Fig. 7 Running time for different k **Fig. 8** Message exchanged for different k

exponentially with the number of partitions, k, and so does the size of the messages being exchanged between sites as shown in Figure 8.

5 Conclusion

We have presented an algorithm for finding dense and closed patterns in horizontally partitioned datasets. Compared with the case of exchanging all patterns found at each local site, our algorithm integrates the minimum density pruning at the early stage to save both the computational cost and the communication cost. Our scheme does not need a centralized site to control the mining process, and each site acts autonomously and cooperatively to find the global patterns. We have presented both analytical validation and experimental verification to show that all globally dense and closed patters will be generated for a δ threshold without any approximation.

References

1. Agrawal, R., Gehrke, J., Gunopulos, D., Raghavan, P.: Automatic subspace clustering of high dimensional data for data mining applications. In: Proceedings of the 1998 ACM SIGMOD international conference on Management of data (SIGMOD 1998), pp. 94–105. ACM Press, New York (1998)
2. Blake, C., Merz, C.: UCI repository of machine learning databases (1998), http://www.ics.uci.edu/~mlearn/MLRepository.html
3. Evfimievski, A., Srikanta, R., Agrawal, R., Gehrke, J.: Privacy preserving mining of association rules. In: Proceedings of ACM SIGKDD International Conference on Knowledge Discovery and Data Mining (KDD 2002) (2002)
4. Ganter, B., Wille, R.: Formal Concept Analysis: Mathematical Foundations. Springer, Heidelberg (1999)
5. Kriegel, H.P., Kroger, P., Pryakhin, A., Schubert, M.: Effective and efficient distributed model-based clustering. In: Proceedings of Fifth International Conference on Data Mining (ICDM 2005), pp. 258–265 (November 2005)
6. Kroger, P., Kriegel, H.P., Kailing, K.: Density-connected subspace clustering for high-dimensional data. In: Jonker, W., Petković, M. (eds.) SDM 2004. LNCS, vol. 3178, pp. 246–257. Springer, Heidelberg (2004)
7. Pasquier, N., Bastide, Y., Taouil, R., Lakhal, L.: Discovering frequent closed itemsets for association rules. In: Beeri, C., Bruneman, P. (eds.) ICDT 1999. LNCS, vol. 1540, pp. 398–416. Springer, Heidelberg (1999)
8. Peeters, R.: The maximum edge biclique problem is np-complete. Discrete Applied Mathematics 131, 651–654 (2003)
9. Zaki, M.J., Hsiao, C.J.: Charm: an efficient algorithm for closed itemset mining. In: Proceedings of the Second SIAM International Conference on Data Mining (SDM 2004) (April 2002)

Improving Distributed Semantic Search with Hybrid Topology and Peer Recommendation

Juan Li

Abstract. In this paper, we propose a novel framework for discovery Semantic Web data in large-scale distributed networks. In this framework, peers dynamically perform topology adaptations to spontaneously create communities with similar semantic interests, so that search requests have a high probability of being satisfied within the local community. For queries which cannot be efficiently solved inside the community, a directory overlay built on Distributed Hash Table (DHT) is used to assist the search. Recommendation from peers of the same community is employed to extract only semantically related results thus improving the precision. Experiments with simulations substantiate that our techniques significantly improve the search efficiency, scalability, and precision.

1 Introduction

Semantic web has been presented as an evolving extension of World Wide Web [1, 2, 3]. With the development of semantic web technologies, more and more semantic web data are generated and widely used in Web applications and enterprise information systems. These data are structured with ontologies [4] for the purpose of comprehensive and transportable machine understanding. To fully utilize the large amount of semantic data, an effective search mechanism customized for semantic web data, especially for ontologies, is needed by human users as well as software agents and services. The unique semantic features and the inherent distributed nature of semantic web data make its discovery highly challenging.

Peer-to-peer (P2P) technology has been used as a solution to distributed resource discovery, since it scales to very large networks, while ensuring high autonomy and fault-tolerance. The recently proposed structured P2P systems in the form of DHTs [5-8] are a promising approach for building massively distributed data management platforms. However, they offer few data management facilities, limited to IR (Information Retrieval) -style keyword search. Keyword search is appropriate for simple file-sharing applications, but is unable to deal with complex semantic queries which have various properties and sophisticated relations with each other.

More recently, a few studies [9, 10] extended the DHT-based P2P to support semantic queries. The basic idea is to map each keyword of a semantic entity to a

Juan Li
Department of Computer Science, North Dakota State University
e-mail: j.li@ndsu.edu

R. Lee, G. Hu, H. Miao (Eds.): Computer and Information Science 2009, SCI 208, pp. 83–92.
springerlink.com   © Springe-Verlag Berlin Heidelberg 2009

key. For example, RDFPeer [9] indexes each RDF [20, 21] triple to support semantic RDF query. A query with multiple keywords then uses the DHT to lookup each keyword and returns the intersection. Systems like [8] avoid this multiple lookup and intersection by storing a complete keyword list of an object on each node. In this way, the DHTs can support multi-keywords queries. However, DHTs still have difficulty to support other richer queries, such as wildcard queries, fuzzy queries, and proximity queries. In addition, most DHT-based applications require all peers in the system sharing a uniform ontology schema, which is impractical in reality. These limitations restrict the deployment of DHTs to semantic web data discovery.

To support flexible complex queries, many P2P systems [11, 12] use flooding or maintain a broadcast structure, such as a tree or a super cube, to propagate the queries to the network. For example, to execute an RDF query, Edutella [11] broadcasts the query to the whole hypercube. However, the overhead of flooding and broadcast may cause scalability issues.

To overcome the shortcomings of existing discovery approaches, we propose a hybrid search mechanism, which integrates structured DHT P2P technology with unstructured P2P technology. Recommendation feedback from semantically similar peers is employed to retrieve the most relevant results thus improving the efficiency and precision of searching. In our system, each node is associated with a semantic summary representing the node's interest. Based on the summary, we design a method to compute the semantic similarity between different nodes. The network topology is reconfigured with respect to nodes' semantic similarity, so that peers with similar semantics are close to each other, forming a semantic community. The semantic community is loosely structured as an unstructured P2P overlay, called community overlay. Because of its unstructured topology, the community overlay is able to handle flexible complex queries. The semantic locality property guarantees that the system's query evaluation can be limited to relevant peers only. A structured DHT-based overlay is used to facilitate the construction of the community overlay and to assist evaluating queries which cannot be effectively resolved by the community overlay.

Members in the same community share similar interests hence are able to make recommendations to each other. Recommendations allow users to disambiguate search requests quickly. Moreover, they can personalize query results for users by ranking higher the results that are relevant to users' semantic properties. Therefore, the search quality in terms of both precision and recall is improved. In addition, peers recommend neighbors for each other according to their query experience to adapt to the evolving network property.

With the assistance of peer recommendation, community overlay and directory overlay complement each other, providing efficient search for the system. Compared to search in pure structured P2P systems, our hybrid search system has inherent support for complex semantic query or partial match; in addition, the retrieved results are more relevant. Compared to search in pure unstructured P2P systems, our community-based structure saves the overhead of flooding the query to unrelated nodes, thus enjoying more scalability.

2 System Overview

This section gives an overview of the system architecture. The proposed system consists of two logical overlays – an unstructured community overlay and a structured directory overlay –taking different roles for efficient operations of the system.

Query evaluation is mainly performed in the community overlay. In a community overlay, peers are connected to those sharing similar semantic interests. As a result, the query propagation tends to first reach those that are more likely to possess the data being searched for. This semantic locality property enables the community overlay to answer most queries originated from the local community. Unlike DHTs, community overlay does not specify any requirements for the query format, hence is able to handle any arbitrary types of complex queries. For the above reasons, a large portion of complex queries can be resolved inside the local community. However, it is still possible that a small portion of queries cannot be answered within the community overlays a peer belongs to, even if the peer may belong to multiple communities. Peers may have more interests which cannot be covered by the community overlays they reside. In this case, the index maintained by the directory overlay can be consulted for hints about where to forward the query for a second try.

The directory overlay is built on top of DHT protocols. It provides a high-level directory service for the system by indexing abstract ontology skeletons. The directory overlay has two main functionalities: (1) It facilitates the construction of community overlay. (2) It resolves queries not covered by the community overlay. Unlike community overlay, directory overlay does not give exact answers of a particular query; instead, it locates all peers possessing semantic keywords of the query. Then the query will be broadcasted to all peers related to the keywords for further evaluation. However, a keyword may have multiple meanings, not all of these meanings match the requestor's intention. Simply forwarding the query to all peers containing the keywords is not accurate and consumes lots of unnecessary network bandwidth.

The directory overlay employs peers' recommendation and feedback to solve the aforementioned semantic ambiguity problem. After receiving results from the directory overlay, the requestor first checks the validity of the results. Then it reports its findings back to the directory overlay nodes. The feedback will benefit future requesters with similar interests.

A physical node may be involved in both of these two overlays. Community overlay and directory overlay benefit from each other: directory overlay facilitate the construction of community overlay, while feedback from communities improves the search precision of directory overlay. Working together, these two overlays improve the search efficiency and accuracy of the system.

3 Community Overlay

The construction of the community is a topology adaptation process, i.e., to make the system's dynamic topology match the semantic clustering of peers. The community topology enables queries to be quickly propagated among relevant peers. In addition, this topology allows semantically related nodes to establish ontology mappings.

3.1 Semantic Similarity

To find semantically similar neighbours, peers should be able to measure semantic similarity between each other. There has been extensive research [17 - 19] focusing on measuring the semantic similarity between two objects in the field of information retrieval and information integration. However, their methods are very comprehensive and computationally intensive. In this paper, we propose a lightweight method to compute the semantic similarity between two nodes.

Our system supports semantic web data represented as OWL ontology. OWL ontology can be divided into two parts: the terminological box (TBox) and the assertion box (ABox) as defined in the description logic terminology [16]. TBox ontology defines the high-level concepts and their relationships. It is a good abstraction of the ontology's semantics and structure. Therefore, we use a node's TBox ontology to represent its semantic interest. In particular, we use keywords of a node's TBox ontology as its ontology summary. However, a semantic meaning may be represented by different keywords in different ontologies, while it is also possible that the same keyword in different ontologies means totally different things. Ontology comparison based on TBox keywords may not yield satisfying results. In order to solve this problem, we extend each concept with its semantic meanings in WordNet [22]. We use two most important relationships in WordNet – synonyms and hypernym – to expand concepts. In this way, semantically related concepts would have overlaps.

After extension, a node's ontology summary set may get a number of unrelated words, because each concept may have many senses (meanings), but not all of them are related to the ontology context. A problem causing the ambiguity of concepts is that the extension does not make use of any relations in the ontology, which are important clues to infer the semantic meanings of concepts. To further refine the semantic meaning of a particular concept, we utilize relations between the concepts in an ontology to remove unrelated senses from the summary set. Since the dominant semantic relation in an ontology is the *subsumption* relation, we use the *subsumption* relation and the sense disambiguation information provided by WordNet to refine the summary. It is based on a principle that a concept's semantic meaning should be consistent with its super-class's meaning. We use this principle to remove those inconsistent meanings. For every concept in an ontology, we check each of its senses; if a sense's hypernym has overlap with this concept's parent's senses, then we keep this sense and the overlapped parent's sense to the ontology summary set. Otherwise, they are removed from the set. In this way we can refine the summary and reduce imprecision.

To compare two ontologies, we define an ontology similarity function based on the refined ontology summary. The definition is based on Tversky's "Ratio Model" [23] which is evaluated by set operations and is in agreement with an information-theoretic definition of similarity [24]. Assume A and B are two nodes, and their ontology summary are S(A) and S(B) respectively. The semantic similarity between node A and node B is defined as:

$$sim(A,B) = \frac{|S(A) \cap S(B)|}{|S(A) \cap S(B)| + \alpha |S(A) - S(B)| + \beta |S(B) - S(A)|}$$

In the above equations, "$\cap$" denotes set intersection, "$-$" is set difference, while "$| \ |$" represents set cardinality, "α" and "β" are parameters that provide for differences in focus on the different components. The similarity *sim*, between A and B, is defined in terms of the semantic concepts common to A and B: $S(A) \cap S(B)$, the concepts that are distinctive to A: $S(A) - S(B)$, and the features that are distinctive to B: $S(B) - S(A)$. Two nodes, node A and node B are said to be semantically related if their semantic similarity measure, *sim(A,B)*, exceeds the user-defined similarity threshold t ($0 < t \leq 1$).

3.2   Community Construction

The construction of an ontology-based overlay is a process of finding semantically related neighbors. A node joins the network by connecting to one or more bootstrapping neighbors. The bootstrapping neighbors try to recommend some other neighbors to this new node according to their semantic. If the bootstrapping neighbors do not have such recommendation information at hand, the new joining node will issue a neighbor-discovery query. The neighbor discovery query contains the new node's ontology summary compressed with a Bloom Filter [25]. It then uses strategies (such as [13-15]) to efficiently propagate the neighbour discovery query over clusters. Nodes receiving the query compute its semantic similarity with the new node based on the semantic summary. Semantically related nodes then return a positive reply to the new node. If there are not enough neighbors discovered within the hops limited by TTL, the new node will turn to the directory overlay for assistance. After the neighbor-discovery process, a new node is positioned to the right community. Inside the community overlay, nodes randomly connect with their neighbors. Queries looking for particular contents can be forwarded inside the community overlay using flooding- or random-walk-based simple forwarding algorithms.

Because of the dynamic property of the large-scale network, and the evolution of nodes' ontology property, neighbor discovery for a node is not once and for all, but rather the first-step of the topology adaptation scheme. Based on the query experiences a node may add or delete neighbors accordingly. At the same time, it recommends new neighbors to its existing neighbors. As a result, the network topology is reconfigured with respect to peers' dynamic semantic properties, and peers with similar ontologies are always close to each other.

4   Directory Overlay

As a facilitator and complement of the community overlay, the directory overlay indexes top-level semantic interests and unpopular semantic concepts. As mentioned,

OWL ontology can be divided into two parts: TBox and ABox. Similar to a database schema, a node's TBox knowledge is more abstract, describing the node's high-level concepts and their relationships. In contrast, ABox includes concrete data and relations, for example, the instances of classes defined in the TBox. Directory overlay indexes TBox and ABox ontology for different purpose: TBox indexing helps nodes locate communities, while ABox indexing assists nodes finding instances which cannot be quickly located in the community overlay.

The directory overlay is constructed according to the mechanism of the corresponding DHT overlay. We employ RDFPeer's indexing method presented by M. Cai *et al* [9]. The basic idea is to divide RDF description into triples and then index the triples in a DHT overlay. We store each triple three times by applying a hash function to its *subject*, *predicate*, and *object*. In this way, a query providing partial information of a triple can be handled. Peers register their top semantic interests in the form of TBox ontology through the *insert(key,value)* operation in the directory overlay. The directory overlay node in charging of that key maintains a Least Recently Used (LRU) cache storing contact information of registered peers. A neighbor discovery query can get contacts of other peers interested in the same ontology through this directory overlay node. Then the new node can connect with these contacts and join their community. At the same time, the new node registers to the directory overlay by adding itself to the cache of the indexing node. A node with multiple interests can register with multiple indexing nodes. The directory overlay also indexes unpopular ABox instances which cannot be quickly located inside the community.

5   Semantic Query Evaluation

The semantic community reduces the search time and decreases the network traffic by minimizing the number of messages circulating between nodes. There are many strategies, such as [13-15] to effectively propagate queries in an unstructured P2P network. Popular data items are more likely to be located quickly since they have more replicas in the community, whereas an unpopular data item cannot be found unless a large number or all of the peers are searched. Also, queries for data in other semantic communities are unlikely to be solved inside the local community overlay. For these cases, nodes turn to the directory overlay to get assistance.

Directory overlay indexes top semantic interests and unpopular instances, thus is able to give hints to queries which cannot be solved by the community overlay. A node can find interested community by lookup its interest in the directory overlay, then connects to all related nodes returned. For unpopular ABox instances, DHT indexing has the semantic ambiguity problem. For example, it is difficult to figure out whether the search term *palm* is a company (company: palm), a technology (operating system: palm), or a product (PDA:palm). We solve the ambiguity problem with community recommendation feedbacks.

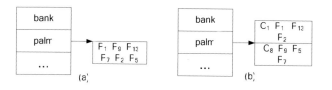

Fig. 1 Example of a data entry stored in an index node

To facilitate query refinement with community feedback, the indexing peers need to perform some additional tasks. Besides storing the ABox keywords, an indexing peer is also responsible for maintaining clusters of peers related to each sense of the keyword. Figure 1 shows an example of a data entry stored in an indexing peer. There are six peers related to the term, *palm*. Initially when a node issues a query related to term *palm* trying to find information about a PDA, all six peers are returned to the requester as shown in Figure 1 (a). The requester will contact each of them, although only three of them (P_1, P_{13}, P_2) are related to PDA. After the requester contacts all these six peers and evaluates their data, it returns its feedback (i.e., which peers are related) to the indexing peer. The indexing peer will link those related three peers with the requester's community, as shown in Figure 1 (b). Next time, a requester from the same community will take advantage of this clustering and be given only the three related peers. In this way, the precision of the query evaluation is improved and the network traffic is reduced.

6 Experiment

As it is difficult to find representative real world ontology data, we have chosen to generate test data artificially. The algorithm starts with generating the ontology schema (TBox). Each schema includes the definition of a number of classes and properties. The classes and properties may form a multilevel hierarchy. Then the classes are instantiated by creating a number of individuals of the classes. To generate an RDF instance triple t, we first randomly choose an instance of a class C among the classes to be the subject: $sub(t)$. A property p of C is chosen as the predicate $pre(t)$, and a value from the range of p to be the object: $obj(t)$. If the range of the selected property p are instances of a class C', then $obj(t)$ is a resource; otherwise, it is a literal. The queries are generated by randomly replacing parts of the created triples with variables.

The directory overlay is implemented as a Pastry [6] virtual network in Java. Each peer is assigned a 160-bit identifier, representing 80 digits (each digit uses 2 bits) with base b=2. After the network topology has been established, nodes publish their TBox knowledge and some unpopular ABox data on the overlay network. Then nodes are randomly picked to issue queries. Each experiment is run ten times with different random seeds, and the results are the average of these ten sets of results.

We examine the system performance in three different aspects, namely scalability, efficiency, and precision by executing the experiment in different network configurations. The performance is measured using two Information Retrieval (IR) standards: recall and precision. Recall refers to completeness of retrieval of relevant items, as defined below:

$$recall = \frac{|\ relevantDocuments \cap retrievedDocuments\ |}{|\ relevantDocuments\ |}$$

Precision measures the purity of the search results, or how well a search avoids returning results that are not relevant. The "document" in the IR definition represents a resource in our experiment.

$$precision = \frac{|\ relevantDocuments \cap retrievedDocuments\ |}{|\ retrievedDocuments\ |}$$

First, we vary the number of nodes from 2^9 to 2^{15} to test the scalability of the system. The results are listed in Figure 2. Our hybrid system gets higher recall in all these different sized networks. In addition, our recall decreases less with the increase in network size.

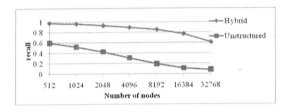

Fig. 2 Recall rate vs. network size

Figure 3 illustrates the system efficiency by showing the relationship between query recall rate and query TTL. With a small TTL, our system gets a higher recall rate, i.e., resolves queries faster.

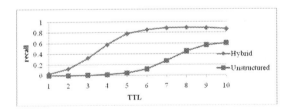

Fig. 3 Recall rate vs. TTL

To testify the effect of community recommendation, we create a special experimental scenario which uses a small-sized dictionary D to generate the ontology data. We randomly pick S words from D, representing polysemy or homonymy (words with multiple meanings); if these words appear in different communities, they represent different meanings. In this experiment, we count the number of

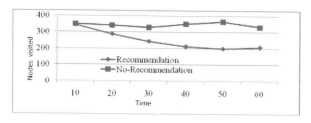

Fig. 4 Effect of recommendation

nodes visited to find 30 results at different time period. As shown in Figure 4, with the time going, using community feedback may reduce the number of nodes to be explored. Because feedback from communities helps eliminating semantic ambiguity of the directory overlay, queries are only forwarded to the most relevant nodes. Consequently, the precision of the search is increased.

7 Conclusion

The main contribution of this paper is to present an effective framework for query evaluation in a large-scale distributed network. Our system combines the structured and unstructured P2P topology to form a hybrid architecture. We organize nodes' topology according to their semantic similarity, so that queries can be focused in semantically related regions only. For queries that cannot be effectively resolved in the semantic community, they can be sent to a structured directory overlay. Recommendations from users are cached for disambiguating future queries. Simulation experiments demonstrate that this framework improves the scalability, efficiency and precision of search in a large semantic heterogeneous network.

References

1. Berners-Lee, T., Fischetti, M.: Weaving the Web, The original design and ultimate destiny of the World Wide Web. Harper (1999)
2. Berners-Lee, T., Hendler, J., Lassila, O.: The Semantic Web: A new form of Web content that is meaningful to computers will unleash a revolution of new possibilities. Scientific American (May 2001)
3. Fensel, D., Musen, M. (eds.): The Semantic Web: A Brain for Humankind. IEEE Intelligent Systems (March/April 2001)
4. Gruber, T.R.: Principles for the Design of Ontologies Used for Knowledge Sharing. International Journal Human-Computer Studies (1995)
5. Zhao, B.Y., Kubiatowicz, J.D., Joseph, A.D.: Tapestry: An Infrastructure for Fault-Tolerant Wide-Area Location and Routing. Technical Report. UCB 2000 (2000)
6. Rowstron, A., Druschel, P.: Pastry: Scalable, Distributed Object Location and Routing for Large-Scale Peer-to-Peer Systems. In: Guerraoui, R. (ed.) Middleware 2001. LNCS, vol. 2218, p. 329. Springer, Heidelberg (2001)

7. Stoica, I., Morris, R., Karger, D., Kaashoek, M.F., Balakrishnan, H.: Chord: A Scalable Peer-to-Peer Lookup Service for Internet Applications. In: ACM SIGCOMM (2001)

8. Ratnasamy, S., Francis, P., Handley, M., Karp, R., Shenker, S.: A Scalable Content-Addressable Network. In: ACM SIGCOMM (2001)

9. Cai, M., Frank, M.: RDFPeers: A scalable distributed RDF repository based on a structured peer-to-peer network. In: Proc. of WWW conference, New York, USA (May 2004)

10. OntoGrid project, http://www.ontogrid.net/

11. Nejdl, W., et al.: EDUTELLA: a P2P Networking Infrastructure Based on RDF. In: Proceedings of the 11th international conference on World Wide Web (WWW) (2002)

12. Arumugam, M., Sheth, A., Arpinar, I.B.: Towards peer-to-peer semantic web: A distribuited environment for sharing semantic knowledge on the web. In: Proc. of the International World Wide Web Conference (2002)

13. Li, J., Vuong, S.: SOON: A Scalable Self-Organized Overlay Network for Distributed Information Retrieval. In: Proceedings of the 19th IFIP/IEEE International Workshop on Distributed Systems (2008)

14. Chawathe, Y., Ratnasam, S., Breslau, L., Lanhan, N., Shenker, S.: Making Gnutella-like P2P Systems Scalable. In: Proceedings of ACM SIGCOMM 2003 (2003)

15. Li, J., Vuong, S.: Efa: an Efficient Content Routing Algorithm in Large Peer-to-Peer Overlay Networks. In: Proceedings of the Third IEEE Peer-to-Peer Computing (2003)

16. Baader, F., Calvanese, D., McGuinness, D.L., Nardi, D., Patel-Schneider, P.F.: The Description Logic Handbook: Theory, Implementation, Applications. Cambridge University Press, Cambridge (2003)

17. Jiang, J., Conrath, D.: Semantic Similarity Based on Corpus Statistics and Lexical Taxonomy. In: Proceeding of the Int'l Conf. Computational Linguistics (1997)

18. Lee, J., Kim, M., Lee, Y.: Information Retrieval Based on Conceptual Distance in IS-A Hierarchies. J. Documentation (1993)

19. M. A. Rodriguez, M. J. c

20. Lin, D.: An information-theoretic definition of similarity. In: Proc. 15th International Conf. on Machine Learning, pp. 296–304. Morgan Kaufmann, San Francisco (1998)

21. Bloom, B.: Space/time tradeoffs in hash coding with allowable errors. Communications of the ACM 13(7), 422–426 (1970)

Spatial Semantic Analysis Based on a Cognitive Approach

Shiqi Li, Tiejun Zhao, and Hanjing Li

Abstract. In order to extract and represent spatial semantics from linguistic expressions, a method is proposed based on an extension of Kinstch's Construction-Integration (CI) model. Compared with the traditional feature-based methods, it can implicitly integrate the contextual and general knowledge into the calculating process. Firstly, we define a propositional spatial semantic representation, and generate a sequence of proposition candidates using Chinese Proposition Bank to construct a text base network. Then the contextually appropriate spatial propositions of the net will be strengthened and inappropriate ones will be inhibited through iteratively spreading activations until the network stabilizes. Finally, the experimental result shows an encouraging performance on two datasets.

1 Introduction

Spatial semantic analysis (SSA) refers to the process of generating a list of semantic units that represent the spatial information including entities, motions and spatial relationships from natural language description. SSA can be used in various intelligent applications, particularly in GIS and language visualization. In this paper, it serves as a semantic parser for a specific Text-to-Animation task that convert Chinese children's story to the corresponding animation.

To achieve the goal of visualizing text of stories, two crucial issues must be addressed. One is how to obtain the spatial semantics that the text conveys, and then recode the semantics into a representation which should be both powerful for expression and understandable for computer. How to effectively interpret the semantic representations into animation is the other issue, which is not the focus of this paper. Recently some work has been done on the spatial semantics. WordsEye system 0 makes use of a rule-based semantic interpreter to achieve spatial semantic frames from the result of dependency parsing. In (Ma and McKevitt 2003), Ma provides a visual semantic representation that can express the procedure of action verbs and facilitate the automatic animation generation.

This paper proposes a novel cognitive solution for the issue of SSA, as shown in Figure 1. Firstly, we define a propositional form of spatial semantic representation

Shiqi Li, Tiejun Zhao, and Hanjing Li
School of Computer Science and Technology
Harbin Institute of Technology, Harbin, China
e-mail: {sqli,tjzhao,hjlee}@mtlab.hit.edu.cn

R. Lee, G. Hu, H. Miao (Eds.): Computer and Information Science 2009, SCI 208, pp. 93–103.

based on Propositional Encoding Language (PEL) (Kapusuz 2001) and Lexical Visual Semantic Representation (LVSR) (Ma 2006). Then we create a sequence of possible propositions from text using some statistical Natural Language Processing tools and linguistic resources including HowNet (Dong and Dong 1999) and Chinese Proposition Bank (CPB) (Xue 2005). Finally on the basis of an extended CI model (Kintsch 1988), we build a propositional text base network, and then iteratively spread activations around the net to gradually strengthen the contextually appropriate propositions and inhibit unrelated and inappropriate ones until the network stabilizes. The final activations of nodes in the network indicate the winning propositions, namely the result of the spatial semantic representation. A detailed description of this method is provided in the following three sections. Finally, we present the evaluation of the method.

2 Spatial Semantic Representation

First, it is necessary to well define the spatial semantics representation as the objective of SSA. Here we employ a propositional form, the atomic unit in which uses a predicate-argument structure, rather than others such as frame, script or semantic network for the reason that it is convenient for both cognitive model and animation generation. And this form has been proved to be powerful enough to represent the spatial semantics by practical applications in (Kapusuz 2001) and (Ma and McKevitt 2003). Table 1 shows the definition of the spatial semantics representation by the Backus-Naur Form.

Table 1 Definition of the spatial semantics representation

No.	Derivation Rules
1.	<text> ::= {<proposition>}<proposition>
2.	<proposition> ::= <prop_info>:<content>
3.	<prop_info> ::= <class><No>
4.	<class> ::= O | H | E | S | PL | PA | PR | A
5.	<No> ::= {<digit>}<digit>
6.	<digit> ::= 0 | 1 | ... | 9
7.	<content>::=<predicate>@<anchor>({<argumentlist>});
8.	<predicate> ::= <verb>
9.	<verb> ::= {<letter>}<letter>
10.	<anchor> ::= <No>-<No>
11.	<argument list> ::= {<argument>}<argument>
12.	<argument> ::= <object>|<proposition>
13.	<object> ::= {<letter>}<letter>
14.	<letter> ::= a | b | ... | z | A | B | ... | Z

It can be seen that a proposition includes five components: class, No, anchor, predicate, and argument list. The anchor which indicates a proposition's prototype in the text consists of two numbers: position and offset. In addition, proposition embedding is allowed here. Referring to the ontological categories in (Ma and McKevitt 2004), we divide propositions into eight categories according to the spatial semantics they conveyed, as shown in Table 2.

Table 2 Example of predicates in each category

Categories	Examples
Object	Obj(tree);
Human	Hum(Tom);
Amount	Amount(Obj(tree), n);
Property	Prop(Hum(Tom), smart);
Place	On(Obj(tree)); Near(Hum(Tom));…
Path	To(Obj(tree)); From(Hum(Tom));…
State	Be(Hum(Tom), On(Obj(tree)));…
Event	Run(Hum(Tom), To(Obj(tree)));…

For example, a Chinese sentence "小猴走进树林，看见一只漂亮的小鸟在树 梢上唱歌" (means `A little monkey walked into the forest, and saw a beautiful bird singing song on a tree`), can be interpreted into the representations as shown in Table 3. It can be clearly seen that entire spatial information of the sentence: the monkey walked into the forest, the bird singing on the tree and the monkey saw the singing bird, are explicitly expressed by the propositions: E10, E11 and E12. After defining the spatial semantics representation, we will discuss how to automatically achieve them using cognitive approach.

Table 3 The propositions of the example sentence

Prop Info	Predicate-Anchor-Arguments
H1, H2	Hum@1-2(□□); Hum@15-2(□□);
O3, O4, O5	Obj@5-2(□□); Obj@18-2(□□); Obj@22-1(□);
A6	Amount@10-1(H2, 1);
PR7,	Prop@12-2(H2, □□);
PA8	Into@4-3(O3);
PL9	On@20-4(H2, O4);
E10, E11, E12	Walk@3-1(H1,PA8); Sing@21-1(H2,O5, PL9); See@8-2(H1,E11);

3 Construction of Proposition Candidates

Among five elements of proposition, predicate and argument are the most important, and the other three are subordinate to them. The predicate is relatively steady and easily available because it is always anchored to an entity or a verb in the text. The argument list is flexible as it depends upon both the predicate and the context. In this paper we use some rules to extract possible proposition candidates rather than the precise proposition to achieve a reasonable trade-off between performance and flexibility. And the improper candidates will be filter out by the CI model in the following phase. The method for extracting proposition candidates varies with the proposition's class.

3.1 The First Six Classes

First, for the Human and Object classes, the proposition of which anchors to an entity in text and insists of only one fixed predicate with an argument, we use a

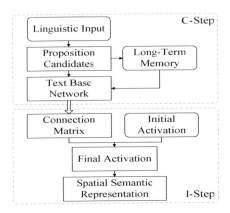

Fig. 1 The framework of the spatial semantic analysis based on CI mode

Chinese Named Entity Recognizer to identify the entities in the text, combining with HowNet which is an computational semantic lexicon to assign them to the right class. In addition, anaphor resolution must be done as preliminary work. Here pronouns are interpreted by simply searching through the context to find the antecedents according to their number, gender and distance.

For the Property and Amount classes, the proposition has a fixed predicate and two arguments. One is a proposition of Human or Object class, and the other is a property or an amount attached to this human or object. We use a dependency parser (Lang et al. 2004) to get the links between entities and their modifiers. The property and amount can be obtained along the "attribute" and "quantity" links of each entity in the parsing result. And if the explicit amount information of an entity is absent, its amount will be set to the default value of 1.

For the Place and Path classes, we find that their propositions stems from preposition phrases in the text, each of which can be decomposed into a preposition, a noun phrase and an optional localizer. And both of preposition and localizer are of a limited set. Therefore, we use several syntactic patterns to extract the corresponding phrases which can be directly converted to propositions from text.

3.2 Event and State Classes

Finally, the emphasis lays on the more complex classes: Event and State. The predicates of these two classes are the verbs in the text which can be achieved by a Part-of-Speech tagger. And the form of argument is decided by the syntactic function of the predicate in the context, namely the verb frame, a descriptive semantic frame containing some slots filled with semantic roles (Fillmore and Baker 2001). Therefore we take advantage of the definition of verb frames in the CPB which is a large annotate corpus of semantic roles. CPB has defined 5796 frames of 4858 Chinese verbs, and most of verbs (4090) only have one frame. There are two types of argument in the verb frames of CPB, general or adjunct-like arguments marked as ArgM, and numbered arguments which represent the verb-specific roles marked as Arg0, Arg1, etc. The precise constrain of argument is also labeled for each particular verb. For example, Arg0 is typically the actor, and Arg1 is typically the thing acted upon.

For the verbs with one frame, we create the proposition candidates by filling its slots with the possible propositions in the same sentence under the constraint of each slot role; For the verbs with more than one frame, we use *spatial valence* to filter inappropriate verb frames by context. Spatial valence of a verb frame is the number of spatial arguments in the text which contain all numbered arguments and ArgMs with the LOC subtype. Spatial valence indicates the basic capacity of a verb frame to accommodate spatial arguments. In actual task, numbered arguments can be mapped to the propositions of Human and Object classes, and LOC-ArgMs can be mapped to the propositions of Place and Path classes. We sum the number of these four kinds of propositions that anchored on both sides of the verbs, and then filter the frames with spatial valence greater than the summation. Finally, for each of the rest frames, we use the same method as for the verb with one frame to generate proposition candidates.

4   Proposition Retrieval by Cognitive Model

According to the cognitive view of text processing, as a text is read, information in the text and any other information already activated in working memory will trigger a spread of activation through the reader's knowledge base, activating associated information (McKoon and Roushey 1990).

The CI Model is a unitary cognitive model that perfectly integrates the above idea. Its computational model which combines the symbolic system with connectionist techniques has been proven to be effective in some comprehension applications such as WSD, Anaphor Resolution (Kinstch 7988) and Story Understanding (Mueller 2002). The comprehension process involves an initial phase of construction, which is chaotic, but attains coherence in the integration phase. Then let's introduce the proposition retrieval from these two phases.

4.1   Construction Phase

The goal of the construction phase is to construct a propositional text base network whose nodes stand for propositions and whose links stand for the degree of relationship between them. First we get nodes of the network from two sources: text and Long-Term Memory (LTM) network, representing contextual knowledge and general or domain knowledge respectively. The textual nodes are composed by the proposition candidates generated in Section 3. And LTM nodes are retrieved using each of the textual nodes as seed through a probabilistic selection of LTM net which is another similar propositional network built up before specific tasks. In this paper the LTM network is set up using annotated propositions by the same method as constructing textbase network. In the probabilistic selection, nodes that are more related to the seed have a larger chance of being chosen. Assume that seed node i is positively associated with n other nodes in the LTM net. Let $l(i, j)$ be the connection weight between nodes i and j. Then the retrieval probability of node j:

$$p(j|i) = \frac{l(i,j)}{\sum_{k=1}^{n} l(i,k)} \tag{1}$$

After achieving these two kinds of nodes, the rest work of construction is to assign connection weights between them. The value is additive and ranges from -1 to 1. We defined six principles for the assignment referring to the guide from (Mannes and Roushey 1990):

(1). Argument overlap. If two nodes, A and B, share a same proposition as their argument, then assign value 0.4 to $l(A,B)$.

(2). Proposition embedment. If two nodes, A and B, the proposition that node A represents acts as an argument of node B, then assign value 0.5 to $l(A,B)$.

(3). Mutual exclusion. If two nodes, A and B, both have the a predicate with the same anchor, then assign value -1 to $l(A,B)$.

(4). Inheritance. If two nodes, A and B, both are derived from the LTM net, then assign their link value in the LTM net $l_{LTM}(A,B)$ to $l(A,B)$.

(5). Inhibition. If two nodes, A and B, are clearly inconsistent or contradictor to each other, then assign value -0.6 to $l(A,B)$.

(6). No relation. If two nodes, A and B, all situations mentioned above all does not happen between, then assign value zero to $l(A,B)$.

The result of the construction process is a text base network consisting of text nodes, plus some LTM nodes that activated by the text nodes, and their connection weights. At this stage, the network is disorderly, redundant, and even contradictory. Therefore it should be integrated into a coherent structure through a spreading activation process to find the appropriate propositions.

4.2 Integration Phase

The integration phase means integrate the network to a coherent one by spreading activation around the text base network until convergence occurs. At the beginning, the text base network should be abstracted into a connection matrix (CM), and each node of the network attaches a numerical activation that denotes its active degree. In the original model, spreading activation is simulated just by repeatedly postmultiplying CM by activation vector until the vector unchanged. Here we extend it further to iteratively calculating the activation as following.

Assume n is the number of nodes in the text base network, namely there are n propositions in all. In the Euclidian space R^n, let a real number a_i be the activation of the ith node. Let real number l_{ij} be the connection weight between the ith and j^{th} nodes. Then the activations of the network can be represented by an activation vector: $A = (a_1, a_2, \cdots, a_n)$. And the connection weights of the net can be represented by a $n \times n$ connection matrix $CM_{n \times n}$. When i equals to j, the l_{ij} as the values in the main diagonal of $CM_{n \times n}$ is set to 1, because we consider that the node has relations with itself.

Then we introduce a decay mechanism of activation into spreading process to simulate the human mind and adjust the equilibrium point. We define a decay function $D(t)$ to express the decrease of activation with the iteration number t.

$$D(t) = 1 - e^{-dt} \tag{2}$$

The first derivative of $D(t)$ is positive and the second derivative is negative. That is consistent with the cognitive process. Parameter d which controls the rate of descent is assigned to 0.5 empirically. In each iteration, new activation a_i' can be figured out by:

$$a_i' = \sum_{j=1}^{n} a_j \cdot CM_{ij} - a_i \cdot D(t) \qquad 0 \leq i, j \leq n \tag{3}$$

According to formula (3) the iterative approach of activation vector can be expressed as:

$$A' = \left(CM_{nxn} - D(t) \cdot E_n \right) \cdot A \tag{4}$$

Then we define three functions to describe the process of spreading activation. f_1 denotes the iterative function of activation vector. f_2 sets all the negative activations to zero. And f_3 is the normalized function to assure the summation of activations:

$$f_1(x) = \left(CM_{nxn} - D(t) \; E_n \right) \cdot x ,$$

$$f_3(x) = \frac{x}{\sum_{i=1}^{n} |x_i|} , \qquad f_2(x) = \begin{pmatrix} \max(x_1, 0) \\ \max(x_2, 0) \\ \vdots \\ \max(x_n, 0) \end{pmatrix}$$

To simplify the description, we defined a composite function $F(x) = f_3(f_2(f_1(x)))$, in which variable x represents an activation vector. Then the algorithm of spreading activation can be described as Figure 2.

```
Algorithm of Spreading Activation:
1: BEGIN input:  A(0). //initial acitivation vector
2:                CT = 10^{-3}. //convergence threshold
3:                i = 1; //iteration numbers
4:         do
5:             A(i) = F(A(i-1));
6:             i = i + 1;
7:         while ‖A(i-1) - A(i-2)‖ > CT
8:         return A(i);
9: END
```

Fig. 2 Algorithm of Spreading activation

In this paper, the initial activations of the textual nodes are all set to $1/m$, where m is the total number of textual nodes, and that of the LTM nodes are all set to zero. A stable state is reached when the change in the activation values after a multiplication is less than the threshold. Finally, we split the proposition candidates into several sets by their anchors. And pick out the one with the maximum activation from each set as the result. These chosen propositions constitute the spatial semantic representation of the text.

5   Evaluation

5.1   Data Set and Evaluation Metric

As far as we know there is no public corpus for the spatial semantic analysis at present. According to the specific task that our method orients to, we build up two datasets by respectively selecting 100 children's stories from Aesop's Fables (Chinese Edition) and Web, marking as D1 and D2. Then manually translate them into the corresponding propositions like Table 3 as the answer of spatial semantic analysis. The experiment is conducted on the most commonly used metrics in the field of information processing: Precision (P), Recall (R) and The F-measure (F).

5.2   Experiment and Result

In order to verify the validity of the method presented by this paper, we set two baselines, random selection and artificial selection, marking as RS and AS respectively. Random selection yields results by randomly selecting a proposition from the candidates with the same anchor. Similarly, artificial selection means manually select a proposition from candidates with the same anchor. CI corresponds to the method proposed in this paper. The average P, R and F of three methods are shown in Table 5.

Table 5 The experiment result on two datasets

	D1			D2		
	P (%)	R (%)	F (%)	P (%)	R (%)	F (%)
RS	66.85	82.32	73.78	64.49	81.49	72.00
CI	83.25	86.85	85.01	77.19	84.83	80.93
AS	88.34	91.14	89.72	87.30	89.61	88.44

Furthermore, we compare the performance of these three methods on the propositional level. The F-measures of each proposition category on D1 and D2 are presented graphically in Figure 3.

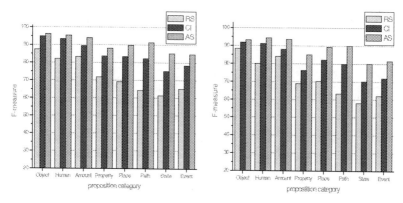

Fig. 3 Propositional level result on D1 and D2

5.3 Discussion

As can be seen from experimental results, the cognitive method achieve an average F-measure over 80%, performs significantly better than the random baseline on both datasets, especially on D1 which is much less complicated than D2. It is due to our method is affected by the deep semantic of dataset, unlike the random selection which is almost independent of the source. The artificial selection still can not achieve the perfect result because of the cumulative error in the processes of natural language processing and proposition candidates constructing.

Comparing with the artificial selection, our method performs slightly worse on the propositions of superficial levels including Object, Human, Amount and Property classes. But the gap becomes more apparent when comes to the deep levels such as State and Event classes. It suggests that proposition candidates of surface classes are of higher quality than the candidates of deep classes. The main reason is that the off-the-shelf NLP tools for creating proposition candidates of the first six categories outperforms the frame-based method used for creating proposition candidates of State and Event classes. The verb frames we used are defined according to the CPB corpus in which most documents are news excerpted from Wall Street Journal. But it can't well support our datasets which are collections of children's stories. The inconsistency originates from two aspects: constrain and number of role between frames. While the results of this paper are significant, there remains much room for improvement.

6 Conclusions and Future Work

This paper presents a cognitive method for a task-oriented spatial semantic analysis. Comparing with traditional semantic analysis methods using machine learning models and surface features, our method implicitly integrates the contextual and general knowledge into analyzing process so that the semantically consistent propositions will be strengthened, and the opposite ones will be excluded. The evaluation has shown that this method can successfully identify spatial semantics on two datasets of stories.

In future, we plan to improve the method in several ways, such as defining more appropriate verb frames, assigning the initial activation of proposition candidates distinctively and establishing a training mechanism for the weights of the text base network. Moreover, we will explore other unitary cognitive models that could solve the issue of spatial semantic analysis better.

Acknowledgments. The work described in this paper was supported by the National Natural Science Foundation of China under grant No. 60575041 and 60803094.

References

Coyne, B., Sproat, R.: WordsEye: an automatic text-to-scene conversion system. In: Proc. of the Siggraph Conference, Los Angeles, CA, USA, pp. 487–496 (2001)

Dong, Z., Dong, Q.: HowNet (1999), http://www.keenage.com

Fillmore, C.J., Baker, C.F.: Frame semantics for text understanding. In: Proc. of WordNet and Other Lexical Resources Workshop. NAACL, Pittsburgh (2001)

Gildea, D., Jurafsky, D.: Automatic labeling for semantic roles. Computational Linguistics 28, 245–288 (2002)

Guha, A., Rossi, J.P.: Convergence of the integration dynamics of the Construction-Integration model. Journal of Mathematical Psychology 45, 355–369 (2001)

Haenggi, D., Kintsch, W., Gernsbacher, M.A.: Spatial Situation Models and Text Comprehension. Discourse Processes 19, 173–199 (1995)

Johansson, R., Williams, D., Berglund, A., Nugues, P.: Carsim: a system to visualize written road accident reports as animated 3D scenes. In: Hirst, G., Nirenburg, S. (eds.) ACL 2004: Workshop on Text Meaning and Interpretation, Barcelona, Spain, pp. 57–64 (2004)

Kapusuz, E.: Refining the representational basis of the construction-integration model of text comprehension with syntactic cues. Master thesis of METU, Turkey (2001)

Kintsch, W.: The Role of Knowledge in Discourse Comprehension: a construction and integration model. Psychological Review 95, 163–182 (1988)

Kintsch, W.: Comprehension: a paradigm for cognition. Cambridge University Press, Cambridge (1998)

Lang, et al.: Language Technology Platform (2004),
http://ir.hit.edu.cn/demo/ltp

Ma, M., Mc Kevitt, P.: Visual semantics and ontology of eventive verbs. In: Su, K.-Y., Tsujii, J., Lee, J.-H., Kwong, O.Y. (eds.) IJCNLP 2004. LNCS, vol. 3248, pp. 187–196. Springer, Heidelberg (2004)

Ma, M.: Automatic Conversion of Natural Language to 3D Animation, PhD Thesis, University of Ulster, Northern Ireland (2006)

Ma, M., Mc Kevitt, P.: Semantic representation of events in 3D animation. In: Proc. of the Fifth International Workshop on Computational Semantics (IWCS-5), Tilburg, The Netherlands, pp. 253–281 (2003)

Mannes, S., Roushey, M.: NETWORK: A computer simulation of the construction-integration model, Technical Report, Institute of Cognitive Science, University of Colorado, Boulder (1990)

McKoon, G., Ratcliff, R.: Memory-based language processing: Psycholinguistic research in the 1990s. Annual Review of Psychology 49, 25–42 (1998)

Mueller, E.T.: Story understanding. In: Encyclopedia of Cognitive Science. Macmillan Reference, London (2002)

Sanjose, V., Vidal-Abarca, E., Padilla, O.M.: A connectionist extension to Kintsch's Construction-Integration Model. Discourse Processes 42, 1–35 (2006)

Xue, N., et al.: Chinese Propbank (2005),
http://verbs.colorado.edu/chinese/cpb

Xue, N., Palmer, M.: Automatic semantic role labeling for Chinese verbs. In: Proc. of the International Joint Conferences on Artificial Intelligence (IJCAI), Edinburgh, Scotland, pp. 1160–1165 (2005)

Formal Semantics of OWL-S with F-Logic

Huaikou Miao[*], Tao He, and Liping Li

Abstract. OWL-S provides Web Service providers with a core ontological framework and guidelines for describing the properties and capabilities of their Web Services in unambiguous and, computer-interpretable form. To support effective verification tools of OWL-S, a formal semantics of the language is highly desirable. In this paper we propose a declarative methodology based on F-logic for modeling OWL-S ontologies, present a formal F-logic semantics of OWL-S and specify their global properties and frame as F-logic formulas. This methodology allows us to bring to bear a body of work for using first order logic based model checking to verify certain global properties of OWL-S constructed service systems.

Keywords: OWL-S, F-logic, Formal semantics, Model checking.

1 Introduction

OWL-S has a model-theoretic semantics that provides a formal meaning for OWL-S ontologies and instance data expressed in them. An OWL-S ontology typically consists of a number of classes, a number of relations (some-times called properties) between these classes, a number of instances and a number of axioms. These elements are all expressed using some logical language. Additionally, F-Logic allows using the same identifier as a class, instance, or property identifier, while still having a standard first-order logic-based semantics.

Huaikou Miao, Tao He, and Liping Li
School of Computer Engineering and Science, Shanghai University, 200072, Shanghai,China
e-mail: hkmiao@shu.edu.cn, he_tao@foxmail.com, liping2000@yahoo.com.cn

Huaikou Miao and Tao He
Shanghai Key Laboratory of Computer Software Evaluating & Testing, 201112, Shanghai, China

[*] This work is supported by National Natural Science Foundation of China (NSFC) under grant No. 60673115, National High-Technology Research and Development Program (863 Program) of China under grant No. 2007AA01Z144 and the National Grand Basic Research Program (973 Program) of China under grant No. 2007CB310800, the Research Program of Shanghai Education Committee under grant No. 07ZZ06,and Shanghai Leading Academic Discipline Project, Project Number: J50103

Antonio Brogi[1] presents a translator from OWL-S descriptions to Petri nets which makes such analyses possible thanks to the many tools available for Petri nets. S. Narayanan and S. A. McIlraith [2] proposed a model based on Petri nets to specify a web service of OWL-S. They suppose that the service is composed of several atomic services, and the supposition is not suitable in open-environment composite service. But the static semantics of OWL-S have not been discussed in their methodology. Jiang Yun-Chen[3] analyzed the formal semantics of semantic Web services description ontology OWL-s and provides reasonable theory foundation for semantic Web services. However, there is just no effective reasoner to support reasoning and transaction.

In this paper, we present a formal denotation model of OWL-S using the F-logic specification language [4]. F-logic has been used to provide one single formal model for the syntax, the static semantics of OWL-S. The advantage would be that of a full-fledged logic that integrates notions of description, reasoning, object-orientation and logic programming can serve as a unifying formalism for current description languages; and can extend their expressivity in the above listed directions. This model not only provides a formal unambiguous model which can be used to develop tools and facilitate future development, but can be used to identify and eliminate errors in the current documentation.

The paper is organized as follows. Section 2 briefly introduces the notion of OWL-S and F-logic. Section 3 is devoted to map OWL DL syntax to F-logic syntax. Section 4 discusses static formal semantics of OWL-S and gives a formal model of OWL-S framework using F-logic specification language. Section 5 discusses some inference problems of OWL-S with F-logic. Section 6 concludes the paper and discusses possible future work.

2 Background

2.1 OWL-S

OWL-S, one of the most significant Semantic Web Service ontologies proposed to date, is an upper ontology for modeling web-service composition which offers a process-based perspective.

OWL-S process model describes formation of services by composition

(i) Service profile –which presents what function the service computes. This information is expressed in terms of the transformation that the service produces.

(ii) Process model – which describes the service behavior providing a view of the service in terms of process compositions. OWL–S defines three types of processes: atomic processes, composite processes, simple processes.

(iii)Service grounding – which offers all details about their invocation.

An atomic process can not be decomposed further and it executes in a single step (similarly to a black box providing a functionality), while a composite process is built up by using a few control constructs: Sequence, Split, Split-Join, Any-Order, Iterate, If-Then-Else, Choice, repeat-while and repeat-until.

2.2 F-Logic

Frame Logic [5] provides a logical foundation for frame-based and object-oriented languages. It has a model-theoretic semantics and a sound and complete resolution based proof theory [4]. In F-Logic, classes and properties are interpreted as objects. This may hamper interoperation with Description Logic-based ontology languages (e.g. OWL DL [6]), in which classes and properties are interpreted as unary and binary predicates, respectively.

Definition 1 (Syntax of F-Logic). The alphabet of an F-Logic language consists of a set F of function symbols, playing the role of object constructors. For convention, function symbols start with lowercase letters whereas variables start with uppercase ones. Id-terms are composed of object constructors and variables and are interpreted as elements of the universe.

In the sequel, let o, c, d, $d_1, \ldots, d_n$, p, v, $v_1, \ldots, v_n$ stand for id-terms or literals. Note that URLs as a subclass of strings can denote objects. [7]

1. An is-a atom is an expression of the form $o : c$ (object o is a member of class c), or $c :: d$ (class c is a subclass of class d).

2. The following are object atoms:

2a. $c[p \Rightarrow (d_1, \ldots, d_n)]$ and $c[p \Rightarrow\!\!\!\!> (d_1, \ldots, d_n)]$: the values of the scalar or multi-valued, respectively, property p of objects of class c belong (simultaneously) to all classes $d_1, \ldots, d_n$,

2b. $o[p \rightarrow v]$: the scalar property p of object o has the value v,

2c. $o[p \rightarrow\!\!\!\!> \{v_1, \ldots, v_n\}]$: $\{v_1, \ldots, v_n\}$ are amongst the values of the multivalued property p of object o,

2d. $c[p \bullet\!\!\rightarrow v]$: for objects of class c, the default value of the scalar property p is v.

2e. $c[p \rightarrow\!\!\!\!> \{v_1, \ldots, v_n\}]$: for objects of class c, the default values of the multi-valued property p are $\{v_1, \ldots, v_n\}$.

A semantic structure F-structure for an F-logic language is tuple $I = <U, \preceq_U, \in_U, I_F, I_\rightarrow, I_{\rightarrow}, I_{\rightarrow}, I_{\rightarrow}, I_{\Rightarrow}, I_{\Rightarrow}>$, where $(U, \preceq_U)$ and $\in_U$ are the partially ordered domain and the membership relations. I_F is the interpretation mapping for object constructors, a standard function mapping. The other six mappings are the associations of domain objects with features, roles, and types.

3 From OWL DL to F-Logic

OWL DL, which is one of the official variants of OWL, is based on description logics and is thus basically a decidable fragment of first-order predicate logic, from which it borrows its formal semantics. OWL and standard logic programming paradigms thus differ in modeling styles and formal semantics in ways which go beyond syntactic issues. The Semantic Web ontology language OWL DL is based on the Description Logic SHOIN. F-Logic is an ontology language which is also based on first-order logic, but classes and properties are modeled as terms rather than predicates.

The given translations use the basic atoms and molecules of F-Logic. More translations of OWL DL constructs are possible with the help of F-Logic rules. This translation is presented in the following section.

Table 1 Mapping from OWL DL abstract to F-logic syntax

OWL Abstract Syntax	F-logic Syntax
Class axioms	
Class(A partial $C_1 \ldots C_n$)	$\wedge A::C_i$ AND $i\Rightarrow\{1,\ldots,n-1\}$
Class(A complete $C_1 \ldots C_n$)	$A::C_1 \wedge \ldots \wedge A::C_n$
EnumeratedClass(A $o_1 \ldots o_n$)	$A\rightarrow\!\!\rightarrow\{o_1,\ldots,o_n\}$
SubClassOf(C_1 C_2)	$C_1::C_2$
EquivalentClasses($C_1 \ldots C_n$)	$C_1=\cdots=C_n$
DisjointClasses($C_1 \ldots C_n$)	$a\rightarrow(\ )\leftarrow a:C_1 \wedge \ldots \wedge a:C_n$
Property axioms	
ObjectProperty(R super(R_1)...super(R_n)	$R::R_i$
domain(C_1) ... domain(C_n)	$D\Rightarrow\{C_1,\ldots,C_n\}$
range(C_1) ... range(C_n)	$R\Rightarrow\{C_1,\ldots,C_n\}$
[inverseOf(R_0)]	$\sim R_0$
[Symmetric]	$X[R\rightarrow\!\!\rightarrow\{Y\}]\leftarrow Y[R\rightarrow\!\!\rightarrow\{X\}]$
[Functional]	$C_1[Func\rightarrow\!\!\rightarrow\{C_2\}]$
[InverseFunctional]	$C_1[invFunc(R)\rightarrow\!\!\rightarrow\{C_2\}]=C_2[R\rightarrow\!\!\rightarrow\{C_1\}]$
[Transitive])	$C_1::C_3\leftarrow C_1::C_2 \wedge C_2::C_3$
Datatype(T)	$D:T$
DatatypeProperty(U super(U_1)...super(U_n)	$U::U_i$
domain(C_1) ... domain(C_n)	$x[C\Rightarrow\{C_1,\ldots,C_n\}]$
range(T_1) ... range(T_n)	$x[R\Rightarrow\{T_1,\ldots,T_n\}]$
[Functional])	$x[funcName@parameter\rightarrow value]$
SubPropertyOf(O_1 O_2)	$O_1::O_2$
EquivalentProperties($O_1 \ldots O_n$)	$O_1=\ldots=O_n$
Individual assertions	
Individual(o type(C_1) ... type(C_n)	$i\Rightarrow\{1,\ldots,n\}$ AND $\wedge o:C_i$
value(R_1 o_1) ... value(R_m o_m)	$i\Rightarrow\{1,\ldots,n\}$ AND $R_i\rightarrow O_i$
value(U_1 t_1) ... value(U_n t_n))	$i\Rightarrow\{1,\ldots,n\}$ AND $U_i\rightarrow t_i$
SameIndividual($o_1 \ldots o_n$)	$O_1=\ldots=O_n$
DifferentIndividuals($o_1 \ldots o_n$)	$O_1\neq\ldots\neq O_n$

Table 2 Mapping from OWL DL axioms to F-logic syntax

OWL Abstract Syntax	F-logic Syntax
	DL Axioms
intersectionOf(C_1,...,C_n)	$i \Rightarrow \{1,...,n\}$ AND $\bigwedge C_i$
unionOf(C_1,...,C_n)	$i \Rightarrow \{1,...,n\}$ AND $\bigvee C_i$
complementOf(C)	$\neg C$
one of(o_1,...,o_n)	$C \Rightarrow \{o_1,...,o_n\}$
restriction(R someValuesFrom(C))	$\leftarrow \exists y : C \ x[R \rightarrow y]$
restriction(R allValuesFrom(C))	$\leftarrow \forall c_i : C \ x[R \Rightarrow \{c_i\}]$
restriction(R value(o))	$\leftarrow x[R \rightarrow o]$
restriction(R minCardinality(n))	$\leftarrow x[R@number->> \{C_1,...C_n,...,C_i\} : number \Rightarrow \{n, n+1,...\}]$
restriction(R maxCardinality(n))	$\leftarrow x[R->> \{C_1,...,C_n\}]$
restriction(U someValuesFrom(T))	$\leftarrow \exists y : T \ x[U \rightarrow y]$
restriction(U allValuesFrom(T))	$\leftarrow \forall t_i : T \ x[U \Rightarrow \{t_i\}]$
restriction(U value(t))	$\leftarrow x[U \rightarrow t]$
restriction(U maxCardinality(n))	$\leftarrow T_1 \neq ... \neq T_n \wedge x[U->> \{T_1,...,T_n\}]$
restriction(U minCardinality(n))	$\leftarrow T_1 \neq ... \neq T_i \wedge x[U@number->> \{T_1,...,T_n,...,T_i\} : number \Rightarrow \{n, n+1,...,i\}]$

4   From OWL-S to F-Logic

In this section, we present F-logic semantics of OWL-S. In this semantic model, all the different aspects of the language. syntax (an OWL-S model is well-formed), static semantics (an OWL-S model is meaningful) and dynamic semantics (how is an OWL-S model interpreted and executed), are defined in one single unified framework, so that the semantics of the language can be more consistently defined and revised as the language evolves. The OWL-S elements are modeled as different F-logic classes. The syntax of the language is captured by the attributes of an F-logic class. The predicates defined as class invariant are used to capture the static semantics of the language. The class operations are used to define OWL-S's dynamic semantics, which describe how the state of a Web service changes. This paper focuses on the first two aspects of OWL-S, i.e. the formal model of syntax and static semantics.

4.1 Upper Level Model

```
Servic [ presents ⟹ ServiceProfile;
    describedBy ⟹ ServiceModel;
    supports ⟹ ServiceGrounding ].
```

Service denotes the OWL-S service and it has three attributes presents, supports and describedBy, which denotes three essential types of knowledge of an OWL-S serivce – the profile, the process model and the grounding. The OWL-S profile describes what the service does. Thus, a Service presents ServiceProfiles. ServiceProfile, ServiceModel and ServiceGrounding will be formally defined later.

4.2   OWLS-S Profile

OWL-S is an OWL-based Web ontology, which is intended to provide Web service providers with a core set of constructs for describing the properties and capabilities of their Web services. OWL-S often refers to externally defined data types using the namespace notation. A Web now give a partial description of OWL-S profile ontology specified using F-logic [8]. Though some implementation of F-logic support URIs and namespaces, our description will omit all namespace definitions and will reference the corresponding external data types and concepts by enclosing them in single quotes, e.g., XMLString.

```
'service:ServiceProfile' : 'owl:Class'.
'Profile' :: 'service:ServiceProfile'[
   serviceName  ⟹ XMLString;
   contactInformation  ⟹≫ 'Participant : owl:Class';
   hasProcess  ⟹ Process;
   serviceCategory  ⟹≫ ServiceCategory;
   serviceParameter  ⟹≫ ServiceParameter;
   hasParameter  ⟹≫ Parameter;
   hasInput  ⟹≫ Input;
   hasOutput  ⟹≫ ConditionalOutput;
   hasPrecondition  ⟹≫ 'expr:Condition';
   hasEffect  ⟹≫ ConditionalEffect ].
```

The frame-based syntax of F-logic enables concise and clear description of the properties of the various classes defined by OWL-S.

4.3   ServiceModel – Modeling Services as Processes

An OWL-S service can be viewed as a process from an interaction point of view. The OWL-S process model is intended to provide a basis for specifying the behaviors of a wide array of services. Before we present the formal definition of Process, we first show some necessary elements for defining Process.

4.3.1   Parameters and Expressions

```
ProcessVar::Variable[parameterType→URI].
```

The class Variable denotes the generic variable type, while ProcessVar, defined as a subclass of Variable, denotes the variables used in an OWL-S service process. It has an attribute parameterType denoting the type of the variable which is specified using a URI. We use the F-logic class URI to denote all possible URI address.

```
ProcessVar ⟹≫ {Parameter, PreCond, Participant, ResultVar, Local}←
Participant::ProcessVar AND Parameter::ProcessVar AND PreCond::ProcessVar
AND ResultVar::ProcessVar AND Local::ProcesVar.
```

In an OWL-S process model, there are five distinct kinds of ProcessVar, as Parameter , PreCond , Participant , ResultVar and Local .

    a→{ } ←(a:Parameter ∨ a:PreCond ∨ a:ResultVar) AND a:Participant.
    theClient ≠theServer← theClient:Participant AND theServer:Participant.

A Participant is a variable used to denote an agent involved in a process. It is disjoint with Parameter, PreCond and ResultVar. There are two predefined Participant agents. One is theClient, representing the agent from whose point of view the process is described, and another is theServer, representing the principal element of the service that the client deals with. theClient and theSever are modeled as instances of Participant.

    Parameter —→ {Input,Output} AND (a:Input AND a:Output∧a→{ }) ← Input:Parameter AND Output:Parameter.
    Parameter is the disjoint union of Input and Output and is used to denote the data transformation produced by the process.

    a→{ } AND b→{ } ← (a:PreCond AND a:Parameter) AND (b:PreCond AND b:ResultVar).
    FORALL a:PreCond. EXISTS e:Expression, a:e[usedVars —→ ()] AND (EXISTS p:Process, e:p[hasPrecondition —→ ()]).

An PreCond is a variable whose value will be bound in a process precondiction to make it true and this value of the existential so obtained can appear in the effects of Results, and, if the Process is composite, can be referred to in its body. The above axiom also ensures that a PreCond will be bound in some preconditions. Expression and Process will be defined later.

    a→{ }←a:ResultVar AND a:Parameter.
    FORALL rv:ResultVar, EXISTS e:Expression, rv:e.usedVars AND (EXISTS r:Result, e:r.inCondition).

ResultVars are analogous to PreConds. Whereas PreConds are variables to be bound in preconditions and then used in the specifying result conditions, outputs and effects, ResultVars are scoped to a particular result, are bound in the result's condition (inCondition), and are used to describe the outputs and effects associated with that condition. The above axiom ensures that a ResultVars will be bound. Result will be defined later.

    Local is the disjoint union of Loc and Link and it used for intermediate results in a composite process. The detail of the usage will be discussed later when we define the data flow in OWL-S.

    a→{ }←(a:Participant OR a:PreCond OR a:ResultVar) AND a:Local.

OWL-S defines a flexible framework which allows users to choose various logic languages to describe services. The key idea is that in OWL-S, expressions are treated as literals (either string or XML). Due to the space limitation, we abstract the XML literals and string literals as F-logic classes. The class Expression has the attribute expressionLanguage to denote which LogicLanguage is used to express the expression. LogicLanguage is referenced by a URI and the attribute expressionBody to denote the actual literal expression.

We also define a secondary attribute [9] usedVars to denote the variables used in expressionBody. Depending on different languages used to describe an expression, there are different ways to retrieve the variables from the expression. We abstract this as the function VarsInExpression, that is given an expression and a logic language name. VarsInExpression returns the set of variables used in the expression. The secondary attribute means that the value of usedVars depends on the values of some other attributes such as expressionLanguage and expressionBody or around environment.

LogicLanguage[refURI→URI].
        Literal —↠ {XMLString, StringLiteral}.
        Expression[VarInExpression ⟹{LogicLanguage,Literal};
                expressionLanguage⟹LogicLanguage:
                expressionBody⟹Literal;
                usedVars ⟹ ProcessVar].
    Condition::Expression [value —↠ {BoolValue,NilValue}].
    usedVar —↠ Vars ← VarsInExpression @ (expressionLanguage,
    expressionBody) —↠ Vars.

Condition is a subclass of Expression and it has a determined truth value. We also define a special kind of value called the nil value. This is used when variables are not bound to any concrete values.

4.3.2 Processes

OWL-S defines three different kinds of processes – atomic process, composite process and simple process. Before formally defining each one in detail, we first specify some of the common attributes of an OWL-S process.

ProcessCom::ServiceModel [
                hasClient::hasParticipant ⟹ Participant;
                performedBy::hasParticipant ⟹ Participant;
                hasParticipant::hasVar ⟹ Participant;
                hasInput::hasParameter —↠ Input;
                hasOutput::hasParameter —↠ Output;
                hasParameter::hasVar ⟹ Parameter;
                hasPrecondition ⟹ Condition;
                hasResult ⟹ Result;
                hasPreCond::hasVar ⟹ PreCond;
                hasVar ⟹ ProcessVar].
    FORALL   re:hasresult,   ob:re.withOutput,   ob.toVar:hasOutput   ←
ProcessCom::ServiceModel.

The attribute hasClient denotes those agents from whose point of views the process is described and performedBy denotes those elements of the service that the client deals with. HasParicipant represents all the parts involved in the process. HasClient and performedBy are subsets of hasParicipant . HasInput denotes the set of data required by the process for execution. HasOutput defines the information provided back by the process to the requester. HasPrecondition denotes the set of conditions that has to be satisfied for the process to perform successfully. HasVar denotes all the variables used in a process. HasResult specifies the results of the service. The last predicate in the invariant ensures that all the output variables used in hasResult are declared in the process.

Result in OWL-S model specifies under what conditions the outputs are generated as well as what domain changes are produced during the execution of the service. InCondition denotes the conditions under which the result occurs. withOutput denotes the output bindings of output parameter of the process to a value form. hasEffect is a set of expressions that captures possible effects to the context. HasResultVar declares variables that bound in the inCondition. Furthermore, these variables are scoped to this particular result. The containment notation 'c' in F-logic can ensure this. OutputBinding is a subset of Binding (which will be defined later) with toVar as an Output.

$$Result [\quad inCondition \Longrightarrow Condition;$$
$$withOutput \Longrightarrow OutputBinding;$$
$$hasEffect \Longrightarrow Expression;$$
$$hasResultVar \Longrightarrow ResultVar;$$
$$resultForm \Rightarrow Literal].$$

FORALL c:inCondition, hasResultVar::c.VarsInExpression $\leftarrow$ inCondition:: Condition AND withOutput::OutputBinding AND hasEffect::Expression AND hasResultVar::ResultVar AND resultForm::Literal.

AtomicProcess is a kind of Process denoting the actions a service can perform by engaging in a single interaction. The invariant shows that an atomic process has the client as theClient and is performed by theServer; and that only those two Participants exist.

$$AtomicProcess::ProcessCom [realizes \Longrightarrow SimpleProcess;$$
$$hasClient \longrightarrow \{theClient\};$$
$$performedBy \longrightarrow \{theServer\};$$
$$hasParticipant \longrightarrow \{hasClient, theServer\}].$$

FORALL s:realizes, self:s.realizedBy $\leftarrow$ realizes::SimpleProcess.

SimpleProcess is another kind of Process. It is mainly used either to provide a view of (a specialized way of using) some atomic processes, or a simplified representation of some composite process (for purposes of planning and reasoning).

$$SimpleProcess::ProcessCom [realizedBy \Longrightarrow AtomicProcess;$$
$$expandsTo \Longrightarrow CompositeProcess].$$

FORALL s:realizedBy, self:s.realizes← realizedBy::AtomicProcess AND expandsTo::CompositeProcess.

FORALL a:expandsTo, self:a.subdivideTo ← realizedBy::AtomicProcess AND expandsTo::CompositeProcess.

Composite processes are processes which are decomposable into other (non-composite or composite) processes and are modeled as CompositeProcess. A CompositeProcess can subdivideTo some SimpleProcsses. HasLocal defines a set of local variables used in the 'Producer-Push' data flow pattern used in a composite processes. We will define this pattern later. The attribute invocable is an optional boolean value which is used to tell whether the CompositeProcess bottoms out in atomic processes. A CompositeProcess must have a composedOf attribute by which it indicates the control structure of the composite, using a ControlConstruct. The secondary attribute bottomProcess denotes the set of bottom processes in a CompositeProcess. Given a controlConstruct, the function getBottomProcess will return the set of bottom processes. The detail definition of getBottomProcess is omitted here.

CompositeProcess::ProcessCom[
        collapsesTo ⟹ SimpleProcess;
        hasLocal::hasVar ⟹ Local;
        composeOf ⟹ controlConstruct;
        bottomProcess::Process[
        getBottomProcess@composeOf→( )];
        hasResult ⟹ Result;
        hasInput ⟹ Input].
FORALL s:realizes, self:s.realizedBy← CompositeProcess.
hasLocal::hasVar ∧ invocableNum⟹{1,…,n} ← CompositeProcess.
invocable→{true} ← bottomProcess::AtomicProcess.
invocable→{false} ←
    bottomProcess [AtomicProcess @ bottomProcess ⟹bottomprocess]
    AND AtomicProcess [bottomProcess@AtomicProcess ⟹AtomicProcess
    ].

Each control construct, in turn, could associate with an additional property called components to indicate the nested control constructs from which it is composed, and, in some cases, their ordering. The secondary attribute definedInProcess defined in controlConstruct denotes the composite process the controlConstruct belonging to.

controlConstruct[defineInProcess ⟹CompositeProcess].
    EXISTS cp:CompositeProcess, cp.composeOf ⟶⟶ self AND definedInProcess→cp ← defineInProcess::CompositeProcess.

Perform is a special kind of ControlConstruct used to refer a process to a composite process. It has two attributes – performProcess which denotes the process to be invoked and hasDataFrom which denotes the necessary input bindings for the invoked process, where InputBinding is a subset of Binding with toVar to

Input. The invariant of Perform ensures that all the inputs of performProcess are provided and if the input is derived from onther process then it can only be derived from the parameters of another Performs in the same composite process. OWL-S also introduces a standard variable, ThisPerform, used to refer, at runtime, to the execution instance of the enclosing process definition.

Perform::controlConstruct [performProcess$\Rightarrow$Process;
                      hasDataFrom $\Longrightarrow$ InputBinding;

          performProcss[hasInput $\longrightarrow$ df:hasDataFrom[toVar$\Rightarrow$ProcessVar]]].
FORALL   df:hasDataFrom,   vs:df.valueSource,   fp:vs.fromProcess, defimeInProcess$\Rightarrow$fp.defimedInProcess
  $\leftarrow$ performProcess::Process $\wedge$ hasDataFrom::InputBinding.
FORALL ib:InputBinding, ib.tovar $\longrightarrow$ Input$\leftarrow$InputBinding::Binding.

In OWL-S, the Service Profile provides a way of describing the services offered by providers, and the services needed by requesters. Three basic types of information about a service is provided, as what organization provides the service, what function the service computes, and a host of features that specify the service characteristics. ServiceProfile denotes the generic OWL-S profile, while Profile defines a subclass of ServiceProfile that denotes the predefined profile. It is used to acknowledge the different ways of profiling services from the default one.

The grounding of a service specifies the details of how to access a service — details such as protocols and message formats, serialization, transport, and addressing. OWL-S does not include an abstract construct for explicitly describing messages. Therefore, we just abstract ServiceGrounding as a simple F-logic class.

5 Inference of OWL-S with F-Logic

In order to permit cross referencing within LNCS-Online, and eventually between different publishers and their online databases, LNCS will, from now on, be standardizing the format of the references. This new feature will increase the visibility of publications and facilitate academic research considerably. Please base your references on the examples below. References that don't adhere to this style will be reformatted by Springer. You should therefore check your references thoroughly when you receive the final pdf of your paper. The reference section must be complete. You may not omit references. Instructions as to where to find a fuller version of the references are not permissible.

In order to verify an F-logic model of OWL-S ontologies design, the consistency, satisfiability, subsumption, equivalence, instantiation of an F-logic model must be verified. When referring to consistency of models, we must check if knowledge is meaningful, and if there is some model I of O, some possible I of C, and so on. Whether two classes denote them same set of instances must be checked by reasoning the equivalence of models, too. To confirm the instantiation of model, we must check if individual i is an instance of class C and retrieve set of individuals that instantiate C. F-OWL, an inference engine basing on F-logic for OWL language, is designed to accomplish this task.

We can prove the Soundness of F-logic Deduction with theorem "if $S \vdash C$ then $S \models C$" and the Completeness of ground deduction with theorem "If a set of ground clauses, S, is unsatisfiable then there exists a refutation of S" [4]. The following theorems are given for the Provenance Model and verification of system.

Let L be a predicate-based ontology language with signature $\Sigma = <F, P>$ and their F-logic counterparts, then the corresponding sorted F-logic language L^F has the signature Σ', the thus obtained translation function is denoted as δ.

Theorem 1. Let P be a positive logic program and KP be the corresponding Horn F-Logic theory, then P has one stable model M and for every ground atom or molecule α, $\alpha \in M$ iff $K_P \models \alpha$.

Theorem 2. Let $K \subseteq L$ be a F-logic theory and let $\kappa \in L$ be a formula. Then $K \models \kappa$ iff $\delta(K) \models_f \delta(\kappa)$.

Theorem 3. There are mapping

    Γ:{F-formuals} {well-formulas of predicate calculus}

    Φ:{F-structures} {semantic structures of predicate calculus}

    such that $M \models_F \psi$ if and only if $\phi(M) \models_{PC} \Gamma(\psi)$ for any F-structure M and any F-formula ψ, where "$\models_F$" and "$\models_{PC}$" denote logical entailment in F-logic and predicate calculus, respectively.

F-OWL inference engine's core responsibilities are to adhere to the formal semantics in processing information encoded in OWL, to discover possible inconsistencies in OWL-S data, and to derive new information from known information.

6 Conclusions

We have introduced a translation from OWL-S ontologies to ontologies in F-Logic and shown that this translation preserves entailment for large classes of predicate-based ontology languages in this paper. We also focus on the formal model for the syntax and static semantics of OWL-S. The dynamic semantics of OWL-S will be discussed in future. The results obtained in this paper can be used for, for example, F-Logic based reasoning with, and extension of, classes of predicate-based ontology languages, and also used for analyzing the modeling paradigms of OWL -S, as well as reasoning tasks supported by OWL-S.

References

1. Brogi, A., Corfini, S., Iardella, S.: From OWL-S descriptions to Petri nets. In: Proceedings of Third international workshop on engineering service-oriented applications: analysis, design and composition (WESOA 2007). Wien, Austria, September 17 (2007); Service-Oriented Computing - ICSOC 2007 Workshops, pp. 427–438. Springer, Heidelberg (2007)
2. Thomas, J.P., Thomas, M., Ghinea, G.: Modeling of Web Services Flow. In: Proceedings of the IEEE International Conference on ECommerce (CEC 2003), p. 391 (2003)

3. Yun-Chen, J., Zhong-zhi, S.: The Formal Semantics of OWL-S. Computer Science 32(2), 5–7 (2005)
4. Kifer, M., Lausen, G., Wu, J.: Logical Foundations of Object-Oriented and Frame-Based Languages. JACM 42(4), 741–843 (1995)
5. Zou, Y., Finin, T., Chen, H.: F-OWL: an inference engine for the semantic web. In: Hinchey, M.G., Rash, J.L., Truszkowski, W.F., Rouff, C.A. (eds.) FAABS 2004. LNCS, vol. 3228, pp. 238–248. Springer, Heidelberg (2004)
6. Dean, M., Schreiber, G. (eds.): OWL Web Ontology Language Reference. W3C Recommendation, February 10 (2004)
7. Kattenstroth, H., May, W., Schenk, F.: Combining OWL with F-Logic Rules and Defaults. In: International Workshop on Applications of Logic Programming to the Web, Semantic Web and Semantic Web Services (ALPSWS 2007), September 13, vol. 287, pp. 60–75 (2007)
8. OWL-S Coalition. OWL-S: Semantic markup for Web services. Release 1.2 (December 2004), http://www.daml.org/services/owl-s/1.2/overview/
9. Davidson, S., Boulakia, S., Eyal, A., Ludäscher, B., McPhillips, T., Bowers, S., Anand, M., Freire, J.: Provenance in Scientific Workflow Systems. IEEE Data Eng. Bull. 30(4), 44–50 (2007)

Cache Offset Based Scheduling Strategy for P2P Streaming

Xunli Fan, Wenbo Wang, and Lin Guan

Abstract. In an application of P2P streaming media system, the performances of playback continuity and startup latency have been regarded as the most important two metrics intuitively. However, the rarity and urgency of data block is not considered when the popularity of data block is calculated in the traditional scheduling strategies. That leads to the inferior playback continuity and startup latency. To overcome the drawbacks of the traditional scheduling strategies, this paper presents a new scheduling strategy based on Cache Offset (BCOP), which deals with the rarity and urgency of data block simultaneously. In the proposed BCOP strategy, the drawbacks mentioned above have been solved well by means of counting and processing the information about the cache offset of playback of requested data block and the two metrics have been improved noticeably. The results of experiments demonstrate the effectiveness of the proposed strategy in terms of higher continuity and lower startup latency.

Keywords: P2P streaming; playback continuity; startup latency; cache offset of playback.

1 Introduction

With the widespread penetration of broadband accesses, multimedia services are getting increasingly popular among users and have contributed a significant amount to today's Internet traffic. Recently, Peer-to-Peer (P2P) has emerged as a promising technique for deploying large-scale live media streaming systems over the Internet, which represents the paradigm shift from conventional networking applications. In a P2P system, peers communicate directly with each other for the sharing and exchanging of data as well as other resources such as storage and CPU capacity. Each peer acts both as a client, who consumes resources from other

Xunli Fan and Lin Guan
Department of Computer Science, Loughborough University, Loughborough,
LE11 3TU, UK

Xunli Fan and Wenbo Wang
School of Information Science & Technology, Northwest University, Xi'an 70127 China

R. Lee, G. Hu, H. Miao (Eds.): Computer and Information Science 2009, SCI 208, pp. 119–129.
springerlink.com   © Springer-Verlag Berlin Heidelberg 2009

peers, and also as a server, who provides service for others. Compared with traditional streaming techniques such IP multicast and CDN (content delivery networks), P2P based streaming system has the advantages of requiring no dedicated infrastructure and being able to self-scale as the resources of the network increase with the number of users [3].

The design of a mechanism for streaming from a distributed group of heterogeneous peers introduces several challenges. First, since each sender peer should perform TCP-friendly congestion control, the available bandwidth from each sender is not prior known and could significantly change during a session. Second, the connections between different senders are likely to exhibit different characteristics, and even worse, each sender peer can potentially leave the session during delivery. More importantly, a multi-source quality adaptive streaming mechanism requires a coordination mechanism among senders [2, 5].

During recent years, streaming applications have become increasingly popular in P2P systems. At present, as for P2P media streaming models based on Gossip protocol, the scheduling strategies are Rarest-first (RF) [8], Weighted Round-robin (WRR) [1], Greedy [9], and Random [4], etc.

RF strategy is a heuristic algorithm on time of fast response and was promoted in CoolStreaming/DONet [8]. The heuristic algorithm first calculates the number of potential suppliers for each segment. Since a segment with less potential suppliers is more difficult to meet the deadline constraints, the algorithm determines the supplier of each segment starting from those with only one potential supplier, then those with two, and so forth. The RF strategy has the advantages of expediting data-block to disperse in the overlay network, boosting the system's total throughput, and gaining the better playback continuity. However, the major drawback of RF is having the worse performance of start-up latency.

WRR strategy is used in delaminated streaming media systems, such as P2P Adaptive Layered Streaming (PALS) [1]. As we know the PALS is a receiver-driven approach that allows a receiver to orchestrate the adaptive delivery of layer-structured streams from heterogeneous sender peers. Firstly given a set of sender peers, the receiver monitors the available bandwidth from each sender and periodically determines the target quality that can be streamed from all senders. Then, the receiver determines the required packets from different layers for each period, properly divides them among senders, and requests a subset of packets from each sender. The PALS achieves this goal with relatively low amount of receiver buffering, and thus it can be used in a spectrum of non-interactive streaming applications ranging from playback to live sessions in P2P networks. The WRR strategy has the advantage of gaining better balance of payload, while the drawback is that it considers less of how the data sharing is to be improved.

Greedy strategy is also called nearest deadline first [9]. This strategy aims to fill the empty buffer location closest to the playback time first. While from a single peer's point of view, Greedy may be the best for playback, and it is often too shortsighted from a system's point of view when the peer population is large.

Random strategy is adopted by PAI in Chainsaw [4]. A P2P overlay multicast system completely eliminates trees. Peers are notified of new packets by their neighbors and must explicitly request a packet from a neighbor in order to receive it. In this way, duplicate data can be eliminated and a peer can ensure it receives all packets. Its performance is unstable, especially in the isomerism network environment.

In this paper, our aim is to study the scheduling strategy used in P2P streaming media and analyze the factors that affect the performance of streaming media scheduling strategy in P2P networks. We propose a new scheduling strategy based on Cache Offset of playback (BCOP). In this strategy, we compute the scheduling priority according to the position of the requested data block at the buffer of requesting node and the offering node, which can better reflect the situation of data requiring and offering. Empirical simulation experiments with various data analysis have demonstrated the improvement of efficiency of the proposed strategy on streaming media scheduling and correctness of the proposed algorithm in P2P networks.

The rest of the paper is organized as follows. Section 2 reviews the details of RF strategy, its drawback, and possible solutions to overcome its drawback, while Section 3 elaborates and implements a new scheduling strategy based on cache offset of playback. Section 4 presents numerical evaluation results for the BCOP algorithm applied to this model, and conclusions and future work follow in Section 5.

2   Problems of RF Strategy in Streaming Media

Among the strategies metioned on the above, RF strategy is one of the most widely used in P2P file sharing and P2P streaming media systems, such as BitTorrent and CoolSteaming, due to its simplicity and stability. Initially, the RF strategy was adopted in P2P file sharing. Its aim is to increase the diversity of data block in the system and keep the whole data blocks composing the file existing for a long time in the overlay network potentially, and reduce the upload payload of seed node.

2.1   RF Strategy in Streaming Media

In the application of P2P streaming media, RF strategy still brings to play some functions. At the same time, compared with the application of P2P file sharing, the RF strategy represents some other characteristics in the application of P2P streaming media.

Suppose in a constructed and static overlay network environment, there are no nodes joining or exiting in the network, and no unsuccessful events happening. Considering a process of data block dispatching from the data source: at the time t, the data block just enters the overlay network, that is, the data block with the max serial number has the minimum copies in the overlay network, which is called the newest data block at time t. According to RF strategy, the requiring node puts the highest priority to this data block. Since all the nodes in the system use the RF strategy to schedule the streaming media data, then the copies of the newest data block at time t will increase after some time, and the priority of the newest data block at time t will reduce accordingly, and eventually all of the nodes have the newest data block at time t. Correspondingly the priority of the newest data block at time t reduces to 0.

During this period, the undergoing time is the time that the newest data block at time t dispatches through the overlay network. In fact, RF strategy schedules the newest data block with the highest priority. However, the deadline of playback of the newest data block is relatively the longest compared with that of playing at current node. Therefore, in P2P streaming media systems, it usually uses RF strategy to quicken the dispatching of data blocks. To do so, it makes the node to have more chances to download all the data blocks dispatched by server in the next period. Consequently, it can obtain better performance of playback continuity.

2.2 Drawbacks of RF Strategy

From the analysis on the above, we find that RF strategy has good behavior in the application of P2P streaming media system, but at the same time it has two major drawbacks.

Firstly the copies of data block cannot exactly reflect the real situation of data block in the system. To illustrate this, we use the example as Fig. 1.

In Fig.1, the requiring node Pr is requesting data block i and j. At the moment, the number of node holding data block j is 1, and that of i is 3. According to RF strategy, node Pr should request j prior to i. On the contrary, according to the buffer position of data block i and j, we find that after 3 playing unit, the data block i on all the nodes will be removed from the buffer, while data block j is still in the buffer. That means, it is reasonable to schedule the data block i prior at the moment. Fig.1 shows that it is insufficient to consider the copies of data block in the system only when determining the priority for the absence of data blocks. Besides, we need to consider the life cycle for each data block copy.

Secondly, RF strategy only deals with the rarity of the data block. However, it fails to consider the urgency of the requested data block. On one hand, if a data block near the current playing site cannot be scheduled timely, it may be unable to be downloaded before its deadline, which may cause the reducing of playback continuity. On the other hand, RF strategy is propitious to the download of newest data block. As for a newly joined node, it must wait more time to play the streaming media, which causes unnecessary start-up latency. Therefore, it is necessary to consider the urgency factor of data block when determining the priority of the data block.

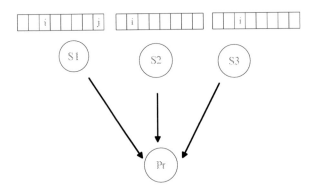

Fig. 1 Position of losing data in the buffer of offer node

2.3 Possible Solutions for RF Strategy in Streaming Media

According to the analysis on the above, the definition of the scheduling priority of the requested data block may affect two important factors, i.e. playback continuity and startup latency, in which user will feel sensitive. Therefore, it is very necessary to improve the system performance through mending the computation method of data block scheduling priority.

Besides, from the views of sending node and requesting node, the cache position of the requested data block respectively has important significance. The details may be described as follows.

(a) For sending nodes, the cache position of the requested data block presents the life cycle of data block at this node. It is said that the longer the distance between the sending node and the current playing data block is, the longer the life cycle of the requested data block exists, and accordingly, the higher the probability of successfully downloading the data block from the sending node is. So the scheduling priority of scheduling data block reduces correspondingly. If considering the life cycle of requested data block on a sending node as a factor of computing the scheduling priority, then, the life cycle of data block on multiple sending nodes can substitute the holding number of data block as a factor of computing the scheduling priority.

(b) For enquiring nodes, the cache position of the requested data block presents the desired-degree of the node enquiring the data block. That is the urgency of data block. It can be said that the further the distance between the enquiring node and the current playing data block, the lower the desire degree of the requested data block from the enquiring node accordingly. Similarly, we can consider the distance between the enquiring data block and the current playing data block as a factor of computing the scheduling priority.

We name the distance between the enquiring data block and the current playing data block as cache offset of playback. Whether the node is enquiring node or sending node, the cache offset of playback is an important factor of computing the data block scheduling priority. So it can exactly reflect the enquiring and offering situation of data block using the cache position of the requested data block of enquiring node and offering node to compute the scheduling priority. Based on the ideas described on the above, we propose a new scheduling strategy based on cache offset of playback (BCOP). The details of the BCOP strategy algorithm are in Section 3.

3 BCOP Scheduling Strategy

3.1 The Computation of Data Block Scheduling Priority

Based on the analysis made in Section 2, the process of computing the scheduling priority of requested data block can be described as follows:

Relevant parameters and hypotheses are described here. Suppose that the data block played at any nodes will not preserve in the cache any more and the buffer size of all the nodes in the system is B, which means that the number of data block preserved in the buffer is B. Each data block contains the streaming data with the same length. Let n be the number of neighbour nodes, $offset_i^k$ presents the cache offset of playback of data block A_i at the k-th neighbour node, $offset_i^R$ refers to the cache offset of playback of requested data block at the buffer of enquiring node, urg_i the urgency of the requested data block A_i, $rarity_i$ the rarity of the requested data block A_i, and $priority_i$ the priority of the requested data block A_i. Where,

$$rarity_i = \prod_{k=1}^{n} (1 - offset_i^k / B), 0 \le offset_i^k < B \qquad (1)$$

$$urg_i = 1 - offset_i^k / B \qquad (2)$$

$$priority_i = rarity_i \times urg_i \qquad (3)$$

Notes: In formula (1), if the cache offset of playback $offset_i^k$ of data block A_i at the k-th neighbour node is 0, it means that the given k-th node does not hold the data block, accordingly, $1 - offset_i^k / B = 1$. If all the n neighbour nodes do not hold the data block, then $rarity_i = 1$, which means the rarity reaches the maximum. As for $1 - offset_i^k / B$, the bigger the $offset_i^k$ is, it implies that the longer life cycle of the data block on the k neighbour node is, and the smaller the value of $1 - offset_i^k / B$ is, which causes the reduction of $rarity_i$.

Formula (2) implies that the bigger the offset $offset_i^R$, the lower the urgency of data block A_i is and the lower the urg_i is.

Formula (3) shows how to compute the scheduling priority using the value of $rarity_i$ and urg_i of data block A_i.

3.2   Implementation of the Proposed Scheduling Algorithms

It is necessary to sort the data block, which needs to be downloaded by scheduling priority descending after computing the scheduling priority of each data block. For the data block A_i, multiple nodes will hold it. Normally, we select the neighbour node, which transmits A_i fastest to send it. However, this may cause the conflict among the download tasks. For example, when we select one node to send two data blocks, one data block must wait or be sent by the other reselected node. That is to say, we need to consider the payload balance of the sending nodes when we assign transmitting tasks to the sending nodes.

Besides, the problem of selecting the proper sending node for the data block can be come down to one of the variation of parallel machine scheduling [8], which is the problem of NP hard. As we know, it is hard to find the optimal solution for this kind of problem. To solve this problem, we present an algorithm to obtain the data block with the highest scheduling priority and also consider the payload balance of the sending nodes. The details of the algorithm are found as follows.

Input

(1) Sort the data block serial by scheduling priority descending, A_1, A_2, A_3, …, A_m;

(2) The data block set serial holding the nodes, S_1, S_2, S_3, …, S_m, accordingly;

(3) The sending speed of all neighbour nodes, $R(1)$, $R(2)$, $R(3)$, …, $R(n)$;

(4) Intending sending waiting time of all neighbour nodes at current scheduling period, wait(1), wait (2), wait (3), …, wait (n), all the initial value is 0.

Output

The sending node $sender_i$ of each data block A_i.

The details of the proposed algorithms:

(1) $get_{max} \leftarrow min(m, \tau \times l)$; //set the maximum number of receiving data block at one scheduling period.

(2) for i = 1 to get_{max} do

(3) $t_{min} = \infty$; // initializing the earliest gaining time of A_i

(4) for j = 1 to k do

(5) $t_{deliver} = \frac{Length}{R(S_i^j)}$; // getting the intending time of A_i from S_i^j

(6) if $t_{deliver} + Wait(S_i^j) < t_{min}$ and $t_{deliver} + Wait(S_i^j) < \tau$

(7) $t_{min} \leftarrow t_{deliver} + Wait(S_i^j)$; $sender_i \leftarrow S_i^j$;

(8) end for j ;

(9) if $sender_i \neq null$

(10) $Wait(sender_i) \leftarrow t_{min}$

(11) end for i.

The computation complexity is $O(n \cdot min(m, \tau \cdot l))$, where n is the number of neighbor nodes, m is the number of requested data block, which is scoped with [0, B](B is the buffer size), $\tau \cdot l$ is the download bandwidth. In our system, n and τ are constants. Obviously, m is related to the data block distribution. So the two factors, which affect the executing time, are the download bandwidth and data block distribution.

The meaning of parameters in data scheduling algorithm is notated in Table 1.

Table 1 Parameters in data scheduling algorithm

Parameter	Description
τ	Period of data scheduling
l	Take-over bandwidth of requesting node
S_i^j	The j-th holding node of block A_i
S_i	The holding node set $S_i = \{S_i^1, S_i^2, \cdots, S_i^k\}$ of block A_i
Length	The length of data block

3.3 The Determination of Sending Speed

In fact, there is a problem when implementing the above scheduling algorithm. That is how to ascertain the sending speed of the neighbour nodes. To do so, we adopt a simple and effective approach to solve it, which is historic information estimation. Before executing a scheduling, estimate the data sending speed that the current neighbour node may offer according to the quantity of the data gained from the node in former M periods. Suppose that $q_k^{(m)}$ is the total data blocks obtained from the k neighbour node at m-th period. Then the sending speed of the k node at (m+1)-th period is

$$R^{(m+1)}(k) = \alpha \cdot \left(\sum_{t=m-M+1}^{m} q_k^{(t)} \right) \bigg/ M\tau \qquad (4)$$

Where, α is modulation coefficient.

We check the percentage of the successful return of requested data of each neighbour node at the last period while executing every new scheduling. If all of the neighbour nodes successfully respond, we say that the given neighbour node can offer uploading service at the speed of $R^{(m+1)}(k)$ at least. If the cumulate number of successful return data reaches the given higher threshold, we say it is conservative that the selected neighbour node sends the assigned transmitting task at the current speed of $R^{(m+1)}(k)$. In this case, we should increase the sending speed in the next period. Accordingly, we should adjust α to a bigger one. Similarly, if the cumulate number of unsuccessful return data reaches the given lower threshold, we say the selected neighbour node cannot offer good service at the current speed of $R^{(m+1)}(k)$. So we should decrease the sending speed so that it can offer better service in the next period. That is, we should adjust α to a smaller one.

4 Algorithm Simulation and Analysis

4.1 Simulation Environment and Parameters Setting

NS-2 simulator is used to simulate the proposed BCOP strategy algorithm, and GT-ITM [7] is used to set up the simulator topologic diagram. The diagram of Transit-Stub is shown in Fig. 2. In this diagram, there are 30 transit domains, which include 6 transmit nodes. Each transmit node is connected with 3 stub domains, each of which in turn includes 3 stub nodes, which is a kind of topology of ts-large.

In the experiments, user node may be located on any stub node, and resource server on any transit node. We randomly select 100 to 1500 stub nodes as the peer node, and put the resource server on a transmit node. The output bandwidth of user node is from 100 to 500kbps, the input bandwidth is from 400 to 1000kbps, and the output bandwidth of the server is 100Mbps. To test our algorithm, we

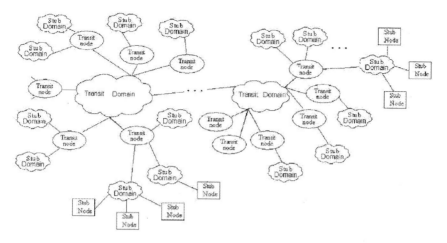

Fig. 2 Diagram of Transit-Stub

select the test stream film with the stream speed of 400kbps and content length of 30 minutes, supposing that the user node joins the network with Poisson distribution and the nodes are active and effective.

4.2 Performance Analysis of Simulation Results

Through the empirical simulation experiments with various data analysis, we found that the playback continuity of BCOP is worse than that of RF if the number of peer nodes is less than 100. While the node number reached about 400, the playback continuity of BCOP had a perfect performance.

To demonstrate the effectiveness of the proposed new algorithm, the comparison results of BCOP algorithm with the existing RF strategy are illustrated in the following sub-section with figures.

4.2.1 Analysis of Percentage of Data Block Losing

In P2P streaming media system, the percentage of data block losing is one of the important factors, which is the embodiment of user playback continuity. Here we adopt the node's average data block losing percentage as evaluating target and analyze the performance of BCOP strategy. Fig. 3 and Fig. 4 demonstrate that as the network size gets bigger, the average data block losing percentage of two strategies increases as well. This can attribute to the limitation of upload bandwidth of the peer nodes, which causes that part of the data block fails to arrive at requiring node before the deadline of playback. Besides this, compared with that of RF strategy, the performance curve of BCOP strategy is increasing slowly and smoothly. It implies that the data block with high rarity and urgency can priority be scheduled under the same restrict condition, which can effectively reduce the losing percentage of data block and improve the performance of playback continuity of the system.

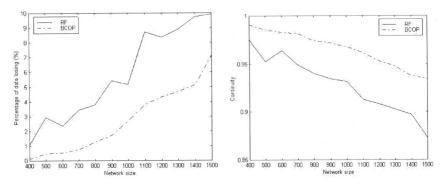

Fig. 3 Percentage of data block losing   Fig. 4 Continuity of the Network

4.2.2 Analysis of Start-Up Latency

The other important factor of assessing the performance of P2P streaming media system is the start-up latency of node, since it is related to the user's feeling when watching the media.

Fig. 5 illustrates the performance curve of BCOP strategy and that of RF strategy. Since we induct the urgency factor of data block while computing the scheduling priority, with the BCOP strategy, we can gain enough data block in the buffer in very short time. The experiment results shown in Fig. 5 verify the conclusion made in Section 2.

Fig. 5 start-up latency of the Network

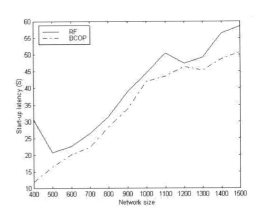

5 Conclusions and Future Work

Data scheduling strategy is an important part of P2P streaming media system, which has great effect on the whole system performance. This paper analyzes the drawbacks of the traditional strategy and provides a new scheduling strategy concerned with the rarity and urgency of data block, which is BCOP strategy. It describes the proposed BCOP strategy algorithm in details and gives the process of how to compute the priority of scheduling. In the BCOP strategy, it schedules the streaming media according to statistical information based on the data block cache offset of playback, which is able to reflect the situation of offering and requiring of data block much more precisely and comprehensively. The simulation experiment results demonstrate that the performances of playback continuity and the start-up latency of BCOP strategy are obviously improved compared with that of traditional RF strategy.

In future work, further investigation is needed to improve the efficiency of scheduling strategy in unstructured P2P networks. We will further research on the scheduling and congestion control mechanisms in order make sense to the Internet.

Acknowledgement. This work was supported by Chinese Postdoctoral Funding (20070410381), Shaanxi Education Bureau Funding (08JK447), and China Scholarship Council.

References

[1] Agarwal, V., Rejaie, R.: Adaptive multi-source streaming in heterogeneous peer-to-peer networks. In: Multimedia Computing and Networking, San Jose, CA, USA, pp. 102–109 (2005)

[2] Floyd, S., Handley, M., Padhye, R., et al.: Equation-based congestion control for unicast applications. In: ACMSIGCOMM, pp. 43–56 (2000)

[3] Gao, W., Huo, L.: Challenges on Peer-to-Peer Live Media Streaming. LNCS, pp. 37–41. Springer, Heidelberg (2007)

[4] Pai, V., Kumar, K.C.: Eliminating trees from overlay multicast. In: Peer-to-peer systems IV 2005, Ithaca, N.Y, pp. 127–140 (2005)

[5] Rejaie, R., Handley, M., Estrin, M.: RAP: An end-to-end rate-based congestion control mechanism for realtime streams in the internet. In: IEEE INFOCOM, pp. 1337–1345 (1999)

[6] Wang, J., Yurcik, W., Yang, Y.: Multiring techniques for scalable battlespace group communications. IEEE Communications Magazine 43(11), 124–133 (2005)

[7] Zegura, E., Calvert, K., Bhattacharjee, S.: How to model an Internet work. In: IEEE INFOCOM, San Francisco, CA, USA, pp. 594–602 (1996)

[8] Zhang, X., Liu, J., Li, B., et al.: Cool streaming/DONet: A data-driven overlay network for peer-to-peer live media streaming. In: IEEE INFOCOM 2005, Miami, Florida, USA, pp. 2012–2111 (2005)

[9] Zhou, Y., Chiu, D.M., Lui, J.C.S.: A simple model for analyzing P2P streaming protocols. In: IEEE ICNP, Beijing, China, pp. 226–235 (2007)

Audio ACP-Based Story Segmentation for TV News

Jichen Yang, Qianhua He, Yanxiong Li, Yijun Xu, and Weining Wang

Abstract. Story segmentation is essential for topic detection and information extraction in TV news. In order to explore to segment TV news in an audio perspective, which has not been researched much yet. This paper proposes an algorithm for story segmentation based on anchor change point from audio, which is different from the traditional story segmentation from video. The proposed algorithm mainly comprises three steps: potential speaker change detection, speaker change refinement and anchor identification. Experimental results show that the proposed algorithm can achieve satisfactory results with total accuracy of 87.84%.

1   Introduction

TV news audio is one of the most important multimedia materials. TV news program consists of news events happened daily all over the world, and is valuable for data analysis for the government, information providers, and television consumers [1]. Generally speaking, a news program consists of more than ten news stories. For example, there are about 14 stories in the TVB (Hong Kong Television Broadcast Ltd) News at 6:30 pm. Segmenting continuous news program into individual news stories is the first step for further audio-video content analysis, which is the key problem for multimedia information retrieval or multimedia understanding.

Many works have been done on the TV news segmentation, in which only from the viewpoint of video processing was adopted [2, 3, 4]. In [2], the temporal locations of the anchor group shots were used to segment the broadcast news program into terms of individual news items. In [3], the news program was first parsed into video shots, and then the anchor shots were identified. Finally, individual news story

Jichen Yang, Qianhua He, Yanxiong Li, Yijun Xu, and Weining Wang
School of Electronic and Information Engineering,
South China University of Technology, GuangZhou, 510640, China
e-mail: NisonYoung@yahoo.cn, eeqhhe@scut.edu.cn,
yanxiongli@163.com, yjz@163.com, weiningwang@scut.edu.cn

R. Lee, G. Hu, H. Miao (Eds.): Computer and Information Science 2009, SCI 208, pp. 131–138.
springerlink.com   © Springer-Verlag Berlin Heidelberg 2009

were constructed based on a simple temporal structural model of news program. In [4], the broadcast TV news was segmented into individual stories based on the locations of the anchor on shots within the program.

Some works on broadcast news story segmentation which have used not only video and audio features, but also features from automatic transcriptions of the programs using speech recognition. For example [14, 15]. But the role of audio in these work is not dominant.

Though the main stream in broadcast news segmentation has been in a visual perspective, we want to explore to segment TV news in an audio perspective, which has not been researched much yet. Maybe the performance of story segmentation can improve much by using more audio features in future.

In this paper, TV news audio is segmented mainly based on anchor change, which is different from general audio stream segmentation based on speaker change [5, 6, 7]. In TV news there are usually two anchors, who report news in turns. In most of the cases, the news story begins with a brief report from the anchor in the studio, followed a detail report by a reporter from the field and a interview by a interviewee in a story unit. In this paper, we define anchor change is from the end of a brief report in a story unit by one anchor to the beginning of another brief report in another story unit by the other anchor. We can see that there is only speaker change in a story unit.For example, the change from anchor to reporter, the change from reporter to interviewee. And anchor change appears in two or more story units. From high-lever semantic of audio, speaker change usually doesn`t stand for story change but anchor change does. Anchor change point (ACP) is often the boundary of between stories.

The remainder of this paper is organized as follows. Section 2 discusses system framework. Section 3 details story segmentation algorithm. Experiment and the evaluations of the proposed algorithm are given in Section 4. Finally, conclusion is given in Section 5.

2   System Overview

The flow diagram of our proposed story segmentation for TV news is illustrated in Fig.1. It is composed of six modules: segmentation speech and non-speech, feature extraction,potential speaker change detection, speaker change refinement, anchor identification and story segmentation.

Audio contains speech, environmental sounds, music and silence in TV news. It is need for us to extract speech from audio firstly. We used method in [5], firstly, the features of high zero-cross-rate, low short-time energy ratios and spectrum flux are extracted from audio. Then a classifier based on K-nearest-neighbor and linear spectral pairs-vector operates on one second window to distinguish speech and non-speech.

In order to seek the potential speaker change, the speech is firstly divided into frames of 32ms with 50% overlap and then the features of 12 MFCCs (Mel-Frequency cepstral coefficients) are extracted for each frame. Finally, computing Kullback-Leibler divergence distance between two adjacent analysis windows (not

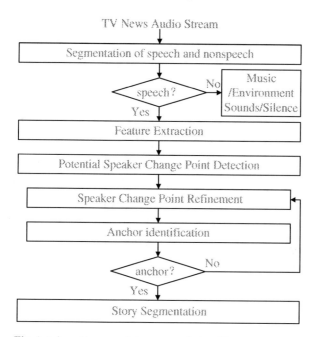

Fig. 1 A flow diagram of the proposed algorithm

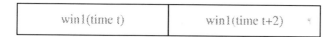

Fig. 2 Two adjacent analysis windows

overlap, but connected, which is illustrated in Fig.2) with the same size (2s in length) shifted by a fixed step (0.16s) along the whole feature vector. Bayesian Information Criterion (BIC) [8] is used in speaker change refinement. Arithmetic harmonic sphericity (AHS) distance [9] is employed for anchor identification.

3 Story Segmentation

Potential speaker change detection and speaker change refinement are described in subsection 3.1 and 3.2 respectively, while anchor identification is discussed in subsection 3.3.

3.1 Potential Speaker Change Detection

This step is a metric-based computation to seek potential speaker change. For each analysis window, a single multi-dimension Gaussian module is estimated from the feature data using maximum likelihood estimation [7]. The Kullback-Leibler divergence distance [10] is chosen for seeking potential speaker change.

$$D = \frac{1}{2}tr[(C_i - C_j)(C_i^{-1} - C_j^{-1})] + \frac{1}{2}tr[(C_i^{-1} - C_j^{-1})(\mu_i - \mu_j)(\mu_i - \mu_j)^T] \quad (1)$$

where C is the estimated covariance matrix and μ is the estimated mean value, tr[·] denotes the trace of a matrix. Speaker model can be expressed as $N(\mu_i, C_i)$ for the i-th analysis window and $N(\mu_j, C_j)$ for the j-th analysis window. The distance is composed of two parts. The first part is determined by the covariance, and the second is determined by covariance and mean. Suppose D(i) equal the Kullback-Leibler divergence distance between the i-th analysis window and its adjacent analysis window. A potential SCP is found if the following conditions are satisfied [10]:
$D(i) > D(i-1), D(i) > D(i+1), D(i) > th_i$.

where th_i is a threshold, defined as

$$th_i = \frac{\alpha}{s} \sum_{n=0}^{s} D(i - n - 1) \quad (2)$$

α is a coefficient as amplifier.

3.2 Speaker Change Refinement

In this step BIC is used to judge whether a potential speaker change is a real speaker change or not. The BIC formula for two speech segments S_i and S_j is as follows:

$$\Delta BIC = \frac{1}{2}((M+N)\log|C| - M\log|C_i| - N\log|C_j|) - \frac{\lambda_1}{2}(d + \frac{1}{2}d(d+1))\log|M+N| \quad (3)$$

where M and N denote the frame number of S_i and S_j respectively and M+N denotes total frame number of the merged speech segments of S_i and S_j. The three speech segments are all supposed to be Gaussian Model. C_i, C_j and C denote covariance of them respectively. d stands for the dimension of the data vector . λ_1 denotes a penalty weight, theoretically equal to 1 and it needs to be tuned in different acoustic and environmental conditions [11].

A positive value of ΔBIC means two speech segments belonging to two different persons, whereas a negative value of ΔBIC indicates two speech segments belonging to the same person. The more negative the ΔBIC is, the closer the two speech segments are [12].

If we know time t_i is a potential speaker change point (SCP), we can use formula (3) to judge whether t_i is a real SCP or not. In the same way, all the wrong candidates can be discarded and the SCP is refined. In the next step we will identify anchor.

3.3 Anchor Identification

In TV news, the first person who reports news is usually an anchor, the following may be a reporter or an interviewee. When we get a real SCP, it may be one of the three. We only care whether it is an anchor or not, a real ACP means a new topic beginning. So we can segment news at ACP; otherwise, we will continue to seek the real ACP. AHS [9] distance has been proven that it can be successfully used for the speaker identification task [13]. In this step, AHS distance is used to identify anchor.

$$d(A,S) = \log[tr(C_A C_S^{-1}) \times tr(C_S C_A^{-1})] - 2\log(D) \qquad (4)$$

where $tr[\cdot]$ represents the trace of the matrix. C_A represents the covariance of an anchor model and C_S represents the covariance of a speech segment after the SCP. D stands for the dimension of the feature vector. For the identical segments, the AHS measure is zero.

For every SCP, it is necessary to calculate d(A,S) for all the anchors in the anchor database. If all d(A,S) exceed a predefined anchor threshold λ_{anchor} (experimental optimal value is used), the SCP is not a ACP. If d(A,S) is less than λ_{anchor}, it belongs to the scope of the anchor and the minimum value related to anchor corresponds the anchor who begins another news story.

Identification of anchor means that anchor change detection is finished. Because ACP is the boundary of story in TV news, so we can segment news program into a sequence of individual news story units.

4 Experiment and Evaluation

In this section, we introduce database and evaluations results of the proposed algorithm.

The database used in the experiments were recorded from TVB news at 6:30 pm. We have recorded six month news program or so, from Jun. 2008 to Nov. 2008. All the data were sampled at 16 KHz, 16 bits and saved as mono channel wav formats. There are three anchors in news program every time, two are main anchors (one male and one female)and the third is a sport anchor (male). The main anchors in the two adjacent stories are often different. The continuous two-minutes or so brief sport news, as a single story, is usually at the end of news program, which is reported by the sport anchor and without reporter in the field.

Generally speaking, the anchors at the news program don't change frequently, for example, there are totally five mainly anchors (two male and three female) and three sport sport anchors (male)in our database, so we can extract anchor data to build anchor model for every anchor.

We divided our database into two parts. Part one contained all the data in Jun., Oct. and Nov.. All the anchors can be found in part one, which was used to build anchor model and train parameters. Part two contained all the data in the other three months, which was used to evaluated our algorithm.

For part one database, we extract two minutes data to build anchor model for every anchor, which is expressed using covariance and every parameter set is as follows:

In order to get better results, we employ a tuning method to seek an optimal value for every parameter. First, every parameter is set an initial value and then is tuned with a step. For amplifier coefficient a, the initial value is given 1.2 [10] and with a tuning step of 0.02. For penalty weight λ_1, its initial value is set as 1.0 [8] and with a tuning step of 0.2. For anchor threshold λ_{anchor}, the initial value is set as 0.4 with a tuning step of 0.001 according to original experimental results. However, tuning step is not absolute, sometimes it can be much bigger or smaller according to the real situation. Through many tuning iterations, finally we got the experimental optimal values for every parameter, which are $a = 1.08, \lambda_1 = 4.2$ and $\lambda_{anchor} = 0.5372$ respectively. Under these conditions, we can seek most SCP, ACP and get a better F1 measure.

For part two database, five days news program was randomly chosen for TV news story segmentation evaluations. The algorithm is evaluated by precision rate (PRC), recall rate (RCL) and F1 measure.

$$PRC = \frac{CFC}{DET} = \frac{CFC}{CFC+FA}, \qquad RCL = \frac{CFC}{GT} = \frac{CFC}{CFC+MD}, \qquad F1 = \frac{2 \times PRC \times RCL}{PRC+RCL} \qquad (5)$$

where CFC denotes the number of correctly detected ACP, which locates within the range of ± 0.3 seconds from the real ACP. FA denotes the number of false alarms. GT stands for the actual number of ACP. MD denotes the number of miss detections. F1 measure admits a value between 0 and 1. The higher the F1 measure, the better the performance is. Story segmentation results are listed in Table 1.

Through analysis the errors in the experiments, the main factor in false alarm is that the voice of some reporter or interviewee voice is very similar to some anchor and maybe there is a short talk between the two main anchors in a story unit. Fortunately, these situations are not too much. The main factor which results in missing error is that some SCP have not been detected and some anchors can not identify because of the set of anchor threshold of λ_{anchor}. This indicates that our algorithm has room to improve.

On the other hand, we found that some audio information may be useful for segmenting news story. For example, at the end of every story, the reporter from the field often says this is reported by TVB reporter his/her name. Before the report of

Table 1 Story segmentation results

News Program	Story Number	PRC	RCL	F1
1	14	87.50%	100%	93.33%
2	15	87.50%	93.33%	90.32%
3	13	80.00%	92.31%	85.72%
4	15	82.35%	93.33%	87.50%
5	13	78.57%	84.26%	81.32%
Total	70	83.33%	92.86%	87.84%

Table 2 Comparison with some published results

	Ref [2]	Ref [3]	Ref [4]	Ours
PRC	84%	90.45%	49.20%	83.33%
RCL	84%	95.83%	47.23%	92.86%
F1	84%	93.06%	48.24%	87.84%

sports news, a main anchor will tell the sport anchor that it is his time to report sport news. After reporting the sport news , the sport anchor will tell the main anchors that sport news is finished. We will utilize these information in the further work to improve the segmentation accuracy.

Table 2 gives a comparison to our experiments and published results from the view of video processing. It shows that our approach could arrive a comparable result against to that from [2] and [3], and much better than that of [4].

5 Conclusion

This paper presented a story segmentation based on ACP approach for TV news, which segments the TV news into individual and meaningful story units only from the viewpoint of audio. A three-step anchor change detection algorithm is proposed, which includes potential Speaker change detection and speaker change refinement, anchor identification. Story segmentation is based on the results of anchor change point. Experiment gave a total accuracy of 87.84%, which shows that the proposed TV news segmentation scheme is very effective.

Because the proposed scheme depends on ACP to segment TV news, it has some inherent drawbacks. For instance, it is a supervised anchor identification, which is not good. It is also necessary to get parameter value by training data and the parameter value influences segmentation story result. Our future research will focus on fusion audio and video to improve performance of story segmentation.

Acknowledgements. The work in this paper is supported by the National Natural Science Foundation of China (NSFC) (Item No. 60572141, 60602014).

References

1. Boykin, S., Merlino, A.: Improving broadcast news segmentation processing. In: Proceedings of IEEE International Conference on Multimedia Computing and Systems, Florence, Italy, Florence, Italy, July 7-11, 1999, vol. 1, pp. 744–749 (1999)
2. O'Connor, N., Czirjek, C., Deasy, S., Marlow, S., Murphy, N., Smeaton, A.: News story segmentation in the Fischlar video indexing system. In: Proceedings of International Conference on Image Processing, Thessaloniki, Greece, October 7-10, 2001, vol. 3, pp. 418–421 (2001)

3. Gao, X., Tang, X.: Unsupervised and model-free news video segmentation. In: Proceedings of IEEE Workshop on Content-Based Access of Image and Video Libraries, Kauai, USA, December 12-14, 2001, pp. 58–64 (2001)

4. O'hare, N., Smeaton, A.F., Czirjek, C., O'Connor, N., Murphy, N.: A generic news story segmentation system and its evaluation. In: Proceedings of IEEE International Conference on Acoustics, Speech, and Signal Processing, Montreal, Canada, May 17-21, 2004, vol. 3, pp. 1028–1031 (2004)

5. Lu, L., Zhang, H., Jiang, H.: Content Analysis for Audio Classfication and Segmentation. IEEE Transactions on speech and audio processing 10(7), 504–516 (2002)

6. Wu, C.-H., Hsieh, C.-H.: Multiple change-point audio segmentation and classification using an MDL-based Gaussian model. IEEE Transactions on Audio,Speech and Language processing 14(2), 647–657 (2006)

7. Du, Y., Hu, W., Yan, Y., Wang, T., Zhang, Y.: Audio Segmentation via Tri-Model Bayesian Information Criterion. In: Proceedings of IEEE International Conference on Acoustics, Speech, and Signal Processing, Honolulu, USA, April 15-20, 2007, vol. 1, pp. I-205–I-208 (2007)

8. Chen, S.S., Gopalakrishnan, P.S.: Speaker,environment and channel change detection and clustering via the Bayesian information criterion. In: Proceedings of DARPA Broadcast News Transcription and Understanding Workshop, pp. 127–132 (1998)

9. Bimbot, F., Mathan, L.: Text-Free Speaker Recognition using an Arithmetic Harmonic Sphericity Measure. In: Proceedings of Eurospeech, pp. 169–172 (1993)

10. Lu, L., Zhang, H.: Speaker change detection and tracking in real-time news broadcasting analysis. In: Proceedings of the tenth ACM international conference on Multimedia, Juan-les-Pins, France, December 1-6, 2002, pp. 602–610 (2002)

11. Kotti, M., Moschou, V., Kotropoulos, C.: Speaker segmentation and clustering. Journal of Signal Processing 88(5), 1091–1124 (2008)

12. Kim, H.-G., Elter, D., Sikora, T.: Hybrid Speaker-Based Segmentation System Using Model-Level Clustering. In: Proceedings of IEEE International Conference on Acoustics, Speech, and Signal Processing, Philadelphia, USA, March 18-23, 2005, vol. 1, pp. 745–748 (2005)

13. Johnson, S.E.: Who Spoke When? - Automatic Segmentation and Clustering for Determining Speaker Turns. In: Proceedings of Eurospeech, vol. 5, pp. 2211–2214 (1999)

14. Besacier, L., Quenot, G., Ayache, S., Moraru, D.: Video story segmentation with multimodal features: experiments on TRECVID 2003. In: Proceedings of the 6th ACM SIGMM international workshop on Multimedia information retrieval, New York, USA, October 15-16, 2004, pp. 221–227 (2004)

15. Hsu, W., Chang, S.-F., Huang, C.-W., Kennedy, L., Lin, C.-Y., Iyengar, G.: Discovery and fusion of salient multi-modal features towards news story segmentation. In: IS&T/SPIE Symposium on Electronic Imaging:Science and Technology-SPIE Storage and Retrieval of Image/Video Database, San Jose, USA, January 18-21 (2004)

Study on Transform-Based Image Sharpening

Ying Liu, Yong Ho Toh, Tek Ming Ng, and Beng Keat Liew

Abstract. The aim of this paper is to investigate how we can make use of Discrete Cosine Transform (DCT) and Discrete Wavelet Transform (DWT) in image sharpening to enhance image quality. The fundamental idea of image sharpening is to make use of image edges or high frequency components to bring out invisible details. Both DWT and DCT can be used to isolate the high frequency components of the original image as they are able to separate the frequency components into high and low portions. An analysis of the results suggests that DWT is more suited to the task. Focusing on DWT, we propose a wavelet-based algorithm for image sharpening. In this algorithm, an image containing the edge information of the original image is obtained from a selected set of wavelet coefficients. This image is then combined with the original image to generate a new image with enhanced visual quality. An effective approach is designed to remove those coefficients related with noise rather than the real image to further enhance the image quality. Experimental results demonstrate the effectiveness of the proposed algorithm for image sharpening purpose.

Keywords: Image sharpening, Discrete Cosine Transform, Discrete Wavelet Transform, Noise-related coefficients.

1 Introduction

The aim of image enhancement is to improve the visual appearance of images through the sharpening of image features such as edge or contrast [1]. To achieve this, many algorithms such as Unsharp Masking [2] and Laplacian filtering [3] have been designed to increase the maximum luminance gradients at the border between different objectives or within textured areas. That is, to makes use of the image edges or high frequency components to bring out invisible details.

Ying Liu, Tek Ming Ng, Beng Keat Liew
School of Information and Communications Technology,
Republic Polytechnic, Singapore,738964
e-mail: {liu_ying,ng_tek_ming,liew_beng_keat}@rp.sg

Yong Ho Toh
Department of Electrical and Computer Engineering,
National University of Singapore, Singapore, 119077
e-mail: u0509543@nus.edu.sg

R. Lee, G. Hu, H. Miao (Eds.): Computer and Information Science 2009, SCI 208, pp. 139–148.
springerlink.com   © Springer-Verlag Berlin Heidelberg 2009

Unsharp Masking (UM) is a well-known image sharpening technique. The fundamental idea of UM is to subtract from the input image a low-pass filtered version of the image itself. The same effect can be achieved by adding to the input image a high-pass filtered version of the image so as to enhance the high-frequency components [2]. In this paper, we refer to the latter formulation expressed as

$$y(m,n) = x(m,n) + \lambda * z(m,n) \qquad (1)$$

where $x(m,n)$ and $y(m,n)$ are the input image and the enhanced image, respectively. The correction image $z(m,n)$ introduces an emphasis on the high-frequency components and makes the image variation more sharp.

In the conventional UM method, $z(m,n)$ is derived by a linear filter, which can be a simple laplacian filter.

$$z(m,n) = 4x(m,n) - x(m-1,n) - x(m+1,n) - x(m,n-1) - x(m,n+1) \qquad (2)$$

The amplitude response of this filter is a monotonically increasing function of the frequency, and it is used in equation (1) to introduce the required emphasis on the image variations. However, this high-pass filter also amplifies noise. To relieve this problem, different methods have been proposed [2,4,5]. Among these, the Cubic Unsharp Masking (CUM) technique modulates the sharpening signal using a simple function dependent on the local gradient of the data and has been proved to be effective for image sharpening with reduced sensitivity to noise [2, 6]. In CUM filter, the correction term $z(m,n)$ can be expressed as

$$z(m,n) = [x(m-1,n) - x(m+1,n)]^2 * [2x(m,n) - x(m-1,n) - x(m+1,n)]$$
$$+ [x(m,n-1) - x(m,n+1)]^2 * [2x(m,n) - x(m,n-1) - x(m,n+1)] \qquad (3)$$

While the above mentioned algorithms work in spatial domain to enhance image quality, some other algorithms try to achieve the same goal by leveraging image information in frequency domain [1,8,9]. Discrete Wavelet Transform (DWT) and Discrete Cosine Transform (DCT) are two widely applicable signal processing tools. Both DWT and DCT can be used to isolate the high frequency components of the original image because they are able to separate the frequency components into high and low portions [7]. In this paper, we first compare DWT and DCT in obtaining high-pass filtered version of the original image (i.e. image details). The results show that DWT is more suitable to this task. Then, making use of wavelet and UM techniques, we propose an effective image sharpening algorithm, referred to as WUM. In WUM, DWT is first applied on the input image to obtain a set of wavelet coefficients with different frequencies. To cope with the noise amplification problem in UM, we design an effective approach to remove those coefficients related with noise rather than the real image. The correction image containing the edge information of the original image is obtained by applying IDWT on the selected set of high-frequency coefficients. This image is then combined with the original image to generate a new image with enhanced visual quality. Experimental results confirm the effectiveness of the proposed algorithm for image sharpening purpose.

The remaining of the chapter is organized as following. Section 2 first compares DCT and DWT in obtaining image details and then describes the proposed algorithm WUM. In Section 3, experimental results on a set of test images are provided and the performance of WUM is compared with that of CUM. Finally, Section 4 concludes this paper.

2 The Proposed Algorithm

2.1 Obtaining High-Pass Filtered Version of the Original Image

In order to obtain the correction signal $z(m,n)$, we explore two widely-used transformations in image processing, DWT and DCT.

DWT decomposes a signal by using scaled and shifted versions of a compact supported basis function which is called the mother wavelet. The decomposition is done line by line and then column by column. In level 1, wavelet filter is convolved with each line of the image and then convolved with each column of the result. Thus, we get 4 subbands, LL_1, LH_1, HL_1, HH_1. Then, same procedure is done to the LL_1 subband, and LL_1 subband turns into four sub-subbands, LL_2, LH_2, HL_2, HH_2 ... This procedure iterates J times, and we get a J levels decomposition of the image, with $3*J+1$ subbands [7]. All the coefficients are arranged in different subbands (numbered as subband 0,1,...,3*J) and at different levels. Fig. 1 shows the 7 subbands in 2-level DWT. Different wavelet filters are available, we use filter Db6 [7]. Fig.2 displays the 'Lena' image and its wavelet coefficients with $J=3$.

DCT expresses a sequence of finitely many data points in terms of a sum of cosine functions oscillating at different frequencies as in equation (4).

$$F(u,v) = (\frac{2}{N})^{\frac{1}{2}}(\frac{2}{M})^{\frac{1}{2}}\sum_{m=0}^{N-1}\sum_{n=0}^{M-1}A(m)A(n)\cos[\frac{\pi u}{2N}(2m+1)]\cos[\frac{\pi v}{2M}(2n+1)].x(m,n) \quad (4)$$

where $A(\xi) = 1/\sqrt{2}$ for $\xi = 0$ and 1 otherwise, the image dimensions are N pixels by M pixels, x(m,n) is the intensity of the pixel in row m and column n and $F(u,v)$ is the DCT coefficient in row u and column v of the DCT matrix. Fig.3 displays the DCT coefficients for Lena image.

Fig. 1 Seven subbands in 2-level DWT

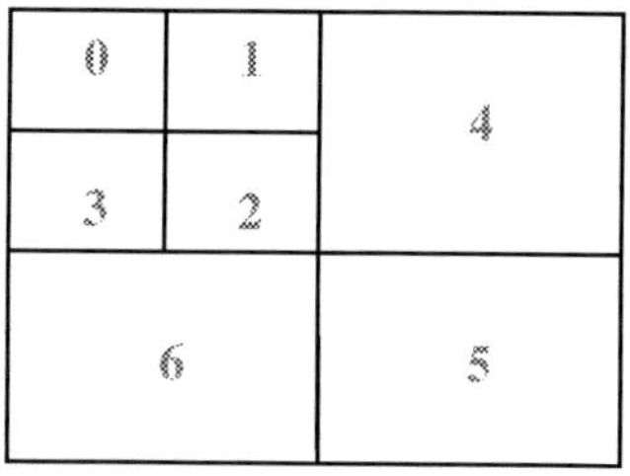

Fig. 2 Example of DWT result
(Left: original Lena image. Right:
3-level DWT result)

Fig. 3 DCT result for Lena image

In DWT domain, the top left rectangle contains the low frequency information. Thus, we can set all the coefficients in this rectangle to 0 to filter out the lowest frequency components. In this way, 25% of the coefficients are removed using 1-level DWT, 6.25% removed using 2-level DWT and 1.5625% removed using 3-level DWT. Then we can perform IDWT to get a high-pass filtered image. In Fig.4, we show the results for Lena image as example. We observe that 3-level DWT is sufficient to obtain the majority of the edge information. Performing DWT with more than 3-levels improves the image details only slightly compared to 3-level DWT.

In DCT, the lowest frequency components are located at the top-left corner and higher frequency components are located diagonally southeast-wards from this corner. In addition, the magnitude of the coefficients becomes generally larger the closer we get to the top left corner, because most of the energy is compacted here. We thus set diagonal sets of coefficients to '0', starting from the top left corner and sweeping southeast-wards. In this way, we keep a certain amount of high frequency coefficients.

Next, we compare the results of DWT and DCT in obtaining high-pass filtered version of an image. We focus on 3-level DWT and its DCT counterpart (with about 1.5625% of the coefficients removed). We observe that DWT does a better job in obtaining details of the original images. For the DCT-based result, we observe the existence of wavelike noise pattern in the correction image that can distort the original image significantly when we add it to the original image. With reference to Fig.5, we can clearly notice the unnatural wavelike patterns in the DCT-based result.

Fig. 4 The high-pass filtered image obtained using DWT of different levels (J)

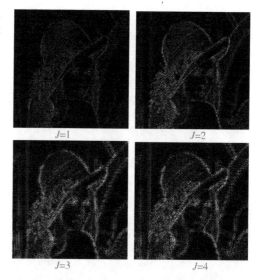

Fig. 5 Example of correction image obtained using DWT and DCT

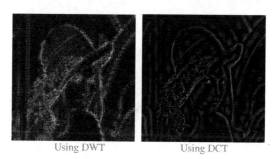

2.2 Suppression of Noise in the Correction Image

After the above discussion, we now settle on wavelet for image sharpening. Given an image $x(m,n), 0 \le m < W, 0 \le n < H$, where W, H are the width and height of the image, we first obtain its wavelet coefficients $c(i,j)$ with $0 \le i < W, 0 \le j < H$. Next, we select a set of high-frequency coefficients from all the above coefficients. Applying IDWT on the selected coefficients, we obtain the correction image $z(m,n)$ with $0 \le m < W, 0 \le n < H$, which describes the edge information of $x(m,n)$.

As described in Section2.1, the high-frequency coefficients are selected from all subbands except subban0 which contains low-frequency information of the image. However, it is improper to take all the high frequency coefficients blindly as some of them may have their source as noise but not the real image. Removing such noise-related coefficients will definitely improve the quality of the result image.

To achieve this, we make use of the degree of correlation between wavelet co-efficients at the same spatial position at successive resolution levels. Due to the fact that noise within an image is normally spatially localized, noise-related coefficients in wavelet domain are usually weakly correlated with the corresponding coefficients on successor planes [10, 11]. Hence, correlation between coefficients at different decomposition levels can be used to discriminate between wavelet coefficients arising from noise within the image and those from image features. In other words, a high degree of correlation in wavelet domain indicates that the corresponding pixels belong to a true image feature. A low degree of correlation instead indicates that the pixels most likely come from noise and should be cleaned.

Taking 3-level DWT as example, we explain how to identify noise-related wavelet coefficients. As shown in Fig. 6, for a coefficient C_{30} at level 3, the corresponding coefficients at level 2 are C_{20}, C_{21}, C_{22}, C_{23}. For C_{20}, the corresponding 4 coefficients at level 1 are C_{10}, C_{11}, C_{12}, C_{13}. We observed that useful edge-related coefficients are usually of larger magnitude. Hence, if $C_{30} \leq T_3$, it is highly noise related. Hence, we set C_{30} and all the corresponding coefficients in successive planes to 0. If $C_{30} > T_3$, we need to check the corresponding coefficients at level 2 and level 1. Here T_3 is a threshold we set based on the average magnitude of the coefficients at level 3. As a coefficient derived from the real image usually has high correction with the corresponding coefficients at next level, if C_{30} is related with real image and has a greater magnitude, then C_{20}, C_{21}, C_{22}, C_{23} should be of reasonably large magnitude. If C_{30} is large, but C_{20} small, then it is highly possible that C_{20} is noise-related.

We tried two methods in identifying noise-related coefficients. Method 1: For a coefficient at level 3 with magnitude greater than T_3, if the average of the magnitudes of all the four corresponding coefficients at level 2 is smaller than a pre-defined threshold, then this coefficient is consider noise-related and all these five

Fig. 6 Identifying noise-related coefficients

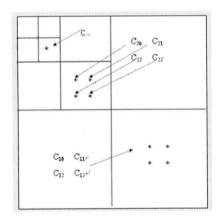

Fig. 7 Correction images obtained before/after removing the noise-related coefficients

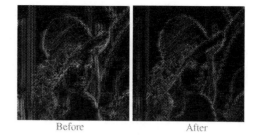

Before   After

coefficients are set to 0. Otherwise, all of them are left untouched. In same way, we examine the coefficients in level 2 and level 3. In this method, we only examine the correlation between coefficients from two neighboring levels. In the second method, we examine the coefficients from all the three levels. Method 2: For C_{20}, C_{21}, C_{22}, C_{23} at level 2, if $(C_{2k} + C_{30}) < T_2$, then C_{2k} is considered to be noise-related. For C_{10}, C_{11}, C_{12}, C_{13} at level 1, if $(C_{1k} + C_{20} + C_{30}) < T_1$, then C_{1k} is considered noise related. Where k=0,1,2,3. T_1, T_2 are the thresholds set based on the average magnitudes of the coefficients at level 1, level 2, respectively. We observe that the second method produces better results. Hence, it will be the noise suppression algorithm adopted in WUM.

In this way, we identify the noise-related coefficients. Setting these coefficients as '0' and keeping all the other high-frequency coefficients, we obtain the entire set of selected coefficients. Applying IDWT on the selected coefficients, we obtain the correction image. Fig. 7 shows examples of the correction images obtained before and after removing the noise-related coefficients. The de-noising effect is obvious. For example, the noise pixels on the face of Lena are removed.

3 Experimental Results

There are different kinds of autofocus functions that can be used to measure the sharpness of images [12]. In our experiments, we choose the famous Tenengrad measure [12], which is a gradient-based sharpness measure and is known for its effectiveness and low computation.

bus   mobile   wall   Lena

Fig. 8 Test images

$J=1, S=0.702$ $J=2, S=0.802$ $J=3, S=0.898$ $J=4, S=0.898$

Fig. 9 WUM with different decomposition level

$$S = \sum_{m,n} [fx^2(m,n) + fy^2(m,n)], \text{ while } \sqrt{fx^2(m,n) + fy^2(m,n)} \geq T \qquad (5)$$

where the horizontal and vertical gradients $f_x(m,n)$ and $f_y(m,n)$ are obtained using the Sobel filters and T is a experimentally set threshold.

In our experiments, we compare the performance of WUM with CUM which is known to be effective for image sharpening with reduced sensitivity to noise [2, 6]. Four test images are used for performance evaluation, as shown in Fig. 8.

WUM is applied to each of the test images in Fig. 8 and λ in equation (1) is experimental set to 0.68. Experimental results show that DWT with $J=3$ is a good choice for image sharpening purpose in our case and decomposition level with $J>3$ does not bring much benefits. As example, Fig. 9 compares the results of WUM with different values of J for the 'mobile' image. We observed that WUM outperforms CUM in obtaining correction image with image details more visible. Fig.10 compares the results for image 'mobile'. We can see that the correction image obtained using WUM is better as there are less noise points at the background. Fig.11 compares the final results of WUM and CUM for all of the four test images. As a side note, we also compare the sharpening results of DCT and DWT for Lena image, as in Fig. 12. The wavelike noise pattern is observed in the DCT-based result. This confirms our conclusion that DWT is more suited to the task of image sharpening.

From the above results, we conclude that WUM is an effective image sharpening algorithm.

Fig. 10 Correction image obtained using WUM and CUM

Using WUM   Using CUM

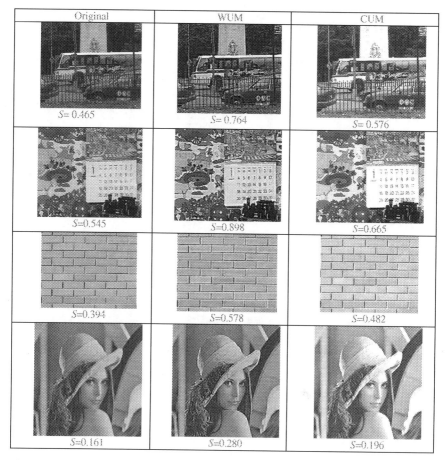

Original	WUM	CUM
S= 0.465	S= 0.764	S= 0.576
S=0.545	S=0.898	S=0.665
S=0.394	S=0.578	S=0.482
S=0.161	S=0.280	S=0.196

Fig. 11 Performance comparison between WUM and CUM

Fig. 12 Image sharpening using DWT and DCT

Using DWT   Using DCT

4 Conclusion

This paper studies the performance of DCT and DWT in obtaining high-pass filtered version of an image to enhance image visual quality. It is fount that DWT is more suited to the task. A wavelet-based image sharpening algorithm is then presented, which is based on the Unsharp Masking technique. The algorithm makes use of the correlation between different wavelet planes to remove those noise-related coefficients and thus selects a set of high frequency coefficients describing the edges in the original image. These selected coefficients are used to obtain the correction image which is combined with the original image to generate a new image with enhanced visual quality. Experimental results demonstrate the effectiveness of the proposed algorithm for image sharpening purpose. In this paper, we rely on our visual perception to determine the value of λ such that we get the image with best visual quality. A method based on equations and optimization algorithms will more accurate since humans may not be able to pick out minute but important details and can only obtain a rough feeling about the image as a whole.

References

[1] Huang, M., Tseng, D., Liu, S.: C Wavelet image enhancement based on Teager energy operator. In: Proceedings of International Conference on Pattern Recognition, vol. 2, pp. 993–996 (2002)

[2] Ramponi, G.: A cubic unsharp masking technique for contrast enhancement. Signal Process 67, 211–222 (1998)

[3] Gonzalez, R.C., Woods, R.E.: Digital Image Processing, 2nd edn. Prentice Hall, Upper Saddle River (2002)

[4] Guillon, S., Baylou, P., Najim, M.: Robust nonlinear contrast enhancement filter. In: Proceedings of IEEE International Conference on Image Processing (ICIP), vol. 1, pp. 757–760 (1996)

[5] Lee, Y.H., Park, S.Y.: A study on convex/concave edges and edge-enhancing operators based on the Laplacian. IEEE Transactions on Circuits and Systems 37, 940–946 (1990)

[6] Yao, Y., Abidi, B., Abidi, M.: Digital image with extreme zoom: system design and image restoration. In: Proceedings of IEEE International Conference on Computer Vision Systems (ICVS), pp. 52–59 (2006)

[7] Daubeches, L.: Ten Lectures on Wavelets, Society for Industrial and Applied Mathematics, Pennsylvania (1992)

[8] Du-Yih, T., Bum, L.Y.: A method of medical image enhancement using wavelet-coefficient mapping functions. In: Proceedings of International Conference on Neural Networks and Signal Processing, vol. 2, pp. 1091–1094 (2003)

[9] Yu-Feng, L.: Image denoising based on undecimated discrete wavelet transform. In: Proceedings of International Conference on Wavelet Analysis and Pattern Recognition, pp. 527–530 (2007)

[10] Zeng, P., Dong, H., Chi, J., Xum, X.: An approach for wavelet based image enhancement. In: Prceedings of IEEE International Conference on Robotics and Biometrics, pp. 574–577 (2004)

[11] Xu, Y., Weaver, J.B., Healy, D.M., et al.: Wavelet transform domain filters: a spatially selective noise filtration technique. IEEE Transactionson Image Processing 3, 747–758 (1994)

[12] Santos, A., Solarzano, C.O., Vaquero, J., Pena, M., et al.: Evaluation of autofocus functions in molecular cytogenetic analysis. Journal of Microscopy 188, 264–272 (1997)

GEST: A Generator of ISO/IEC 15408 Security Target Templates

Daisuke Horie, Kenichi Yajima, Noor Azimah, Yuichi Goto, and Jingde Cheng

Abstract. This paper presents a generator of ISO/IEC 15408 security target templates, named "GEST," that can automatically generate security target templates for target information systems from evaluated and certified security targets. GEST can choose certified security targets resembling target system according to keywords inputted by users, and detect and correct discordance of templates. By using GEST, designers with a little experience can easily get security target templates as the basis to create security targets of target systems.

1 Introduction

ISO/IEC 15408 (Common Criteria) is an international criterion for evaluating whether security facilities of information systems are appropriately designed and implemented [5]. Now, evaluation and certification system according to ISO/IEC 15408 is established in the world. Designers have to prepare a design specification document (called Security Target, ST) if they want to acquire the certification according to ISO/IEC 15408. However, it takes much of cost to prepare a ST. In addition, designers must search for requirements from ISO/IEC 15408 documents (over 300 pages in total) and write the requirements on a ST. Designers may consult certified STs for reducing of cost to prepare a ST, but they must search for informative STs from certified STs (now over 500 documents) and search for requisite information from each of found STs (over 100 pages in some cases). Corporations must pay employment cost to designers, and might expend a million dollars and five years for creating a ST in some case [12].

Furthermore, it is not enough even if information systems are certified according to ISO/IEC 15408 once. Spiteful crackers are active persons who can get knowledge and skills day after day, and then continuously attack target systems always

Daisuke Horie, Kenichi Yajima, Noor Azimah, Yuichi Goto, and Jingde Cheng
Graduate School of Science and Engineering, Saitama University, Saitama, Japan 338-0825
e-mail: {horie,yajima,azim,gotoh,cheng}@aise.ics.saitama-u.ac.jp

R. Lee, G. Hu, H. Miao (Eds.): Computer and Information Science 2009, SCI 208, pp. 149–158.
springerlink.com

with new techniques. Hence, maintenance of information systems is important from viewpoint of security. Maintainers should repeatedly design to add security facilities against attacks. It takes additional cost to remake systems and STs to acquire the certification again. It is obviously desirable to reduce cost to design systems with high security and create STs. Moreover, rich experience is desirable to design information systems according to the format of STs. Expert designers who have rich experience are demanded to create STs. However, small corporations may not employ the expert designers.

In software engineering, some tools for automatically generating specification documents [6, 11] have been developed and used. However, there was no tool that can generate STs. We have developed ISEDS (An Information Security Engineering Database System) [7] that manages the data of specifications described on certified STs. It can manage data described on STs certified according to ISO/IEC 15408. However, it cannot directly support to prepare STs by using the data.

This paper presents a generator of ISO/IEC 15408 security target templates, named "GEST," that supports creating STs of information systems. It automatically generates templates of STs for designers with a little experience to inexpensively, easily and rapidly create STs of information systems with high security requirements. It is the first tool that can automatically generate security specifications documents for information systems.

2   ISO/IEC 15408

Fig. 1 shows the evaluation and certification system according to ISO/IEC 15408. An applicant who is a designer of an information system have to prepare a ST of the system and a corroboration document indicating that the system is appropriately implemented according to the ST. An evaluation agency evaluates them according to ISO/IEC 15408. Then, a certification agency certifies appropriateness of the security facilities according to the result of the evaluation, and publishes certified STs. ISO/IEC 15408 assures that the system is appropriately designed. Orderers of information systems can view the certified STs, and comfortably use the certified systems.

ISO/IEC 15408 defines the security requirements that security facilities should satisfy: the requirements about function (security functional requirement, SFR), the requirements about assurance (security assurance requirement, SAR), and Evaluation Assurance Level (EAL) as reliability of security functions. Designers must design a TOE (Target of Evaluation) as an information system or its category, define design specifications of it, and write them on a ST. They should write the following design specifications:

- **ST introduction**: A title, an overview, a description and references of a TOE.
- **Security Problem (SP)**: A problem or a task including organizational security policies, threats and assumptions at the environment of a TOE.

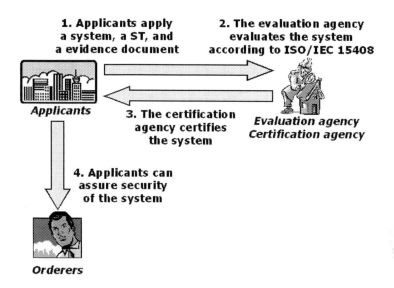

1. Applicants apply a system, a ST, and a evidence document

2. The evaluation agency evaluates the system according to ISO/IEC 15408

Applicants

3. The certification agency certifies the system

Evaluation agency
Certification agency

4. Applicants can assure security of the system

Orderers

Fig. 1 The Evaluation and certification system according to ISO/IEC 15408

- **Security Objective (SO)**: A security countermeasure that promises conditions in the environment or oppose attacks.
- **SFR**: A functional requirement that a TOE should satisfy for achievement of SOs.
- **SAR**: An assurance requirement that a TOE should satisfy for achievement of SOs.
- **TOE Summary Specification (TSS)**: A security facility that TOE should provide for satisfaction of SFRs.
- **Rationale**: A relationship and a rationale of SPs, SOs, SFRs, SARs, and TSSs.

More than 800 certified STs are published on the official web-site of ISO/IEC 15408 in various nations [5, 9]. ISO/IEC 15408 also defines other specification documents, called "Protection Profiles (PPs)." A PP is a specification document for a classification of a TOE. PPs consist of introduction, SPs, SOs, SFRs, SARs, and rationales. PPs do not include TSSs, because they do not depend on an implementation of a particular TOE. Designers can deal with PPs as templates of STs.

3 GEST: A Generator of ISO/IEC 15408 Security Target Templates

3.1 Outline of GEST

GEST is a tool that generates templates of STs for supporting to create STs of information systems. The templates generated by GEST consist of specifications in

TOEs that resemble a designers' target system. The templates adhere a format of
STs. The templates do not always satisfy requirements that designers require, because they are mechanically generated. Therefore, designers should analyze and
correct the templates. The templates are not same as PPs. PPs are good templates
for STs, however. PPs depend on a classification of TOEs. Therefore, PPs do not
include TSSs. The templates adhere a format of STs that include TSSs and depend
on a particular TOE. Moreover, GEST can generate templates for STs that belong
to classifications covered by existing PPs.

We defined two requirements for GEST. Its requirements are as follows;

- **R1:** GEST should generate templates of STs from STs of the existing TOEs
  resembling target systems by only inputting keywords.
- **R2:** GEST should generate templates of STs that provide appropriateness assured
  according to ISO/IEC 15408.

We devised the method to calculate resemblance between a target system and certified TOEs for R1. The method calculates an inner product [10] between a vector
of the keywords inputted by users and a vector of the ST introduction of certified
STs that is defined with tf-idf weight [13], and adopts STs with high inner products. The method can adopt the TOEs that resemble the target system according to
the keywords and weights, if users input appropriate keywords and weights of the
keywords.

We decided that it adopts all of specifications of each adopted TOE as specifications of a template for R2. However, if there is no consistency among adopted TOEs,
the templates cannot provide appropriateness assured according to ISO/IEC 15408.
Therefore, we devised the method that detects and corrects discordance among specifications to keep quality of the templates. We investigated IEEE Std. 830 [8] that is
a standard of recommended practice for software specification documents and CEM
3.1 [4] that rules evaluation methodology for ISO/IEC 15408. We defined unneces-
sary/insufficient specifications, unrealizable specifications, incongruous specifications with surrounding environment of a system, and terminological inexactitude as
discordance of the templates according to IEEE Std. 830 and CEM 3.1. We devised
the detection/correction methods for the discordance. The method detects the discordance with pattern matching of name and definition text of specifications, and
corrects them by eliminating unneeded specifications and renaming specifications
or alerts users to the discordance.

3.2   Design and Implementation of GEST

We designed three functions of GEST.

- **The resemble TOE retrieval function**
- **The quality maintenance function**
- **The specification documentation function**

Fig. 2 shows the construction of the functions. Firstly, users input parameters that are keywords of the target system, weights of the keywords, a number of adopting TOEs, EAL, language of a template, and profiles of the system (TOE name, TOE version, ST title, ST version, authors, developers and category). The resemble TOE retrieval function retrieves the data of certified TOEs from ISEDS. Current ISEDS is a database system that has specification data of 124 TOEs and their SPs, SOs, SFRs, SARs, and TSSs as entities, and rationales as relationships. Next, the function decides to adopt TOEs by comparing the certified TOEs and keywords with the resemble TOE retrieval method. The quality maintenance function detects and corrects discordance among specifications with the discordance detection method. The specification documentation function transcribes the data of specifications whose discordance are resolved, the profiles inputted by users, and the tabular that illustrates the rationale into a template according to the format of STs. Then, the function outputs the template with text, HTML, XML, or LaTeX format.

We designed GEST as a web application for users to ubiquitously create STs without depending on a particular environment. We also implemented it with PHP 5.2. Fig. 3 shows the user interface of GEST, and Fig. 4 shows a template generated by GEST.

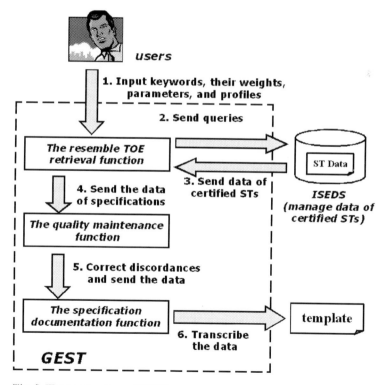

Fig. 2 The construction of GEST

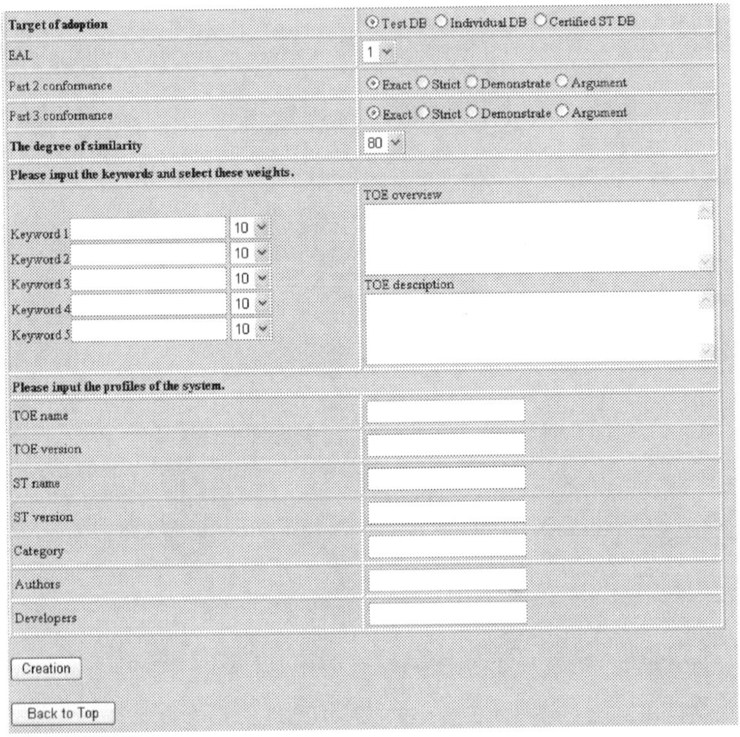

Target of adoption	⊙ Test DB ○ Individual DB ○ Certified ST DB
EAL	1 ∨
Part 2 conformance	⊙ Exact ○ Strict ○ Demonstrate ○ Argument
Part 3 conformance	⊙ Exact ○ Strict ○ Demonstrate ○ Argument
The degree of similarity	80 ∨

Please input the keywords and select these weights.

		TOE overview
Keyword 1	10 ∨	
Keyword 2	10 ∨	
Keyword 3	10 ∨	TOE description
Keyword 4	10 ∨	
Keyword 5	10 ∨	

Please input the profiles of the system.

TOE name	
TOE version	
ST name	
ST version	
Category	
Authors	
Developers	

Creation

Back to Top

Fig. 3 Overview of the interface of GEST

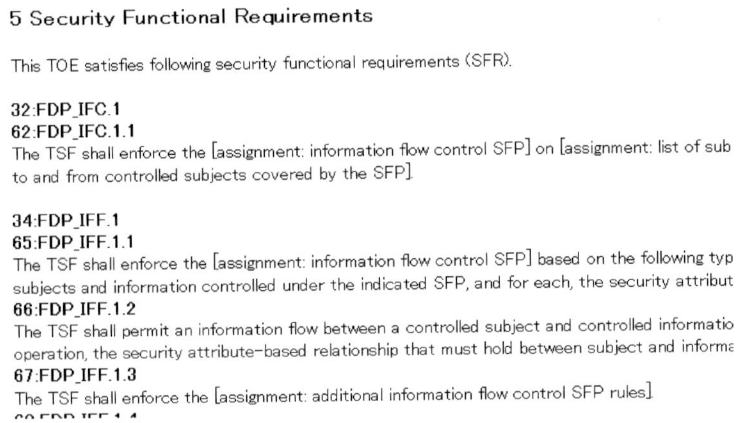

5 Security Functional Requirements

This TOE satisfies following security functional requirements (SFR).

32:FDP_IFC.1
62:FDP_IFC.1.1
The TSF shall enforce the [assignment: information flow control SFP] on [assignment: list of sub
to and from controlled subjects covered by the SFP]

34:FDP_IFF.1
65:FDP_IFF.1.1
The TSF shall enforce the [assignment: information flow control SFP] based on the following typ
subjects and information controlled under the indicated SFP, and for each, the security attribut
66:FDP_IFF.1.2
The TSF shall permit an information flow between a controlled subject and controlled informatio
operation, the security attribute-based relationship that must hold between subject and informa
67:FDP_IFF.1.3
The TSF shall enforce the [assignment: additional information flow control SFP rules]
CO FDB IFF 1 4

Fig. 4 A template generated by GEST

3.3 Evaluation

Practicality of GEST depends on appropriateness of templates generated by GEST.
GEST writes all of specifications described on certified STs on a template, so

appropriateness of certified STs is not lost on the template. When there is discordance between specifications on some STs, GEST detects and/or corrects them. Thus, GEST keeps appropriateness of templates. However, appropriateness of a template depends on appropriateness of keywords and correcting the template. GEST might output a template that is not desirable even if they are generated from certified STs, when the user input inappropriate keywords for the target systems. Therefore, users should carefully choose keywords and analyze the templates with their empirical rules or some tests. It is easier than creating a ST from the top for users with a little experience.

We compared templates generated by GEST with certified STs for ensuring practicality of it. We concretely measured appropriateness and completeness between a certified ST and a template. The definition of appropriateness and completeness defined by IEEE Std. 830 [8] are the below.

$$\text{Appropriateness} = \frac{\text{a number of appropriate specifications}}{\text{a number of all specifications}}. \quad (1)$$

$$\text{Completeness} = 1 - \frac{\text{a number of insufficient specifications}}{\text{a number of all specifications}}. \quad (2)$$

Now, GEST can generate templates from 114 certified STs stored into ISEDS. Firstly, we collected 20 certified STs (the half is in English and another half is in Japanese) from the official web-site of ISO/IEC 15408 [5, 9] and defined them as an ideal STs. Secondly, we automatically selected keywords and their weights from each of the ideal ST by using tf-idf weight. The keywords are selected from sentences of the ST introduction of the ideal STs. Thirdly, we generated templates by inputting the keywords and their weights. We generated a template from 5 certified STs in this experiment. The ideal STs provide specifications that the templates must provide. Fourthly, we compared the templates with each of the ideal ST and measured overlapping, insufficient, and unnecessary SPs, SOs, and SFRs. We judged overlapping specifications whether two specification sentences are scored more than 0.6 with 3-gram [2]. Finally, we calculated appropriateness and completeness of the templates, summarized in each language and SPs, SOs, and SFRs.

Table 1 is the summary of all comparisons: specifications in the ideal STs, specifications in the templates, overlapping specifications between the ideal STs and the templates, appropriateness, and completeness.

Completeness is considerably higher than appropriateness in all of the table. This summary indicates that templates generated by GEST provide 90 % specifications in all of specifications that the templates must provide, but 82 % specifications are unnecessary. This also means that designers of information systems must define 10 % specifications in themselves. Moreover, a certified ST contains 29 specifications and a template contains 104 specifications on average. That is, they must define 3 specifications and delete 85 specifications in themselves. In each specification, appropriateness of SFRs is higher than one of SPs and SOs, however, completeness is lower. In each language, the templates in Japanese have high appropriateness, however, the templates in English antithetically have high completeness.

Table 1 Comparison between 20 templates generated by GEST and 20 ideal STs

	number of specifications in the ideal STs	number of specifications in the templates	overlapping specifications	appro- priateness	comp- leteness
specifications					
SP	206	834	144	0.172	0.925
SO	184	760	118	0.155	0.913
SFR	194	496	118	0.237	0.846
language					
Japanese	317	482	157	0.325	0.668
English	267	1608	223	0.138	0.972
all STs	584	2090	380	0.181	0.902

4 Discussion

We think that practicality of GEST is effective but much needs to be improved.
The comparison in section 3.3 indicates that designers of information systems must
delete 85 unnecessary specifications in themselves. However, it is obviously more
easily to delete unnecessary specifications than to define appropriate specifications.
A rich experience is needed for this difficult task to define appropriate 10 % specifications. However, cost of defining 10 % specifications will obviously decrease rather
than defining all of specifications. Thus, we think that GEST is sufficiently effective for creating STs. However, it is fact that almost of specifications in a template
is unnecessary. We also must devise more effective method for judging similarity
between target systems and certified STs and for detecting unneeded specifications.

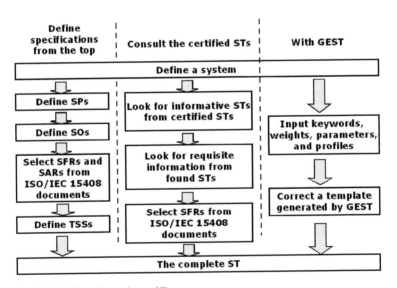

Fig. 5 The flow of creating a ST

Fig. 5 shows comparison of the flow of the creating STs in three kinds of cases, designing all specifications, consulting certified STs, and using GEST. In the case of left in Fig. 5, designers must define target information systems, SPs, SOs, SFRs, SARs, TSSs, and rationales. In this way, they must search for appropriate SFRs and SARs from ISO/IEC 15408 documents. Sometimes, designers might expend a million dollars and five years for creating a ST in some cases [12]. In the case of middle in Fig. 5, designers can consult certified STs for reducing of cost, but they must search for informative STs from certified STs, and search for requisite information from the STs in this case. It is hard to search for requisite information from over 500 STs. In the case of right in Fig. 5, with GEST, designers must define their system, choose and input keywords, and correct templates. However, they can easily generate the templates by only choosing the keywords by deleting 85 specifications and defining 3 new specifications on average. Thus, they can easily and rapidly create STs more than designing all specifications only with their own ability. GEST can reduce cost for creating STs. For example, in our trial, we can create a ST from data about two ISO/IEC 15408 documents and two certified STs about 10 minutes on average. If we try to create the ST by only our own ability, it takes a great deal of time to search for appropriate STs and read the STs. Additionally, at corporations, it takes a great deal of money for employment cost for engineers to create it.

5   Concluding Remarks

We have presented GEST that automatically generates templates of STs from STs certified according to ISO/IEC 15408. By using GEST, designers with a little experience can easily and rapidly create STs of information systems with high appropriateness. GEST is the first tool that can automatically generate security specifications documents for information systems. We publish GEST on Web [1].

We will construct the information security engineering environment that supports all of a life cycle of security facilities [3]. The environment provides tools that standardly, consistently, and formally supports tasks from design to development, management, and maintenance of security facilities. GEST is one of tools that the environment should provide.

References

1. Advanced Information Systems Engineering Laboratory, Saitama University: GEST: Security Target Template Generator,
   http://iseds.aise.ics.saitama-u.ac.jp/posta/
2. Bahl, L., Jelinek, F., Mercer, R.: A Statistical Approach to Continuous Speech Recognition. IEEE Trans. on PAMI (1983)
3. Cheng, J., Goto, Y., Morimoto, S., Horie, D.: A Security Engineering Environment Based on ISO/IEC Standards: Providing Standard, Formal, and Consistent Supoprts for Design Development, Operation, and Maintenance of Secure Information Systems. In: Proceedings of the 2nd International Conference on Information Security and Assurance (ISA 2008), pp. 350–354. IEEE Computer Society Press, Busan (2008)

4. Common Criteria Project: CEM v3.1, http://www.commoncriteriaportal. org/thecc.html
5. Common Criteria Project: Common Criteria Portal, http://www.commoncriteriaportal.org/
6. Goldstein, H., Shannon, D., Bolling, R., Rustici, E.: Document Generation Apparatus and Methods. European Patent (1997)
7. Horie, D., Morimoto, S., Azimah, N., Goto, Y., Cheng, J.: ISEDS: An Information Security Engineering Database System Based on ISO Standards. In: Proceedings of the 3rd International Conference on Availability, Reliability and Security (ARES 2008), pp. 1219–1225. IEEE Computer Society Press, Barcelona (2008)
8. IEEE: IEEE 830-1998 Recommended Practice for Software Requirements Specifications, http://ieeexplore.ieee.org/xpl/freeabs_all.jsp
9. Information-Technology Promotion Agency, Japan: JISEC (Japan Information Technology Security Evaluation and Certification Scheme), http://www.ipa.go.jp/security/jisec/jisec_e/index.html
10. Leacock, C., Towell, G., Voorhees, E.: Corpus-based Statistical Sense Resolution. In: Proceedings of the ARPA Workshop on Human Language Technology pp. 260–265 (1993)
11. Reiter, E., Mellish, C., Levine, J.: Automatic Generation of Technical Documentation. Applied Artificial Intelligence 9(3), 259–287 (1995)
12. Robert Lemos: Linux Makes a Run for Government, http://www.news.com/ 2100-1001-950083.html
13. Salton, G., Buckley, C.: Term-weighting Approaches in Automatic Text Retrieval. Information Processing and Management: an International Journal 6(4), 205–219 (2002)

A General Performance Evaluation Framework for Streaming Service over 3G/B3G

Fangqin Liu, Chuang Lin, and Hao Yin

Abstract. Public wide-area wireless networks are now migrating towards 3G systems. In Europe, the Universal Mobile Telecommunication System (UMTS) developed by 3GPP has been widely adopted as an evolution of the GSM system. One of the most appealing features in 3G and Beyond 3G systems is the introduction of many new multimedia services, which often demand performance guarantees. However, there lacks general performance analysis tools that enable service providers to conveniently obtain performance measurements. This paper develops a general performance metrics framework for streaming services and proposes a modelling method that can effectively map such metrics for real measurement usages. The framework and modelling method presented in this paper have been accepted as part of the 3GPP standards for performance evaluation.

1 Introduction

Public wide-area wireless networks are now migrating towards 3G systems. In Europe, the UMTS developed by 3GPP as an evolution of GSM has been widely adopted. Among all the new services introduced to UMTS, the multimedia services feature prominently, such as Packet-switched Streaming Service (PSS) [2], Multimedia Broadcast Multicast Service (MBMS) [3], Video Telephony (VT), Push-to-talk Over Cellular (PoC) [13] and so on. In order to promote 3G as a global all-purpose communication tool for millions of people, it is essential to guarantee a high quality of these new 3G multimedia services experienced by the user.

However, there lacks general performance analysis tools to conveniently obtain performance measurements. First the performance metrics are disorganized though have a considerable large number, and most of them are service dependent.

Fangqin Liu, Chuang Lin, and Hao Yin
Department of Computer Science and Technology, Tsinghua University,
Beijing, 100084, China
e-mail: {fqliu,clin,hyin}@csnet1.cs.tsinghua.edu.cn

R. Lee, G. Hu, H. Miao (Eds.): Computer and Information Science 2009, SCI 208, pp. 159–168.
springerlink.com   © Springer-Verlag Berlin Heidelberg 2009

Second statistical analysis is often used as the evaluation method [7, 14]. Even if some analytical models (e.g., queueing systems) are employed [10, 11, 6, 12], they just focus on some parts of the network and can not be used to get the service performance of the end-user. Using the disorganized metrics, and the statistical analysis not showing the inherent laws of them clearly, it is quite difficult to evaluate and improve the service performance. In this paper, we aim to sort the metrics out and propose a general performance evaluation framework for streaming services, such as PSS, MBMS, VT, PoC and so on, and make the framework can be used by all the operators, device and service providers to evaluate and improve service quality more conveniently.

First, we introduce a special classification to the performance metrics and unify them into a systematic framework. The framework is requirement-driven. It is built from the users' viewpoints. In addition, due to the classification, it is very suitable for performance analysis from different points of view. All the network operators, device and service providers can use it as a performance analysis tool. We also propose an analytical modelling method to get the mapping relationships of these metrics. Based on the mapping relationships, the operators, device or service providers can obtain the service performance and know how to adjust the metrics to improve the performance conveniently. In [4], PSS performance is taken as an example in detail.

As 3G systems are being deployed, many researches pay attention to advanced Beyond 3G (B3G) communication systems which would provide greatly increased flexibility in deployment and service provisioning. The two predominant features of B3G are: heterogeneity network structure and multi-type applications. The metrics framework and modelling method we propose also match the two features, and can be extended to B3G systems very easily.

The remainder of this paper is organized as follows: Sect. 2 represents the related work. The building of the metrics framework including metrics classification and choosing is presented in Sect. 3. Then Sect. 4 describes the analytical modelling analysis based on the metrics framework. Sect. 5 gives an example of how to use the analytical models. Finally, conclusions are given in Sect. 6 which summarizes the main theme of the paper, reiterates the main contributions and presents some thoughts on our future work.

2 Related Work

European Telecommunications Standards Institute (ETSI) also built a Quality of Service (QoS) metrics framework for all the popular services in 3G networks [7]. Their framework is suitable for measuring and testing. The metrics are classified into two classes: service independent QoS metrics (e.g., network accessibility) and service dependent QoS metrics (e.g., service accessibility). Measurement for each metric is also given.

Oumer Teyeb et al. make a subjective evaluation of packet service performance in UMTS [14]. Their metrics include the basic parameters of the networks (that is bandwidth, delay, jitter and packet loss ratio) and the service performance metrics from *usability testing*. *Usability testing* is a subjective performance evaluation method. It combines many factors such as learnability, efficiency, ease of use, error proneness and overall satisfaction.

To summarize, the metrics in ETSI focus on the service. And the subjective evaluation of Oumer Teyeb et al. just focuses on the basic network transport metrics. Their metric classifications are both very simple and not fit for performance evaluation from the viewpoints of all the operators, device or service providers. Besides, they both build their evaluation frameworks using statistical methods, which can't reveal the inherent laws of these metrics clearly and would cost lots of human and material resources. Our metrics framework is suitable to be used in another way — analytical modelling.

Some works have been done using analytical modelling methods. Kishor S. Trivedi et al. analyze the Base Station to get the accessibility using a Markov Chain [10]. Stochastic Well formed Petri Nets (SWN) [11] has been used to solve the same problem. Irfan Awan and Khalid Al-Begain use a G/G/1/N censored queue with single server and $R(R \geq 2)$ priority classes to evaluate a buffer management scheme to get the performance with different class traffics in 3G Mobile Networks [6]. Majid Ghaderi and Srinivasan Keshav focus on the MMS relay and server to get the performance of MMS using queueing model [12]. However these models just focus on some parts of the network. Considering all the units in the networks, our analytical modelling analysis uses a systematic way to combine these models to get the mapping relationships between the metrics.

3   Metrics Framework

Many performance metrics for streaming services have been proposed. We introduce a special classification to these metrics and unify them into a systematic framework for service performance evaluation.

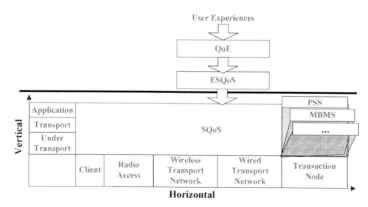

Fig. 1 The overall model of our framework

The main impact of service quality is on the end-user, therefore a detailed study of service performance should involve the end-user. We adopt a top-down requirement-driven approach from the users' viewpoints to build the framework as shown in Fig. 1.

Quality of Experience (QoE), End-to-end Service QoS (ESQoS) and System QoS (SQoS) are all subsets of the conventional QoS, which is defined in ITU Telecommunication Standardization Sector (ITU-T) Recommendation E.800 [9] as "the collective effect of service performances which together determines the satisfaction of a user of a service". The exact definitions and the differences of the three are described below.

3.1 QoE

QoE is the overall performance of a service, as perceived subjectively by the end-user. And it is stated irrespective of its measurability. We have launched a user investigation presented in [4], then choose the metrics that are more important from the users' viewpoints as our QoE.

In general, the network operators, device and service providers should set service requirements in line with the end-user's expected QoE. And QoE needs to be translated into metrics that they can measure and control. Thus, we introduce ESQoS and SQoS metrics.

3.2 ESQoS

ESQoS also indicates the overall performance of a service. It is obtained by mapping QoE into metrics that more relevant to operators, device and service providers. QoE, ESQoS and their mapping relationships are as shown in Fig. 2.

Compared to QoE, ESQoS is more suitable to be measured and analyzed for the mapping between ESQoS and SQoS. Initial Connecting Time and Initial Buffering Time would be mapped to different SQoS, so the Service Setting-up Time is divided. Re-Buffering is divided for measurability. It is difficult to measure these metrics which indicate audio and video quality, however in the mapping, they are both related to the same SQoS, so we aggregate them to three metrics temporally: Audio Quality, Intra-Frame Video Quality and Inter-Frame Video Quality. The further study is our future work.

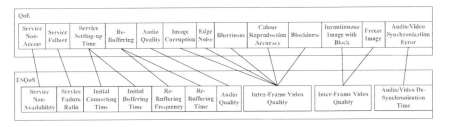

Fig. 2 The mapping from QoE to ESQoS

3.3 SQoS

SQoS indicates the metrics that express the system's capability to support services, such as the bandwidth of networks, transaction capacity of servers and so on. Compared with ESQoS, SQoS denotes the point-to-point QoS, which is specifically related to the units and links of network systems, rather than the whole service. SQoS metrics are some QoS metrics that the network operators, device and service providers can control directly or indirectly.

Based on the network structure, SQoS metrics are transversely divided into five classes: client, wireless transport, radio access, wired transport and transaction node as shown in Fig. 3. This classification covers the networks completely and it is a good fit for our modelling analysis.

The transaction node means the entities where the signalling and data traffic are processed on the network. They are scattered around the whole network, including Base Stations, SGSNs, GGSNs, servers and so on. Because we do not focus on the internet, the routers are not included in the transaction node class.

Each class also has three layers which reflect the protocol layers vertically. The first one is the application layer. The second one is the transport layer. And the third one consists of data link and physical layers.

The difference of the services is reflected by the different transaction nodes. We take two new services in 3G — PSS and MBMS for examples. According to the reference architectures of PSS and MBMS, we can get the transaction nodes of them. Meanwhile in 3G, IP Multimedia Subsystem (IMS) provides a signaling platform for multimedia services [1]. It is optional for some multimedia services' implementation. We also give the metrics to evaluate IMS's performance. Their reference architectures are shown in Fig. 4. The transaction nodes of different services are represented using different shapes. The specific SQoS metrics of each service and IMS are list in Table 1. We just choose the typical functional entities in each service. For example, in the MBMS service, CSE is optional. Whether CSE is used depends on the exact network deployment. And if the service is across more than one network domain, some additional entities may be used, such as Border Gateway (BG) and so on. All these entities are not included in Table 1. Which entities should be used

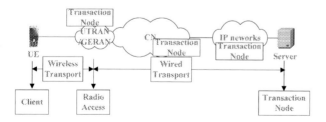

Fig. 3 Five classes of SQoS
UTRAN=UMTS Terrestrial Radio Access Network, GERAN=GSM/EDGE Radio Access Network

Table 1 Transaction nodes of PSS, MBMS and IMS

	PSS	MBMS	IMS
Special	Web Server,	BM-SC,CBC,MBS,	IP-CAN,HSS,DHCP,DNS,IM-MGW,AS,
Transaction Node	Media Server.	Content Provider.	BGCF,MGCF,MRFC,MRFP,CSCF,SLF.
Common Transaction Node		BS,SGSN,GGSN,VLR,HLR,AuC.	

IP-CAN=IP-Connectivity Access Network, DHCP=Dynamic Host Configuration Protocol, DNS=Domain Name System, MBS=Multicast Broadcast Sources, BS=Base Station.

in the framework depends on the specific circumstances. Among these transaction nodes, there are also some used by PSS, MBMS and IMS in common listed as the Common Transaction Node.

The SQoS metrics for each service are all the same, and they are typically chosen as shown in Table 2. The definitions of all the QoE, ESQoS and SQoS metrics can be found in detail in [4].

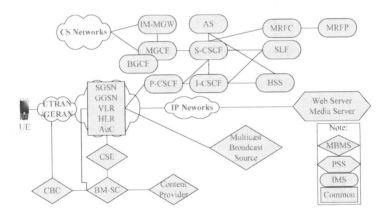

Fig. 4 Services' reference architecture
UE=User Equipment, SGSN=Serving GPRS Support Node, GGSN=Gateway GPRS Support Node, HLR=Home Location Register, VLR=Vistor Location Register, AuC=Authentication Server, CBC=Cell Broadcast Centre, CSE=Camel Server, BM-SC=Broadcast-Multicast Service Centre,CS=Circuit Switched, IM-MGW=IP Multimedia-Media Gate-way, BGCF=Border Gateway Control Function, MGCF=Media Gateway Control Function, CSCF=Call Session Control Function, P-CSCF=Proxy CSCF, I-CSCF=Interrogating CSCF, S-CSCF=Serving CSCF, AS=Application Server, HSS=Home Subscriber Server, SLF=Subscription Locator Function, MRFC=Media Resource Function Controller, MRFP=Media Resource Function Processor

Table 2 SQoS metrics of each class

	Application	Transport	Under Transport
Client	Frame Rate, Frame Loss Ratio, Frame Error Ratio.	– –	– –
Wireless Transport	– –	– –	Block Ratio, Access Delay, Soft Handoff Rate.
Radio Access	– –	Bandwidth, Transfer Delay, Jitter, Packet Loss Ratio.	Frame Loss Ratio, Residual Frame Error Ratio, Frame Discard Ratio, Residual Bit Error Ratio.
Wired Transport	– –	Bandwidth, Transfer Delay, Jitter, Packet Loss Ratio.	– –
Transaction Node	Transaction Rate, Transaction Delay, Transaction Capacity.	– –	– –

The proposed framework has several features. First, it is a top-down requirement-driven one. QoE is from users, and ESQoS is got from QoE. Then we choose the SQoS metrics affecting these ESQoS metrics. So all the metrics are user-centered.

Second, we choose the same SQoS metrics for different streaming services. The difference of the services just lies on the different transaction nodes due to their different reference architectures. So based on the framework, it is much easier to analyze the performance of different services simultaneously.

Third, the framework is very suitable for performance analysis from different points of view. From our framework, all the network operators, device and service providers can choose the related metrics easily to analyze their responsible part of the network based on the SQoS classification.

Finally, our framework can be extended to B3G easily. In B3G heterogenous networks, we only need to add more new transaction nodes or classes transversely to SQoS according to the new network structure.

4   Modelling Analysis

Based on the metrics framework proposed above, we introduce a modelling method to get the mapping relationships from SQoS to each ESQoS metric.

There are lots of analytical models to analyze networks, typically such as Markov Chain modelling Radio access [10, 11], Jackson queueing network modelling the

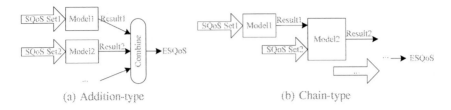

(a) Addition-type   (b) Chain-type

Fig. 5 Analysis combine methods

IP network [8, 5] and queueing system modelling the transaction nodes [6, 12]. First, we model each class of the network in details using these typical models. Then we combine these models to get the ESQoS metrics. There are two ways to combine them: addition-type as shown in Fig. 5(a) and chain-type as shown in Fig. 5(b). Addition-type means that we can analyze each part in isolation, and then just combine the results of them to get the ESQoS metrics. Chain-type is more complicated. The models of all the parts are interrelated and these interrelationships can not be ignored. Like the chain, one model's analysis results are other's input. We divide these metrics into two classes: Service Non-Availability, Service Failure Ratio, Initial Connection Time and Initial Buffering Time belong to the addition-type; Re-Buffering Frequency, Re-Buffering Time and Audio/Video De-Synchronization Time belong to the chain-type. Finally, a quantitative function between SQoS metrics and each ESQoS can be obtained. The mapping results of PSS can be found in detail in [4]. We take Service Non-Availability of PSS as an example in Sect. 5.

The metrics framework we proposed is not only fit for the analytical modelling analysis, but also can be used for statistical modelling. We can substitute these Markov chain or queue models in each class of the network with statistical models, then combine them, or we can just use the metrics in the framework to build a whole statistical model.

Based on the mapping relationships, knowing the SQoS metric values, the operators, device or service providers can get the service's performance. And in order to improve service's performance, they can adjust the relative SQoS to change the ESQoS metric conveniently. Moreover before a new service is implemented, they can estimate ESQoS using the existent SQoS values to check whether the new service should be launched. Based on the framework, they can also adjust the common SQoS for configuring a most suitable network to satisfy all the users of different services.

5 Modelling Service Non-availability of PSS

Take the Service Non-Availability of PSS as an example. First the client sends out the service request data. The data goes through the Base Station, HLR, VLR, AuC, several SGSNs and GGSN, and traverses the IP networks, then arrives at the web server. In the radio access network, Markov Chain can be used to get the probability of access failure; HLR, VLR, AuC and web server are just used for signalling, so they can be simply modelled as M/M/1 queues to get the probability of server failure; In SGSN and GGSN, different kinds of traffic with different process priorities go through, so they are modelled as priority queueing systems.

The Service Non-Availability is an addition-type metric. Suppose that the service request data goes through one SGSN, then the Service Non-Availability can be expressed as: $1 - (1 - P_{access\_failure})(1 - P_{HLR\_failure})(1 - P_{VLR\_failure})$ $(1 - P_{AuC\_failure})(1 - P_{SGSN\_failure})(1 - P_{GGSN\_failure})(1 - P_{WebServer\_failure})$. $P_{access\_failure}$ is the access failure probability, and can be got using Markov Chain model with Block Ratio, Soft Handoff Rate and other setting parameters as the input. $P_{*\_failure}$ $(* : HLR, VLR, AuC, SGSN, GGSN, WebServer)$ are the probabilities that the corresponding nodes fail. They can be got using Queue model with Transaction Rate and Capacity as the input.

The details of these model results are present in [4]. Based on the model results, the operators know how to adjust the block ratio or soft handoff rate to decrease the Service Non- Availability. And the device providers know which device's quality must be improved and how to adjust the transaction rate and transaction capacity to balance the service quality and cost.

6 Conclusion

In this paper, we propose a general performance metrics framework for the streaming services in 3G. The proposed framework is a top-down requirement-driven one. All the metrics are user-centered, so the framework is more practical. Besides, the framework is very suitable for performance analysis from different points of view due to our SQoS metrics' classification. All the network operators, device and service providers can use it as a performance evaluation tool. This metrics framework can also be extended to B3G easily.

On the metrics framework, we also give an analytical modelling analysis to get the mapping relationships from SQoS to ESQoS. Based on the mapping relationships, and knowing the SQoS metric values, the operators, device or service providers can get and improve the performance of a service conveniently. Moreover before a new service is implemented, they can estimate the ESQoS of it using the existing SQoS values to show whether this new service should be launched. Based on the framework, they can also adjust the common SQoS to get a most suitable network for different services. The work presented in this paper has been accepted as part of the 3GPP standards for performance evaluation. Based on the metrics framework, statistical modelling methods can also be used, and it is the next step of our work.

Acknowledgements. This work is supported in part by the National Natural Science Foundation of China (NSFC) under Grant No.60773138, the National Grand Fundamental Research Program of China (973) under Grant No.2009CB320504 and the China Mobile Research Institute. Thanks for the great help.

References

1. 3GPP TS 23.228, IP Multimedia Subsystem (IMS); Stage 2, v8.2.0 (September 2007)
2. 3GPP TS 26.234, Transparent end-to-end Packet-switched Streaming Service (PSS); Protocols and codecs, v7.4.0 (September 2007)
3. 3GPP TS 26.346, Multimedia Broadcast/Multicast Service (MBMS); Protocols and codecs, v7.5.0 (September 2007)
4. 3GPP TR 26.944, End-2-End Multimedia Services Performance Metrics, v7.0.0 (March 2007)
5. Borodin, A., Kleinberg, J., Raghavan, P., Sudan, M., Williamson, D.: Adversarial queueing theory. Journal of the ACM, 13–38 (Janauary 2001)
6. Awan, I., Al-Begain, K.: Performance modelling of differentiated services in 3G mobile networks. In: ESM 2003, pp. 302–307 (2003)
7. ETSI TS 102 250-1, QoS aspects of popular services in GSM and 3G networks; Part 1: Identification of Quality of Service aspects, v1.1.1 (Octobor 2003)
8. Harchol-Balter, M., Black, P.E.: Queuing analysis of oblivious packet-routing networks. In: Proc. of ACM, pp. 583–592 (Janauary 1994)
9. ITU-T Recommendation E.800, Terms and definitions related to quality of service and network performance including dependability (August 1994)
10. Trivedi, K.S., Dharmaraja, S., Ma, X.: Analytic modeling of handoffs in wireless cellular networks. Information Sciences 148(1-4), 155–166 (2002)
11. Mokdad, L., Sene, M.: Performance measures of a call admission control in mobile networks using SWN. In: ACM International Conference Proceeding Series, vol. 180(64) (2006)
12. Ghaderi, M., Keshav, S.: Multimedia Messaging Service: System Description and Performance Analysis. In: Proc. of IEEE/ACM Wireless Internet Conference (WICON) (July 2005)
13. Open Mobile Alliance, Push to talk over Cellular (PoC) - Architecture Approved Version 1.0.1, OMA-AD_PoC-V1_0_1-20061128-A (November 2006)
14. Teyeb, O., Sørensen, T.B., Mogensen, P., Wigard, J.: Subjective Evaluation of Packet Service Performance in UMTS and Heterogeneous Networks. In: Q2SWinet, pp. 95–102 (October 2006)

The Stochastic Network Calculus Methodology

Deah J. Kadhim, Saba Q. Jobbar, Wei Liu, and Wenqing Cheng

Abstract. The stochastic network calculus is an evolving new methodology for backlog and delay analysis of networks that can account for statistical multiplexing gain. This paper advances the stochastic network calculus by deriving a network service curve, which expresses the service given to a flow by the network as a whole in terms of a probabilistic bound. The presented network service curve permits the calculation of statistical end-to-end delay and backlog bounds for broad classes of arrival and service distributions. The benefits of the derived service curve are illustrated for the Exponentially Bounded Burstiness (EBB) traffic model. It is shown that end-to-end performance measures computed with a network service curve are bounded by O(H logH), where H is the number of nodes traversed by a flow. Using currently available techniques that compute end-to-end bounds by adding single node results, the corresponding performance measures are bounded by O(H^3).

Keywords: Stochastic process and computer network performance.

1 Introduction

The network calculus is a framework for analyzing delays and backlog in a network where the traffic, and sometimes also the service, is characterized in terms of envelope functions. Pioneered as deterministic network calculus in the early 1990s for the computation of worst-case performance bounds in packet networks, it has played an important role in the development of algorithms that support Quality-of-Service guarantees in packet networks [1], [2].

The elegance of the network calculus becomes evident in the min-plus algebra formulation, where service guarantees to a flow at a node (switch) are expressed in terms of service curves [3], [4]. In this formulation, bounds for single nodes can be easily extended to end-to-end bounds. More concretely, suppose a flow is assigned a service curve S^h at the *h-th* node on its route *(h = 1, . . .,H)*. Then the service given to the flow by the network as a whole can be expressed in terms of a network service curve S^{net} as.

Deah J. Kadhim, Saba Q. Jobbar, Wei Liu, and Wenqing Cheng
Electronics and information Department
Huazhong University of Science and Technology
Wuhan, China
e-mail: deya_naw@yahoo.com

R. Lee, G. Hu, H. Miao (Eds.): Computer and Information Science 2009, SCI 208, pp. 169–178.
springerlink.com   © Springer-Verlag Berlin Heidelberg 2009

$$S^{net} = S^1 * S^2 * \ldots * S^H \qquad (1)$$

Where $*$ is a convolution operator. With this remarkable property, bounds for the output burstiness, backlog and delay or the entire network are computed in the same fashion as single node results. The resulting end-to-end delay bounds are generally tighter than the sum of the per-node delay bounds. For example, if the service curve at the h-th node is given as a constant rate function, $S^h(j) = cj$, one obtains from Eq. (1) that $S^{net}(j) = cj$. As a result, the end-to-end backlog and delay bounds are identical to the bounds at the first node. In this way, the min-plus version of the network calculus provides simple end-to-end estimates for delay and backlog.

A drawback of the worst-case view of traffic in the deterministic network calculus is that it does not reap the benefits of statistical multiplexing, which can result in an overestimation of the actual resource requirements and a low utilization of network resources. This has motivated the search for a stochastic network calculus which describes arrivals and service probabilistically while preserving the elegance and expressiveness of the original framework. By allowing even a small fraction of traffic to violate its traffic description or performance guarantees, one can achieve significant resource savings.

Most work on extending the network calculus to a probabilistic setting has been concerned with deriving statistical performance bounds for a single node [5], [6]. In a stochastic network calculus framework, traffic arrivals and sometimes also service at network nodes are random processes which are bounded by probabilistic envelope functions. The first and probably most widely known envelope function is the exponentially bounded burstiness (EBB) characterization for traffic arrivals. The EBB model, which has been generalized and has been shown to imply delay and backlog bounds at simple traffic multiplexers. Probabilistic arrival envelopes were used to derive schedulability conditions for a variety of scheduling algorithms. The authors of some our references have established a link between envelope functions and the theory of effective bandwidth, which estimates bandwidth requirements to satisfy given performance guarantees. Probabilistic envelope functions that specify the amount of service made available to a flow at a network node have appeared.

A number of studies have used probabilistic single node bounds on delay, backlog, or the burstiness of traffic departing from a node to derive multi-node performance bounds. Indeed, by relating output descriptions of traffic at a node to corresponding input descriptions, one can obtain end-to-end bounds by adding the per-node bounds. However, such results tend to degrade rapidly as the number of nodes traversed by a flow is increased.

One particular challenge is the formulation of the multi-node convolution expression of a network service curve within a probabilistic context. It was shown that a straightforward probabilistic extension of deterministic concepts yields a network service curve that deteriorates with time. As a solution, it formulated a probabilistic service curve that takes the form of Eq. (1), however, this service curve is difficult to apply in numerical examples. A probabilistic network service curve was derived under the assumption that each node drops traffic that locally violates a given delay guarantee. This dropping policy requires that packets in each buffer are sorted according to a deadline computed from the arrivals envelope and the service curve.

We illustrate the benefits of the network service curve for a statistical end-to-end analysis of multiplexed EBB traffic. By contrasting end-to-end delay bounds obtained with our service curve with bounds obtained by iterating single node results, we show the improvements attainable through our stochastic network calculus approach. We will show that the calculus approach with network service curve renders bounds for delay, backlog, and output burstiness of the order $O(H \log H)$ in the number of nodes H in the network, as opposed to $O(H^3)$ bounds obtained by adding per-node results. Thus, this paper, for the first time, quantifies the benefits of using network service curves in a probabilistic setting. This presents a significant step forward towards the goal of developing the stochastic network calculus into a practical methodology for the analysis of networks. As a remark, in the deterministic calculus, a network service curve leads to end-to-end bounds that scale with $O(H)$, while summing up single-node results gives bounds that scale with $O(H^2)$. Thus, network service curves have comparable benefits in a deterministic and a stochastic setting.

2 Network Service Curves

The input and output traffic of a flow at a network node is described by two stochastic processes $A = A(t)$ for $t \geq 0$ and $D = D(t)$ $t \geq 0$ that are defined on a joint probability space. $A(t)$ represents the cumulative arrivals, and $D(t)$ the cumulative departures in $[0, t)$. We require that A and D are non decreasing, left continuous functions with $A(0) = D(0) = 0$, and that $D(t) \leq A(t)$ for $t \geq 0$. In this paper, we use a continuous-time framework. Extensions to a discrete time setting are discussed in remarks.

In a packet-switching network, the service available to a flow at a node is determined by a scheduling algorithm (e.g., FIFO, Fair Queuing) which sets the order of packet transmissions [7]. A service curve, first presented is an alternate method to describe the service received by a flow in terms of a function which specifies a lower bound on the service. In the min-plus algebra formulation of the deterministic calculus, a service curve is a function $S(\cdot)$, such that $D(t) \geq A * S(t)$ for all $t \geq 0$, where the convolution of two real-valued functions f and g is defined as,

$$f * g(t) = \inf_{0 \leq s \leq t} \{f(s) + g(t - s)\} \text{ for all } t \geq 0 \qquad (2)$$

Next we define our measure of a probabilistic service guarantee for a flow. We adopt a variation of the definition of a statistical service curve, where we add a positivity requirement. We use the notation $[x]_+ = max(x, 0)$ for the positive part of a real number x.

Definition 1: (Statistical Service Curve)
A function $S(t)$ is a statistical service curve for an arrival process A if for every choice of σ and for all $t \geq 0$,.

$$P(D(t) < A^* \, [S - \sigma] + (t)) \leq \varepsilon(\sigma) \qquad (3)$$

Where $\varepsilon(\sigma)$ is a non-increasing function. We refer to the bound $\varepsilon(\sigma)$ on the violation probability as the error function. So the condition is avoid whenever $\varepsilon(\sigma) \geq 1$, and that Eq. (2) for $t = 0$ implies that $\varepsilon(\sigma) \geq 1$ for all $\sigma < S(0)$. We frequently require that the function $\varepsilon(\sigma)$ satisfies the integrability condition,

$$\int \varepsilon(u) \, du < \infty \qquad (4)$$

Comparing Definition 1 to probabilistic service descriptions in the literature, we see that for each choice of σ, the function $[S - \sigma]_+$ is an effective service curve. Here, choosing σ large amounts to increasing the latency and decreasing the violation probability of the service guarantee. Compared to the service curve, the enforced positivity of the statistical service curve can lead to tighter performance bounds on backlog and output Burstiness. Lastly, if $\varepsilon(\sigma) = 0$ for some value of σ, then $[S - \sigma]_+$ defines a deterministic service curve almost surely. In the continuous-time setting, we find it convenient to replace Eq. (2) by the stronger requirement that the service curve satisfies,

$$P\,(D(t) < A^* \, [S - \sigma]_+ \, (t + j_0)) \leq \varepsilon(\sigma) \qquad (5)$$

For all $t \geq 0$, and that $\varepsilon(\sigma) \geq 1$ for all $\sigma < S(j_0)$. Here, $j_0 > 0$ is a parameter that specifies a discretization of the time scale. If $S(t)$ is a statistical service curve in the sense of Eq. (2), then Eq. (5) holds for $S(t - j_0)$.

We now state the main result of this paper. Consider a flow with a network path through $H > 1$ nodes, as shown in Fig. 1. At each node we assume that the flow receives a probabilistic service guarantee in terms of a statistical service curve. The following theorem provides an expression for an end-to-end statistical network service curve in terms of the per-node service curves. In the theorem, we use the notation $f_{\delta(t)} = f(t) + \delta t$ for a real function f and a real number δ.

In a discrete time setting, there is no need for the parameter j_0 appearing in Eq. (5), and we use the definition of the statistical service curve in Eq. (2). If each node provides a service curve S^h in the sense of Eq. (2) with an error function $\varepsilon(\sigma)$, then for any choice of $\delta > 0$, so,

$$S^{net} = S^1 * S^2_{-\delta} * \cdots * S^H_{-(H-1)\delta} \qquad (6)$$

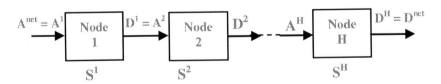

Fig. 1 Traffic of a flow through a set of H nodes

It is a network service curve which again satisfies Eq. (2), with error function,

$$\varepsilon^{net}(\sigma) = \inf_{\sigma_1 + \ldots + \sigma_H = \sigma} \{\varepsilon^H(\sigma^H) + \sum_{h=1}^{H-1} \sum_{k=0}^{\infty} \varepsilon^h(\sigma^h + k\delta)\} \qquad (7)$$

3 Performance Bounds

The derivation of the statistical network service curve in definition 1 does not make assumptions on the arrival functions A at a node, and holds for all deterministic or probabilistic descriptions of traffic. However, applying the network service curve to compute performance bounds for a traffic flow requires a characterization for the arrivals from the flow. In the deterministic network calculus [8], [9], it is generally assumed that the arrivals from a traffic flow A are bounded or regulated by an arrival envelope A^*, such that $A(t + j) - A(t) \le A^*(j)$ for all $t, j \ge 0$. A frequently used envelope is $A^*(t) = \sigma + \rho t$, which corresponds to a leaky bucket with rate ρ and burst size σ. In a stochastic network calculus, traffic arrivals are usually described in terms of probabilistic extensions of this envelope concept. The following definition specifies such an extension.

Definition 2: (Statistical Envelope)
A non decreasing function $G(t)$ is said to be a statistical envelope for an arrival process A if, for all σ and all $0 \le s \le t$,

$$P(A(t) - A(s) > G(t - s) + \sigma) \le \varepsilon(\sigma) \qquad (8)$$

Where $\varepsilon(\sigma)$ is a nonnegative, non increasing function. Note that Eq. (8) is formulated for negative as well as positive values of σ. If Eq. (8) is only known to hold for nonnegative values of σ, we set $\varepsilon(\sigma) = max\{1, \varepsilon(0)\}$ for $\sigma < 0$.

Definition 2 is inspired by the Exponentially Bounded Burstiness (EBB) model, which is the special case where $G(t) = \rho t$ is a constant-rate function and the error function decays exponentially. The Stochastically Bounded Burstiness (SBB) model is a generalization of the EBB model where the condition that the error function decays exponentially is relaxed and replaced by the assumption that it decays faster than polynomially. These decay conditions are used to bound the violation probability of events involving entire arrival sample paths.

We state the following results for completeness, since there are technical differences between our definitions of a statistical service curve in Eqs. (3) and (5) and those used in the literature. The theorem uses the deconvolution operator $\varnothing$, which is defined for two real functions f and g as,

$$f \varnothing g(t) = \sup_{s \ge 0} \{f(t + s) - g(s)\} \quad \text{for all } t \ge 0 \qquad (9)$$

4   An Application with EBB Arrivals

We now present an application that relates the network service curve and the
performance bounds developed in this paper to the literature on statistical
service guarantees from the early 1990s. We demonstrate that statistical network
service curves can faithfully reproduce the single node results which predate the
service curve concept. In a multi-node setting, we show the benefits of the statistical network service curve by comparing statistical end-to-end performance bounds
computed with the techniques (without a network service curve) to those obtained
with the results of this paper.

The network scenario that we consider is shown in Fig. 2. We will refer to the
flows which traverse the network as the through flows and the flows which transit
the network as the cross flows. We are interested in statistical multi-node performance measures for the through flows, such as an output envelope at the last
node in the network and bounds on the total delay experienced along the path
through the network. Our performance bounds hold for all work-conserving
scheduling algorithms that serve traffic from the same flow in the order of arrival.

Network arrivals are described in terms of the exponentially bounded burstiness
(EBB) model defined which is given for the arrival process A by the condition that.

$$P(A(t) - A(s) > \rho(t - s) + \sigma) \leq M\, e^{-\theta\sigma} \qquad (10)$$

For any $0 \leq s \leq t$. Here, ρ represents a bound on the long term arrival rate, θ is
the decay rate of the error function, and $M \geq 1$ is a constant.

By choosing the EBB traffic model we can compare our stochastic network calculus results with existing performance bounds, specifically. Moreover, the EBB
model can be used to describe arrival processes that have relevance in practice,
e.g., Markov-modulated On-Off processes and multiplexed regulated arrivals.
Finally, and most importantly, the EBB model lends itself to simple, closed-form
performance bounds that permit us to gain insight into the scaling properties of
network-wide bounds obtained with the stochastic network calculus.

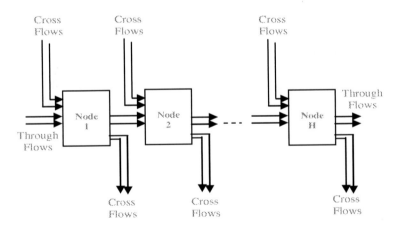

Fig. 2 A network with cross traffic

5 Numerical Results

We next give a numerical example that illustrates the benefits of using network service curves for the computation of statistical end-to-end delay bounds. We consider the network and arrival scenario shown in Fig. 2. Arrivals of cross flows and arrivals of through flows at the first node are each described as an aggregate of independent Markov Modulated On-Off processes. This type of process, which has been used for modeling voice channels, falls into the category of the EBB traffic model.

The Markov Modulated On-Off arrival process of an arrival flow, illustrated in Fig. 3, is a continuous time process with support given by a homogeneous two-state Markov chain $X(t)$ which is described in terms of the generator matrix,

$$G = \begin{bmatrix} -\mu & \mu \\ \lambda & -\lambda \end{bmatrix} \qquad (11)$$

Here, μ denotes the transition rate from the 'On' state to the 'Off' and λ to denote the transition rate from the 'Off' state to the 'On' state. In the 'On' state, the arrival process transmits at the peak rate P, and no arrivals occur in the 'Off' state.

Fig. 3 On-Off Traffic model

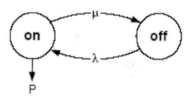

We assume that there are N through flows and N_c cross flows at each node. Through flows at the first node and cross flows are stochastically independent. For the sake of simplicity, we assume that all arrival processes are homogeneous and that there is an equal number of through and cross flows ($N = N_c$). We want to emphasize that computing examples for heterogeneous flows does not pose a problem, other than increasing notation.

Next, following, we quickly derive an EBB characterization of N independent On-Off flows. The moment generating function of a single On-Off flow is bounded by the following $E[e^{\theta A(t)}] \le e^{\theta \rho(\theta)}$, where,

$$\rho(\theta) = 1/2\theta \, (P_\theta - \mu - \lambda + ((P_\theta - \mu + \lambda)^2 + 4\lambda\mu)^{1/2} \qquad (12)$$

The quantity $\rho(\theta)$ is called 'effective capacity' and has the property that $\rho(0) \le \rho(\theta) \le P$. The rate $\rho(0) = \lambda/\lambda + \mu P$ represents the average rate of the flow.

For each choice of $\theta > 0$, this provides us with an EBB characterization for the through flows and the cross flows in the network from Figure 2. Delay bounds for the through flows are provided by [1]. Finally, we numerically optimize the resulting delay bounds over θ.

In the example, the capacity of each node in the network is set to $C = 100\ Mbps$ and time is measured in milliseconds. The parameters of the flows are given in Table 1.

Table 1 Parameters of On-Off sources

Burstiness	T (ms)	P (Mbps)	$\rho(0)$ (Mbps)	λ (ms−1)	μ (ms−1)
Low	10	1.5	0.15	0.11	1.0
High	100	1.5	0.15	0.01	0.1

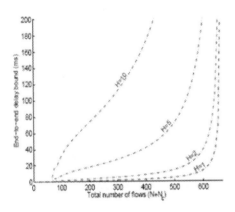

Fig. 4.a: T=10 (Adding per node bounds)   Fig. 4.b: T=10 (Network Service Curve)

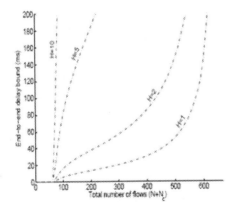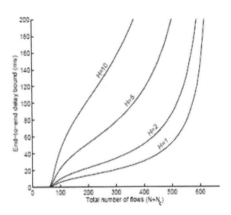

Fig. 4.c: T=100 (Adding per node bounds)   Fig. 4.d: T=100 (Network Service Curve)

We consider two types of flows, with identical peak rate $(P = 1.5\ Mbps)$ and average rate $(\rho(0) = 0.15\ Mbps)$. We introduce a parameter $T = 1/\mu + 1/\lambda$ to describe the Burstiness of a flow. T is the expected time for the Markov chain to change states twice. For flows with given peak rate P and given mean rate $\rho(0)$, a larger value of T indicates a higher degree of burstiness. In Table 1, we use $T = 10$ for flows with low burstiness and $T = 100$ for flows with high burstiness. Lastly, the violation probability for the end-to-end delay bounds is set to $\varepsilon = 10-9$.

In Fig. 4 we show the probabilistic end-to-end delay bounds of flows as a function of the number of flows $N + N_c$. We consider networks where the number of nodes traversed by the through flows is set to $H = 1, 2, 5,\ and\ 10$. The maximum number of flows at each node is given by $\lfloor C/\rho \rfloor = 666$ flows. In Figures 4(a) and (b) we show the delay bounds obtained by adding per-node bounds. For a single node $(H = 1)$, both techniques yield the same delay bounds. The benefits of network service curves become pronounced when the number of nodes H traversed by the through flows is increased. Figures 4(c) and (d) show similar plots for flows that are more bursty.

Here, the delay bounds are higher, underlining that bursty flows have a lower statistical multiplexing gain. Fig. 4 shows End-to-end delay bounds of Markov Modulated On-Off arrivals as a function of the number of flows $N + N_c\ for\ H = 1, 2, 5, 10, \varepsilon = 10-9$ with at first $T = 10ms$ (low burstiness), and then $T = 100ms$ (high burstiness).

6 Conclusions

We have extended the state-of-the-art of the stochastic network calculus by deriving a network service curve formulation that is applicable to a broad class of traffic and service characterizations. The formulation of such a service curve in the presented general form has been a long-standing research problem. Using the network service curve, we calculated statistical end-to-end delay and backlog bounds which are vastly superior to bounds obtained by adding delay bounds of single nodes. For EBB traffic arrivals, we showed that for a flow that traverses H nodes and encounters cross traffic at each node, our network calculus with statistical network service curves gives statistical end-to-end delays that are bounded by $O(H\ logH)$, as opposed to $O(H^3)$ bounds rendered by the method of adding per-node bounds. An immediate research problem suggested by this paper relates to the tightness of the $O(H\ logH)$ bounds.

References

1. Kurose, J.: Open Issues and Challenges in Providing Quality of Service Guarantees in High-Speed Networks. ACM Computer Communication Review (1993)
2. Cruz, R.L.: A calculus for network delay, Part I: Network elements in isolation. IEEE Transaction of Information Theory 37(1), 114–121 (1991)
3. Cruz, R.L.: A calculus for network delay, Part II: Network analysis. IEEE Transactions on Information Theory 37(1), 132–141 (1991)

4. Chang, C.S.: Performance Guarantees in Communication Networks. Springer, Heidelberg (2000)
5. Thiran, P., Le Boudec, J.-Y.: Network Calculus. LNCS, vol. 2050, p. 3. Springer, Heidelberg (2001)
6. Cruz, R.L.: Quality of service guarantees in virtual circuit switched networks. IEEE Journal on Selected Areas in Communications 13(6), 1048–1056 (1995)
7. Jiang, Y.: Stochastic Network Calculus and its Application to Network Design and Analysis. Norwegian University of Science and Technology (NTNU), Department of Telematics, Trondheim, Norway (2007)
8. Ridouard, F., Scharbarg, J.L., Fraboul, C.: Stochastic network calculus for end-to-end delays distribution evaluation on an avionics switched ethernet. In: Proceedings of INDIN. IEEE, Los Alamitos (2007)
9. Jiang, Y., Emstad, P.J.: Analysis of stochastic service guarantees in communication networks: A server model. In: de Meer, H., Bhatti, N. (eds.) IWQoS 2005. LNCS, vol. 3552, pp. 233–245. Springer, Heidelberg (2005)

Anti-spam Filter Based on Data Mining and Statistical Test

Gu-Hsin Lai, Chao-Wei Chou, Chia-Mei Chen, and Ya-Hua Ouv

Abstract. Because of the popularity of Internet and wide use of E-mail the volume of spam mails keeps growing rapidly. The growing volume of spam mails annoys people and affects work efficiency significantly. Most previous researches focused on developing spam filtering algorithm, using statistics or data mining approach to develop precise spam rules. However, mail servers may generate new spam rules constantly and mail server will then carry a growing number of spam rules. The rules might be out-of-date or imprecise to classification as spam evolves continuously and hence applying such rules might cause misclassification. In addition, too many rules in mail server may affect the performance of mail filters. In this research, we propose an anti-spam approach combining both data mining and statistical test approach. We adopt data mining to generate spam rules and statistical test to evaluate the efficiency of them. By the efficiency of spam rules, only significant rules will be used to classify emails and the rest of rules can be eliminated then for performance improvement.

Keywords: Spam mail, Data Mining, Statistical Test.

1 Introduction

Spam, by the simplest definition involves that e-mail users do not want to receive, the contents of the letter is usually commercial advertisement, pornography or malicious mail. Spam has great impact on individuals or enterprises. For individuals, users waste time on spam elimination. There are a lot of malicious codes, internet fraud and phishing through spam. Spam makes people more vulnerable to attacks by malicious code, phishing or fraud. As for the impact on enterprises, they spend money and manpower combating the problem.

Most studies have been focused on spam filtering algorithm which is based on the features of the content and titles of the message. Usually artificial intelligence

Gu-Hsin Lai and Chia-Mei Chen
Department of Information Management, National Sun Yat-Sen University,
Kaohsiung, Taiwan

Chao-Wei Chou and Ya-Hua Ou
Department of Information Management, I-Shou University, Kaohsiung, Taiwan
e-mail: choucw@isu.edu.tw

R. Lee, G. Hu, H. Miao (Eds.): Computer and Information Science 2009, SCI 208, pp. 179–192.
springerlink.com   © Springer-Verlag Berlin Heidelberg 2009

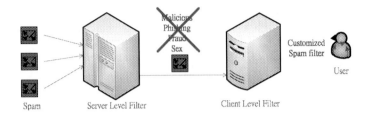

Fig. 1 Structure of a multiple spam mails filters

as well as data mining is used to generate the rules to classify the spam, but the biggest problem in spam filtering arises because the definitions of spam to different users are all different. For example, a "call for paper" mail for the scholars is the extremely important information, but the kind of mail will be classified as spam for White-collar workers. It is the foremost problem in current spam mail filter. In this paper, multiple spam mails filter is considered to be an effective one. Figure 1 illustrates the structure of a multiple spam mails filters system.

As shown in Figure 1, an effective spam mails filter is divided into two layers. The first layer is the server-side e-mail filter which filters out the spam mails that are not uncontroversial. Malicious mails (Worms, viruses, Trojans or backdoor), phishing, Internet fraud or pornographic e-mail are filtered out by this filter. On the other, controversial mails (some commercial mail) are filter out by client-side spam filter, which is mainly based on user's behavior. The difference between the designs of the two layers is very large. In this work we focus on the server-level.

The previous studies usually focused on the generation of laws for e-mail filtering, while few focused on the spam management rules. Also, the current spam rules management policies are lack in theory-based automatic management mechanism. Therefore, in this paper we will propose an automatic anti-spam method combining both data mining and statistical test approach. A mail server constantly learns new rules by using data mining method, and when there is new mail coming in, the mail server will look for corresponding rule. In this paper, what is distinct from the previous studies is that when the mail server finds the corresponding rule, it does no adopt immediately the rule to filter e-mail. Instead, the mail server will use statistical test methods to test the significance for the rule. Only after the approval that the rule is significant, it can be used to filter spam. After receiving e-mails, users can send their feedback to the mail server based on the classification results in order to strengthen accuracy of the rule. This paper uses statistical test and user feedback as well as data mining methods to solve the server-side spam filtering. The rest of this paper is organized as follows. In Section 2 we explore the literature on spam; in Section 3, generation of the rules is investigated; the management model and adoption rules are discussed in Section 4; in Section 5 we prove the theoretical part for the rule management model; Section 6 display system for authentication and the conclusions and the future research are discussed in Section 7.

2 Literature Review

Most anti-spam researches focused on developing a precise filter to classify mails. Several statistic or machine learning approaches are used. Chouchoulas and Shen proposed a rough set-based approach for text classification to filter out spam [1]. Zhao and Zhang proposed a rough set based model to classify emails into three categories - spam, non-spam, and suspicious, rather than two classes (spam and non-spam) [9]. Zhao and Zhu proposed a classification scheme based on decision-theoretic rough set to filter out spam mails. It seems that rough set theory is suitable for classifying data with vagueness and uncertainty that arises from imprecise, inconsistent or incomplete information [10]. However, these researches only used keyword frequency to generate spam rules, while some header information is also important for filtering out spam mails.

Sahami et al. proposed an email filter based on an enhanced Naive Bayes classifier. Recall and precision were improved when phrases and header specific information were added as features [7]. Carreras and Marquez proposed an Adaptive Boosting algorithm (AdaBoost) with confidence-rated predictions to filter out spam mails [2]. However, these researches only develop a precise spam filter.

Drucker et al. used SVM to classify emails, in addition, a measure of confidence is associated with that classification so that the message can be ranked in order [4]. Clark et al. proposed a neural network based system for automated e-mail filing into folders and anti-spam filtering [3]. Li and Hunag used SVM to classify emails [5]. Woitaszek et al. used SVM approach to construct an automated classification system to detect unsolicited commercial e-mail. Several sets of sample messages were collected to build dictionaries of words found in e-mail communications, which were processed by the SVM to create a classification model for spam or non-spam messages [8]. Neural network and SVM seem good for spam filter. However, both neural network and SVM cannot generate manageable spam rules.

3 Spam Rules Generation

The purpose of this study is to put forward a spam mails rule based on management theory by the data mining to manage the information. Theoretically, the following approach can be used to generate spam rule: such as the decision tree, AdaBoosting with Decision Stumps, Naïve Bayes, Ripper, Rough set theory or AD Tree. In this study rough set theory is used to produce the spam rules. Rough Sets Theory is the first propose by Poland logician Prof. Pawlak [6]. The information we get is often inaccurate, incomplete. In order to obtain the conclusion to the information, some operations to these inaccurate and (or) incomplete information must be carry out. At present, there are a number of scholars who have been using rough set theory to generate spam rules and get a good performance (Zhao and Zhang [9]; Zhao and Zhu [10]). Therefore, this study uses rough set theory to generate spam rules, too. Due to the space limitations, the paper only extracts the characteristics of the e-mail features for description.

Table 1 The attributes of the mail

Attributes	Description
From	The sender's name and email address.
Reply to	If this mail specifies an address for replies to go to
CC	If this mail has carbon copy
Received	It means where the message originated and what route it took to get to you.
Subject	The subject of this mail.
Body	The content of this mail.
Length	The length (# of byte) of this mail
Domain	The domain name of sender's mail server
Multi part	Does this mail be multi part?
Text/Html	Is the format of the content of mail.
Hasform	Does this mail have form?
Table	Does the content of mail have tables

In this study, three different types of attributes are adopted to filter e-mail, they are keyword attributes, header attributes and the letters format attributes. Table 1 illustrates the attributes used in this study.

4 Spam Mail Management and the Model Used

Many spam-related researches focused on spam rules generation. They empathized on the precision of the trained rules. The majority of the researches focusing on the classification of client's e-mail do not take into account the special needs of server. In addition to the focuses on the management and choice of the spam rules, the focuses of this paper are also on the model management of the spam rules.

In this study, we take use of reward function to assess the rule. The classification of different acts has different levels of reward. For example, the rewards of misclassification on one normal mail and on one spam mail are different. The four different rewards are illustrated in Table 2.

Table 2 Contingency reward table

Judgment / Truth	Spam	Non-Spam
Spam	R_{ss}	R_{sn}
Non-Spam	R_{ns}	R_{nn}

R_{ss} : The reward of one correct classified spam mail .

R_{sn} : The reward of one incorrect classified spam mail .

R_{ns} : The reward of one incorrect classified normal mail .

R_{nn} : The reward of one correct classified normal mail .

The rewards are shown in inequality 1 .

$$R_{ss} > R_{nn} > 0 > R_{sn} > R_{ns} \qquad (1)$$

5 Statistical Model

In the following, we first give some notation:

Table 3

Symbol	Quantity
N	Total number of mails in the server
R_i	Rule i for spam mail classification
M_i	Mail i , $i = 1, 2, \cdots, N$
P	The probability that one mail is a spam mail.
$\hat{P}_N$	Unbiased estimator for P when there are N mails in the server.
β_i	The probability that one spam mail is incorrect judged as a normal mail by R_i .
$\hat{\beta}_i$	Unbiased estimator for β_i .
α_i	The probability that one normal mail is incorrect judged as a spam mail by R_i .
$\hat{N}_{ss}(i)$	The number of correctly classified spam mails by R_i .
$\hat{N}_{sn}(i)$	The number of wrongly classified spam mails by R_i .
$\hat{N}_{ns}(i)$	The number of wrongly classified normal mails by R_i .
$\hat{N}_{nn}(i)$	The number of correctly classified normal mails by R_i .
$\hat{N}_{ss}$	The number of correctly classified spam mails.
$\hat{N}_{sn}$	The number of wrongly classified spam mails.
$\hat{N}_{ns}$	The number of wrongly classified normal mails.
$\hat{N}_{nn}$	The number of correctly classified normal mails.
$N(i)$	The number of mails filtered by R_i .
$P(i)$	The probability that one mail is judged as a spam mail by R_i .
$\hat{P}(i)$	Unbiased estimator of $P(i)$, $\hat{P}(i) = \dfrac{\hat{N}_{ss}(i) + \hat{N}_{sn}(i)}{N(i)}$
$R(i)$	Reward function of R_i .

where $R(i)$ is stochastic and can be proved to be asymptotically normal distributed as (2):

$$R(i) \sim N(\mu, \sigma^2),$$

where

$$\mu = R_{ss} \cdot P + R_{nn} \cdot (1-P) + P \cdot (R_{sn} - R_{ss}) \cdot \beta_i + (1-p) \cdot (R_{ns} - R_{nn}) \cdot \alpha_i$$

and

$$\sigma^2 = P^2 \cdot (R_{sn} - R_{ss})^2 \cdot (\frac{\beta_i}{N(i)P}) + (1-P)^2 \cdot (R_{ns} - R_{nn}) \cdot (\frac{\alpha_i}{N(i)(1-P)}) \qquad (2)$$

To statistically compare R_i and R_j, we may take use of the distribution of $R(i) - R(j)$ as in (3):

$$R(i) - R(j) \sim N(\mu_{i-j}, \sigma^2_{i-j})$$

with

$$\mu_{i-j} = P \cdot (R_{sn} - R_{nn}) \cdot (\beta_i - \beta_j) + (1-P) \cdot (R_{ns} - R_{nn}) \cdot (\alpha_i - \alpha_j),$$

and

$$\sigma^2_{i-j} = P^2 \cdot (R_{sn} - R_{ss})^2 \cdot (\frac{\beta_i}{N(i)P} - \frac{\beta_j}{N(j)P}) + (1-P)^2 \cdot (R_{ns} - R_{nn})^2 \cdot (\frac{\alpha_i}{N(i)(1-P)} - \frac{\alpha_j}{N(j)(1-P)})$$

$$(3)$$

The testing hypothesis can be as follows:

H_0 : $E[R(i)] \leq E[R(j)]$ (or $\mu_{i-j} \leq 0$)

H_1 : $E[R(i)] > E[R(j)]$

Here we use $\hat{\mu}_{i-j}$ as the statistic for hypothesis testing.

$$\hat{\mu}_{i-j} = P \cdot (R_{sn} - R_{nn}) \cdot (\hat{\beta}_i - \hat{\beta}_j) + (1-P) \cdot (R_{ns} - R_{nn}) \cdot (\hat{\alpha}_i - \hat{\alpha}_j) \qquad (4)$$

Under the case that $\hat{\mu}_{i-j} > Z_\alpha \cdot \hat{\sigma}_{i-j}$, we reject H_0 and conclude that R_i is better than R_j. Here Z_α stands for the Z-value in Normal distribution.

Among all the possible rules, we have possibly two pre-set rules: Non-Spam Rule, R_N and All-Spam Rule, R_A :

R_N : The rule classifies all mails as normal. Hence $\alpha_N = 0$ and $\beta_N = 1$.

R_A : The rule classifies all mails as spam. Hence $\alpha_A = 1$ and $\beta_A = 0$.

Their reward functions are respectively

$$R(N) = R_{nn} \cdot (1-P) + R_{sn} \cdot P \qquad (5)$$

$$R(A) = R_{ns} \cdot (1 - P) + R_{ss} \cdot P \tag{6}$$

Rule R_i can be reasonably applied in a mail server under the conditions that $E[R(i)] > R(N)$ and $E[R(i)] > R(A)$, i.e. under the following hypothesis testing conditions:

$$\hat{\mu}_i - R(N) > Z_\alpha \cdot \sigma_i \tag{7}$$

$$\hat{\mu}_i - R(A) > Z_\alpha \cdot \sigma_i \tag{8}$$

where

$$\hat{\mu}_i = R_{ss} \cdot P + R_{nn} \cdot (1 - P) + P \cdot (R_{sn} - R_{ss}) \cdot \hat{\beta}_i + (1 - P) \cdot (R_{ns} - R_{nn}) \cdot \hat{\alpha}_i \tag{9}$$

Passing the above tests convinces us that applying the rule R_i is better than not applying the rule. After giving the above statistical model, we will give some theoretical results to justify the model.

First we give and prove Lemma 1. The result is independent of the rules.

Lemma 1

$$\lim_{N \to \infty} \Pr(|\hat{P}_N - P| \geq \varepsilon) = 0 \qquad \forall \varepsilon > 0 \tag{10}$$

Proof 1: In the first step we prove that $\hat{P}_N = \dfrac{\hat{N}_{ss} + \hat{N}_{sn}}{N}$ is an unbiased estimate of P: Since

$\hat{N}_{ss} + \hat{N}_{sn}$ can be written as the sum of N i.i.d. random variables, $M_i, i = 1, 2, \cdots, N$, where $M_i, i = 1, 2, \cdots, N$, are indicator functions on whether mail i is a spam or not (if mail i is a spam mail, then $M_i = 1$, else $M_i = 0$), i.e.

$\hat{N}_{ss} + \hat{N}_{sn} = M_1 + M_2 + \cdots + M_N$

and

$$P(M_i = 1) = 1 - P(M_i = 0), i = 1, 2, \cdots, N$$

Hence $E(\hat{N}_{ss} + \hat{N}_{sn}) = NP, E(\hat{P}_N) = \dfrac{1}{N} E(\hat{N}_{ss} + \hat{N}_{sn}) = P$ and we have the result

that $\hat{P}_N$ is a unbiased estimate for P.

Accordingly, by taking use of the Law of Large Numbers we have

$$\lim_{N \to \infty} \Pr(|\hat{P}_N - P| \geq \varepsilon) = \lim_{N \to \infty} \Pr\left(\left|\frac{\sum_{i=1}^{N} M_i}{N} - E(\frac{\sum_{i=1}^{N} M_i}{N})\right| \geq \varepsilon\right) = 0, \forall \varepsilon > 0 \tag{11}$$

and Lemma1 is proved.

According to the result in Lemma 1, we substitute P with $\hat{P}_N$ and take it as a constant in the following. Next, we give and prove Lemma 2:

Lemma 2. $\hat{\alpha}_i = \dfrac{\hat{N}_{ns}(i)}{\hat{N}_{ns}(i) + \hat{N}_{nn}(i)}$ is an unbiased estimate of α_i.

Proof 2: Given the condition that there are $(\hat{N}_{ns}(i) + \hat{N}_{nn}(i))$ normal mails in the server, the conditional expected value of $\hat{N}_{ns}(i)$ is

$$E(\hat{N}_{ns}(i) \mid \hat{N}_{ns}(i) + \hat{N}_{nn}(i)) = (\hat{N}_{ns}(i) + \hat{N}_{nn}(i)) \cdot \alpha_i. \qquad (12)$$

This yields the result that $E(\dfrac{\hat{N}_{ns}(i)}{\hat{N}_{ns}(i) + \hat{N}_{nn}(i)} \mid \hat{N}_{ns}(i) + \hat{N}_{nn}(i)) = \alpha_i \qquad (13)$

By taking use of the Law of Total Probability, we have

$$E(\dfrac{\hat{N}_{ns}(i)}{\hat{N}_{ns}(i) + \hat{N}_{nn}(i)}) = E\left[E(\dfrac{\hat{N}_{ns}(i)}{\hat{N}_{ns}(i) + \hat{N}_{nn}(i)} \mid \hat{N}_{ns}(i) + \hat{N}_{nn}(i)) \right] = \alpha_i \qquad (14)$$

and we prove Lemma 2.

After proving that $\hat{\alpha}_i$ is an unbiased estimate of α_i, we must further find the variance of $\hat{\alpha}_i$, by taking use of some approximating technique.

$$Var(\dfrac{\hat{N}_{ns}(i)}{\hat{N}_{ns}(i) + \hat{N}_{nn}(i)}) = E(\dfrac{\hat{N}_{ns}(i)}{\hat{N}_{ns}(i) + \hat{N}_{nn}(i)})^2 - E^2(\dfrac{\hat{N}_{ns}(i)}{\hat{N}_{ns}(i) + \hat{N}_{nn}(i)}) = E(\dfrac{\hat{N}_{ns}(i)}{\hat{N}_{ns}(i) + \hat{N}_{nn}(i)})^2 - \alpha_i^2. \qquad (15)$$

What is left to be computed or approximated is the second moment $E(\dfrac{\hat{N}_{ns}(i)}{\hat{N}_{ns}(i) + \hat{N}_{nn}(i)})^2$

Among the mails filtered by R_i, the number of normal mails is $\hat{N}_{ns}(i) + \hat{N}_{nn}(i)$. Under the above condition, given $\hat{N}_{ns}(i) + \hat{N}_{nn}(i)$, the conditional distribution of $\hat{N}_{ns}(i)$ is Binomial, i.e.

$B(\hat{N}_{ns}(i) + \hat{N}_{nn}(i), \alpha_i)$ and therefore the conditional second moment is

$$E\left[(\hat{N}_{ns}(i))^2 \mid \hat{N}_{ns}(i) + \hat{N}_{nn}(i) \right]$$
$$= Var\left[\hat{N}_{ns}(i) \mid \hat{N}_{ns}(i) + \hat{N}_{nn}(i) \right] + E^2\left[\hat{N}_{ns}(i) \mid \hat{N}_{ns}(i) + \hat{N}_{nn}(i) \right] \qquad (16)$$
$$= \alpha_i(1 - \alpha_i)(\hat{N}_{ns}(i) + \hat{N}_{nn}(i)) + \alpha_i^2(\hat{N}_{ns}(i) + \hat{N}_{nn}(i))^2$$

For small α_i we have

$$E\left[(\dfrac{\hat{N}_{ns}(i)}{\hat{N}_{ns}(i) + \hat{N}_{nn}(i)})^2 \mid \hat{N}_{ns}(i) + \hat{N}_{nn}(i) \right] \cong \dfrac{\alpha_i}{\hat{N}_{ns}(i) + \hat{N}_{nn}(i)} + \alpha_i^2 \qquad (17)$$

For further approximation, we replace $\dfrac{1}{\hat{N}_{ns}(i)+\hat{N}_{nn}(i)}$ by $\dfrac{1}{N(i)(1-P)}$ since for

large $N(i)$, $\dfrac{N(i)(1-P)}{\hat{N}_{ns}(i)+\hat{N}_{nn}(i)} \approx 1$ with very high probability

According to the above inference and by taking use of the Central Limit Theorem, the distribution of $\hat{\alpha}_i$ can be approximated by a Normal distribution,

$\dfrac{\hat{N}_{ns}(i)}{\hat{N}_{ns}(i)+\hat{N}_{nn}(i)} \sim N(\alpha_i, \dfrac{\alpha_i}{N(1-P)})$. In a similar way, the distribution of $\hat{\beta}_i$ can

also be approximated by another Normal distribution,

$\dfrac{\hat{N}_{ss}(i)}{\hat{N}_{ss}(i)+\hat{N}_{sn}(i)} \sim N(\beta_i, \dfrac{\beta_i}{NP})$. After we approximate distributions of $\hat{\beta}_i$ and $\hat{\alpha}_i$,

the distribution of the reward function, $R(i)$, is therefore approximated by the Normal distribution.

$$R(i) \sim N(\mu, \sigma^2)$$

where

$$\mu = R_{ss} \cdot P + R_{nn} \cdot (1-P) + P \cdot (R_{sn} - R_{ss}) \cdot \beta_i + (1-P) \cdot (R_{ns} - R_{nn}) \cdot \alpha_i$$

and

$$\sigma^2 = P^2 \cdot (R_{sn} - R_{ss})^2 \cdot (\dfrac{\hat{\beta}_i}{N(i)P}) + (1-P)^2 \cdot (R_{ns} - R_{nn})^2 \cdot (\dfrac{\hat{\alpha}_i}{N(i)(1-p)}) \qquad (18)$$

This justifies (2). Through (2), the hypothesis testing can be realized. Further, by the results of the hypothesis testing, automatic administration of rules can be achieved by continuous feedback.

6   The Experimental Results and Verification

In this section, experiment is performed to evaluate the performance of proposed approach. The experimental group system have additional management module for the rules, while the control group system takes direct use of rough set theory without management module.

A.   *Measurements for spam filter performance analysis*
    In this paper, two types of measurements for system verification are proposed. The first type of indicators is of judgmental rate, among them we use spam precision, spam recall, accuracy, miss rate, as well as the whole reward of the system.

B.   *Data source*
    The data in this study comes from the mail server of a college and is shown in Table 4. The ratio of spam and normal mails is about 9:1. Table 4.

Table 4 Experimental data of mails

kind	Total Number of mail (Duration: 3/1-4/25 08')	percentage
spam	54,602	0.9
normal	5,921	0.1

The values of the five important measurements are compared and analyzed in this section. In Figure 2 the data of the spam recall is given.

The experiment shows no statistically significant difference between the experimental group and the control group, therefore we can conclude that the method proposed in this paper has the same spam filtering ability with the traditional way. Another measurement to determine the effectiveness of a spam mail is Spam precision and we show the results reflected in Figure 3.

In figure 3, it can be concluded that the spam precision is also of no significant difference between the experimental groups and control groups. The proposed method in this study seems to be a little higher than the control groups. The reason that it is of no significant difference can be explained as: when the number of the spam mail is very large compared with the normal mails. The ratio of normal mails is very small and thus the number of misclassified normal mails is still smaller. Hence in this paper we use Miss Rate system as another measurement for the ratio of misclassified normal mails. The results of the comparison are shown in Figure 4 as follows.

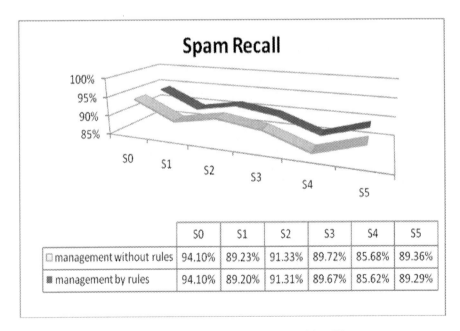

	S0	S1	S2	S3	S4	S5
☐ management without rules	94.10%	89.23%	91.33%	89.72%	85.68%	89.36%
■ management by rules	94.10%	89.20%	91.31%	89.67%	85.62%	89.29%

Fig. 2 The comparison of spam recalls during six time stages S0 to S5

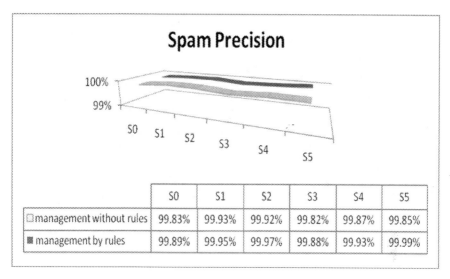

Fig. 3 The comparison of spam precision during six time stages S0 to S5

	S0	S1	S2	S3	S4	S5
☐ management without rules	99.83%	99.93%	99.92%	99.82%	99.87%	99.85%
■ management by rules	99.89%	99.95%	99.97%	99.88%	99.93%	99.99%

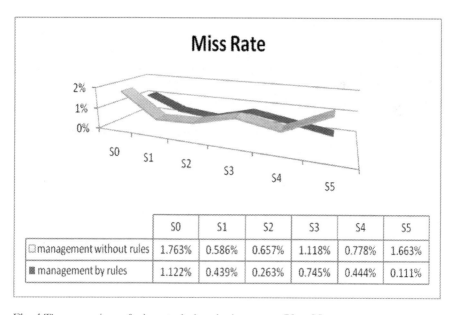

Fig. 4 The comparison of miss rate during six time stages S0 to S5

	S0	S1	S2	S3	S4	S5
☐ management without rules	1.763%	0.586%	0.657%	1.118%	0.778%	1.663%
■ management by rules	1.122%	0.439%	0.263%	0.745%	0.444%	0.111%

As shown in Figure 4, the experimental group of Miss Rate is indeed lower than that of the control group. The Miss Rate means the ratio that a mail is a normal mail and is judged as spam. On a standard system, this is a very important measurement, because with a higher Miss Rate, the system has a higher probability that

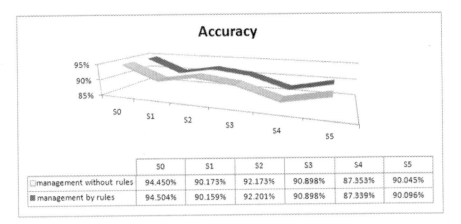

	S0	S1	S2	S3	S4	S5
☐management without rules	94.450%	90.173%	92.173%	90.898%	87.353%	90.045%
■ management by rules	94.504%	90.159%	92.201%	90.898%	87.339%	90.096%

Fig. 5 The comparison of accuracy during six time stages S0 to S5

a normal mail will be judged as spam. The results show that our proposed method can effectively reduce the Miss Rate. In the following Figure 5 shows the rate of accuracy on the whole system.

In this paper we show that our proposed method has a better overall performance than the traditional method. When the number and the ratio of spam mails is much higher than normal mails, the main standard to determine a spam filter is good or bad can be the Miss Rate. Experiments above can convince us on the advantage of the management model of this paper. Besides, another important indicator in this paper is on the Reward of the system. The comparison of the Reward is shown in Figure 6.

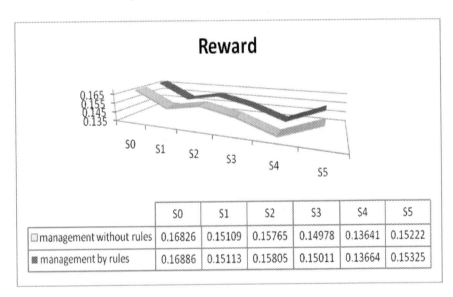

	S0	S1	S2	S3	S4	S5
☐management without rules	0.16826	0.15109	0.15765	0.14978	0.13641	0.15222
■ management by rules	0.16886	0.15113	0.15805	0.15011	0.13664	0.15325

Fig. 6 The comparison of reward during six time stages S0 to S5

In Figure 6 we can obtain more compensation on each mail with the proposed method in this paper, but the data shows that there is no statistically significant effect though the proposed approach gets higher reward. This is because that the majority (more than 90%) of the mail is spam mail, and as long as Miss Rate is not of great difference, then the Reward will be of little difference. But from the experimental data, the proposed method can improve on the judgmental rates of the whole system.

Based on the above experiments, the conclusion obtained in this paper can be that the proposed method has a significant improvement in Miss Rate with a small increase on the effectiveness of spam filtering.

7 Conclusions and Future Research

This paper provides a method to solve the spam mail in server. In addition to the use of data mining to produce spam rules, we also put forward a theory of statistics with automation of management spam rule to manage the increasing number of rules. By way of statistical testing it can be ensured that the reward of applying the law is higher than no management of rules or by using of other rules. The contribution of this study is as the following: (1) This study provides a spam rule by use of the keywords, the headers and formats in the mail. (2) The management of spam rule can reduce the rate of misclassifying. (3) Server administrators can reduce their burden of management by automatic applying rules based on the theory. (4) By ruling out the bad rules can reduce the number of bad rules, so the efficiency of the filter of mail server can be increased.

This paper points out new possibilities for future research. An important area for future research in the years to come will be in the refinement of approaches to the analysis of all filtering rules. A more rigorous test of these filtering rules could be performed. We intend to continue pursuing this line of investigation in a series of experimental studies.

Acknowledgements. This work was partially supported by I-Shou University ISU97-01-12 and National Science Council NSC 97-2221-E-214-043.

References

1. Androutsopoulos, I., Paliouras, G., Karkaletsis, V., Sakkis, G., Spyropoulos, C.D., Stamatopoulos, P.: I Learning to filter spam e-mail: a comparison of a Naive Bayesian and a memory-based approach. In: 4th PKDD's Workshop on Machine Learning and Textual Information Access, pp. 1–13 (2000)
2. Carreras, X., Marquez, L.: Boosting Trees for Anti-Spam Email Filtering. In: 4th International Conference on Recent Advances in Natural Language Processing (2001)
3. Clark, J., Koprinska, I., Poon, J.: A Neural Network Based Approach to Automated E-mail Classification. In: IEEE/WIC International Conference on Web Intelligence, pp. 702–705 (2003)

4. Drucker, H., Wu, D., Vapnik, V.N.: Support Vector Machines for Spam Categorization. IEEE Transactions on Neural Networks 10(5), 1048–1054 (1999)
5. Li, K., Huang, H.: An Architecture of Active Learning SVMs for Spam. In: 6th International Conference on Signal Processing, vol. 2, pp. 1247–1250 (2002)
6. Pawlak, Z.: Rough Sets and Intelligent Data Analysis. Information Sciences 147(1-4), 1–12 (2002)
7. Sahami, M., Dumais, S., Heckerman, D., Horvitz, E.: A Bayesian Approach to Filtering Junk E-mail. In: Proceedings of Workshop on Learning for Text Categorization (1998)
8. Woitaszek, M., Shaahan, M., Czernikowski, R.: Identifying Junk Electronic Mail in Microsoft Outlook With a Support Vector Machine. In: Symposium on Applications and the Internet, pp. 66–169 (2003)
9. Zhao, W., Zhang, Z.: An Email Classification Model Based on Rough Set Theory. In: Proceedings of the International Conference on Active Media Technology, pp. 403–408 (2005)
10. Zha, W., Zhu, Y.: An Email Classification Scheme Based on Decision-Theoretic Rough Set Theory and Analysis of Email Security. In: IEEE TENCON, pp. 1–6 (2005)

vPMM: A Value Based Process Maturity Model

Jihyun Lee, Danhyung Lee, and Sungwon Kang

Abstract. A business process maturity model provides standards and measures to organizations in assessing, defining and improving business processes, thereby guiding them to achieving business goals and values. In this paper, we present a business process maturity model called Value based Process Maturity Model (vPMM) that overcomes the limitations of the existing models. The vPMM is a model that can be used to determine the maturity of an organization's current business process practices by considering an organization's business value creation capability as well. It helps an organization set priorities for improving its product production and/or service provisioning using a proven strategy and for developing the capability required to accomplish its business values.

1 Introduction

Companies today strive to improve and innovate their business processes in real-time to cope with market dynamics. The Business Process Management System (BPMS) is considered to be a technical enabler that helps realize such needs of enterprises [6, 11, 15, 21]. Standard processes help an organization's workforce achieve business objectives by helping them work not harder, but more efficiently and with improved consistency. Effective processes also provide a vehicle for introducing and using new technology in a way that best meets the business objectives of an organization [3]. A business process should not focus on a project or process only; it should also focus on enterprise values because the ultimate purpose of a business process is to achieve business values so that a business process that doesn't contribute to business value achievement should be improved or adjusted to contribute business value creation. Therefore, a business process maturity model should be able to provide not only a roadmap for process improvement but also guidance for business value creation.

There are several existing researches on business process maturity models [4, 7, 10, 20]. However, business process maturity has not been sufficiently defined and standardized to be applied to an organization's business value achievement. The existing Process Maturity Model (PMM) and Process Management Maturity Model (PMMM) are not elaborated to the detailed level to apply in practice. For

Jihyun Lee, Danhyung Lee, and Sungwon Kang
School of Engineering Information and Communications University
e-mail: {puduli, danlee, kangsw}@icu.ac.kr

R. Lee, G. Hu, H. Miao (Eds.): Computer and Information Science 2009, SCI 208, pp. 193–202.
springerlink.com   © Springer-Verlag Berlin Heidelberg 2009

example, P. Harmon, D. M. Fisher, and M. Rosemann [7, 10, 20] only define business process maturity concepts or architectures. The Business Process Maturity Model (BPMM) [4, 5] was adapted as an OMG standard and is defined to the detailed level but leaves much room for improvement because it doesn't consider business value creation that should be achieved through a business process.

In order to overcome these limitations of the existing business process maturity models, we developed the vPMM that has the following characteristics:

- Guidance on how to control and improve an organization's business process competence
- Focusing on an enterprise for aligning with their business values through the business processes
- Conceptual processes that compare the maturity of an organization's current practices against an industry standard
- Reviewing the 'as-is' processes and performing vPMM-based gap analysis for developing 'to-be' processes.

The vPMM stands for value-based business Process Maturity Model. The vPMM is a model that can be used to determine the maturity of an organization's current business process practices by considering an organization's business value creation capability as well. It helps the organization set priorities for improving its operations for products/services (P/Ss) using a proven strategy and developing the capability required to accomplish its business values. Through the vPMM, an organization can efficiently and effectively manage their business processes while trying to achieve and realize its business objectives and values.

This paper is organized as follows: Section 2 presents the vPMM design principles and Section 3 introduces the vPMM by giving an overview, comparing it with its early version and discussing its distinctive characteristics. In Section 4, we describe related works and lastly, in Section 5, we discuss our contributions, conclusions, and future works.

2   The vPMM Design Principles

The vPMM has a five-level structure like CMM/CMMI and PMM. The maturity level of the vPMM reflects two aspects. The first aspect is maturity based on predictability control and effectiveness aspects as CMM/CMMI maturity model did provide [10]. Figure 1 illustrates the meanings of the process maturity in this aspect. Especially, the vPMM defines the maturity levels that include concepts such as the scope of influence of PAs, measurement & analysis, monitoring & control, and organizational process improvement activities.

The second aspect is the capability for embedding an organization's business values into its business processes and measuring the performance and results. Figure 2 shows the meanings of the maturity level from this aspect. In the Managed level, business processes are not coordinated enough to achieve business values and performance for business value accomplishment is measured from the black-box point of view. However, business processes are aligned and coordinated

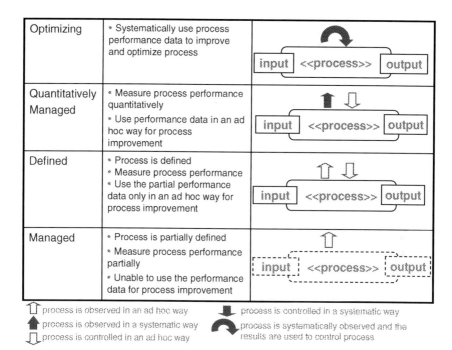

Optimizing	* Systematically use process performance data to improve and optimize process	
Quantitatively Managed	* Measure process performance quantitatively * Use performance data in an ad hoc way for process improvement	
Defined	* Process is defined * Measure process performance * Use the partial performance data only in an ad hoc way for process improvement	
Managed	* Process is partially defined * Measure process performance partially * Unable to use the performance data for process improvement	

process is observed in an ad hoc way   process is controlled in a systematic way
process is observed in a systematic way   process is systematically observed and the
process is controlled in an ad hoc way   results are used to control process

Fig. 1 Process maturities of the vPMM

in accordance with business values as processes are getting mature. Furthermore, a mature organization achieves capabilities to measure and control their business value achievement from the white-box perspective and also an organization can control its business values proactively.

The vPMM identifies generic business process areas with the viewpoint that any business process essentially belongs to one of the four generic kinds of activities – Input, Mechanism, Control, and Output (IMCO) [16]. All companies perform the functions of IMCO to produce P/S. Companies produce P/S using man power, money, and material via organizational processes and control for improving productivity and P/S quality. In order to produce a product or provide a service, a company consumes various resources. A company also requires a mechanism that turns the resources into products or services that will then be provisioned to the customer. For ensuring that these three kinds of activities are performed effectively and efficiently, a mechanism, inputs, and outputs should be monitored and controlled. From these four quadrants, essential KPAs of the vPMM are derived.

In addition, for deciding KPAs of each maturity level, core process and support process principles of value chain [17] is applied. Among core processes, processes with higher priority are deployed to the lower level and thereafter an organization has capability to perform the processes.

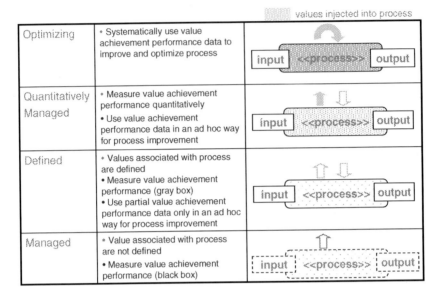

		values injected into process
Optimizing	• Systematically use value achievement performance data to improve and optimize process	input <<process>> output
Quantitatively Managed	• Measure value achievement performance quantitatively • Use value achievement performance data in an ad hoc way for process improvement	input <<process>> output
Defined	• Values associated with process are defined • Measure value achievement performance (gray box) • Use partial value achievement performance data only in an ad hoc way for process improvement	input <<process>> output
Managed	• Value associated with process are not defined • Measure value achievement performance (black box)	input <<process>> output

Fig. 2 Business values in process maturity

3 vPMM

The vPMM is a maturity model that considers an organization's business values within an organization's business processes and makes possible to guarantee that an organization's maturity level means the capabilities for achieving organization's business values.

3.1 An Overview of the vPMM

Figure 3 illustrates major components of the vPMM. Maturity level consists of applicable guideline, KPAs, and the result of process area categorization. Each KPA has specific and generic goals, normative components of the vPMM, and purpose, introductory notes, and related process areas, informative components. The vPMM categorizes process areas into four categories; organizational management, P/S work management, process management, and organizational support.

Each specific and generic goal has practices (sub-practices), typical work products, examples, and guideline reference amplifications. Guideline reference amplification contains guidelines for embedding business values within business processes, business value measurement and analysis, and process map analysis. Moreover, the vPMM supports both staged and continuous representation.

The vPMM is a process maturity model that provides guide on what capabilities should be embodied in an organization for implementing business values. As an organization achieves goals in each maturity level of the vPMM, an organization is viewed as having the capability for accomplishing the corresponding business value.

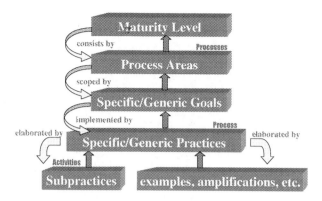

Fig. 3 Structural components of the vPMM

The vPMM guides an organization's processes toward achieving its business values by structuring an organization to consider its business values and measure their achievement.

The vPMM consists of 5 maturity levels, 23 KPAs, 52 specific goals and 5 generic goals for institutionalization. Each specific goal has 3-5 specific practices, sub-practices, amplifications, and examples to guide goal achievements. Especially, the vPMM has processes embedding, measuring and analyzing business values, and includes those related amplification.

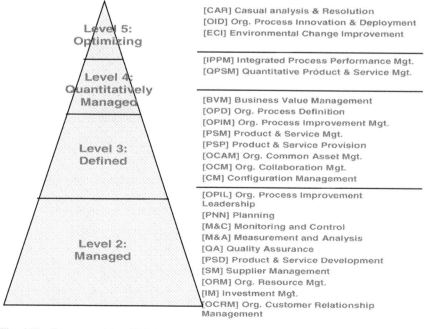

Fig. 4 The Structure of the vPMM

Figure 4 shows KPAs for each maturity level of the vPMM. The vPMM guides measuring, monitoring, and controlling business values achievement in KPAs, goals, and practices. In Fig. 4, the KPAs for measuring, monitoring, and controlling business value achievement are OPIL and BVM. In addition, the vPMM guides processes and practices to take care of related business values in accordance with their roles and responsibilities. For example, in Planning KPA of Level 2 of our vPMM, when a practitioner makes a plan for an assigned business work she is compelled to consider business value by making several plans first and then performing a tradeoff analysis based on the value achievement degree and necessary resources. As another example, with the Measurement & Analysis KPA of the vPMM, she has to find measures (KPIs) that would make significant impact on business value achievement. Especially, the vPMM considers explicitly the achievement of associated business values through accomplishing each goal of KPAs (i.e. Generic Goal 1, Generic Practice 1.2). Namely, accomplishing goals means achievement of associated business values.

3.2   Differences from the Early Version of vPMM

After we developed the early version of vPMM in [26], we conducted a survey to validate and revise the structure of the initial model. Survey questionnaires were completed by 12 representatives of the companies. The survey, which we called Voice Of Customer (VOC), was designed to obtain information such as the organization's business practices, the organization's priority of the KPAs for each maturity level of the initial vPMM, missing elements in the initial vPMM, the organization's needs for managing business process, and directions for improving business processes.

In addition, we conducted a survey on engineers and consultants who are responsible for developing business process management system and for consulting process improvement. The aims of the survey are to obtain information such as priority for each KPA in a provider's perspective, the missing elements on a provider's perspective, and gaps that exist in between markets and the providers.

The survey revealed important market needs, such as the necessity for improved leadership, customer relations, and separation of P/S from process management. There were some differences between VOC and VOE about KPAs. For example, engineers think that the configuration management is needed and should cover defined processes and common asset management to ensure strict compliance of rules in addition to the organizational process definition and the process management. These surveys also revealed the needs to develop capabilities, such as organizing Business Process Management Group (BPMG), P/S life cycle management, and adaptability management to cope with environmental changes.

In addition to incorporating the survey results, we also conducted formal reviews of our initial vPMM with practitioners and as a result of the reviews, some concepts and practices are adjusted. Requirements management processes are included in P/S Development KPA because managing requirements are not so complex in P/Ss works. And changes in requirements for P/S are not requested frequently. And we moved P/S Development and Customer Relationship Management KPAs from Level 3 to Level 2 because these are located in core processes at value chain and because it is better to implement core processes in advance. In addition, Supplier Agreement Management KPA is renamed as Supplier Management because it also includes practices for integrated supplier management and supply chain management.

After constructing the vPMM structure through survey and formal review, we compared the model with CMM/CMMI, IS12207, and IS15288, which are all widely accepted and implemented in various industry sectors [3,13,14]. Additionally, we compared the existing researches of BPMM, PMM, and PMMM to check its compliance with the existing practices.

3.3 The Distinctive Characteristics of the vPMM

Process maturity in the traditional sense meant a degree of how specific processes are accurately and effectively defined, managed, measured, and controlled [27]. However, it does not indicate the degree of business value achievement. an ideal process maturity model should guide an organization so that it can measure and control its process capabilities how they are performing toward achieving an organization's business values. The vPMM is a maturity model in which maturity is an indicator of an organization's ability to control business values. The vPMM presented in this study has the following distinctive characteristics:

* The vPMM guides what capabilities that organization should have in order to implement its business values
* An organization obtains capabilities to implement business values by achieving the goals of each KPA of the vPMM.
* The vPMM guides an organization to measure and control its business processes and improve its business processes steadily in accordance with an organization's maturity level.
* The vPMM makes an organization to implement it processes toward achieving business values.
* The vPMM provides Value-Based Agile Assessment (VBAA) for assessing the vPMM conformance of an organization's business processes. The VBAA guides an assessor how to derive weaknesses, strengths, recommendations, and action plans from assessment.

4 Related Works

Among the business process models that were proposed in the past, BPMM presented by Curtis [4, 5] is the only comprehensive model. It is also an OMG standard. Curtis borrowed well-defined and verified CMM/CMMI concepts and introduced them into the field of business process studies.

Fisher considered business process as having multi-dimensional and non-linear characteristics, unlike the software project/system life cycle [7]. Fisher defined actions on the ground that PMM is represented as five levers of change and five states of process maturity. Due to its high level abstraction on actions, Fisher's model only provides ends, with no means to these ends.

PMM developed by Harmon regards all the core and support processes as a value chain, starting from the resource right up to the final product [10]. It also provides a checklist for accessing organization/process maturity. Although Harmon's approach shows the need for including values in PMM, his maturity model does not provide the means to achieve values.

When Smith introduced PMMM [20, 21], he insisted that "business process maturity should be taken into consideration because process management maturity has an orthogonal relationship with process maturity" [20]. Business process maturity should include business process management maturity for keeping up with the rapid changes of the business environment through continuous process improvement [25]. vPMM includes business process management maturity like BPMM and M. Hammar's PEMM (Process and Enterprise Maturity Model) [23].

Rosemann describes a BPM Maturity Model as a three dimensional structure that consists of Factor (IT/IS, Methodology, Performance, Accountability, Culture and Alignment), Perspective (Align, Design, Execute, Control and Improve), and Organization Scope (which includes time and area, the entity to which the model is applied, one dimension location, a division, a business unit or a subsidiary) [18, 19]. Rosemann's model has an advantage over other models in that it is supported by surveys and case studies. However, Rosemann's model is not well organized and has a complex three dimensional structure.

The 8 Omega Framework of BPMG encompasses the four dimensions of Strategy, People, Process, and Systems; the framework applies Discovery, Analysis, Design, Validate, Integrate, Implement, Control, and Improve (DADVIICI action) to all four dimensions [1]. It is simple and intuitive but it does not have principles or guidelines for its application. Further, the possibility of utilizing CMM model to business process maturity model development is currently being studied and tested by experts in industry and academia [12].

However, the existing maturity models and the associated methods did not directly consider business value achievement, which is the ultimate objective of implementing business processes.

5   Conclusions and Future Works

The ultimate goal of the vPMM is to be a business process maturity model that guides organizations in achieving their business values. In this paper, we reported our vPMM structures, and design principles. Business value achievement, which is not treated in other current business process maturity models including PMM, PMMM and BPMM, has been covered in the vPMM. In this paper, our main focus was to introduce the vPMM that makes an organization's business processes be aligned to a company's declared values. The vPMM was structured through the following procedures:

- Defining KPAs derived from IMCO functional model presented in IDEF0 and not from empirical intuition obtained through best practices
- Reflecting the opinions of experts such as practicing BPMS engineers and consultants
- Verifying whether KPAs are fully defined and value-adding process
- Verifying KPAs that are compliant with the exiting verified principles and models.

With the help of the vPMM, companies can analyze the strengths and the weaknesses of their 'as-is' business processes and develop 'to-be' models to achieve the organization's business values.

As Spanyi [22] said, a business process maturity model has to focus on competence. The maturity level should agree with the organization's competence for achieving their business values such as market share, revenue, and time to market improvement. The vPMM presented in this paper is the first business process maturity model that was constructed with that goal in mind and is carefully engineered such that design of levels, goals and sub-goals, practice areas and practices are all aligned towards the goal.

Acknowledgments. This work was partially supported by the Industrial Technology Development Program funded by the Ministry of Commerce, Industry and Energy (MOCIE, Korea).

References

[1] The 8 Omega Framework, http://www.bmpg.org
[2] Burlton, R.: In search of BPM excellence. In: BPM: From Common Sense to Common Practice. Meghan-Kiffer Press (2005)
[3] Chrissis, M.B., Konrad, M., Shrum, S.: CMMI: Guidelines for Process Integration and Product Improvement. Addison-Wesley, Reading (2003)
[4] Curtis, B.: Overview of the Business Process Maturity Model. San Antonio SPIN (April 2004)
[5] Curtis, B., Alden, J.: The Business Process Maturity Model: An overview for OMG members (2006)
[6] Elzinga, D.J., Horak, T., Lee, C., Bruner, C.: Business Process Management: Survey and Methodology. IEEE Transactions on Engineering Management 42(2), 119–128 (1995)
[7] Fisher, D.M.: The Business Process Maturity Model: A Practical Approach for Identifying Opportunities for Optimization., BP Trends (September 2004)
[8] Fisher, D.M.: Getting Started on the Path to Process-Driven Enterprise Optimization. BP Trends (February 2005)
[9] Garretson, P.: How Boeing A&T Manages Business Processes. BP Trends (November 2005)
[10] Harmon, P.: Evaluating an Organization's Business Process Maturity. BP Trends (March 2004)

[11] Harrington, H.J.: Business Process Improvement: The Breakthrough Strategy for To-
     tal Quality, Productivity, and Competitiveness. McGraw-Hill, New York (1991)
[12] Ho, D.: Business Process Maturity Model — A CMM-Based Business Process Reen-
     gineering Research. Center of Excellence in Business Process Innovation, Stevens In-
     stitute of Technology (2004)
[13] ISO/IEC 12207 (ISO/IEC 12207) Standard for Information Technology Software Life
     Cycle Processes (1995)
[14] ISO/IEC 15288 (ISO/IEC 15288) Standard for Information Technology System Life
     Cycle Processes (2002)
[15] Khan, R.N.: Business Process Management: A Practical Guide. Meghan-Kiffer Press
     (2004)
[16] NIST, Integration DEfinition for Function Modeling (IDEF0), Draft Federal Informa-
     tion Processing Standards Publication 183, National Institute for Standards and Tech-
     nology, Washington D.C. (1993)
[17] Porter, M.E.: Competitive Advantage: Creating and Sustaining Superior Performance.
     Free Press, NY (1985)
[18] Rosemann, M., de Bruin, T.: Application of a Holistic Model for Determining BPM
     Maturity. BP Trends (February 2005)
[19] Rosemann, M., de Bruin, T.: Towards a Business Process Management Maturity
     Model. In: Proc. 13th European Conference on Information Systems (2005)
[20] Smith, H., Fingar, P.: Process Management Maturity Models. BP Trends (July 2004)
[21] Smith, H., Neal, D., Ferrara, L., Hayden, F.: Business Process Management. Business
     Process Management Summit (September 2001)
[22] Spanyi, A.: Beyond Process Maturity to Process Competence. BP Trends (June 2004)
[23] Harmon, P.: BPM Methodologies and Process Maturity. BP Trends (May 2006)
[24] Hammar, M.: The Process Audit. Harvard Business Review 85(4), 111–123 (2007)
[25] Sinur, J.: A Strategist's Perspective. BP Trends (May 2007)
[26] Lee, J.H., Lee, D.H., Kang, S.W.: An Overview of the Business Process Maturity
     Model (BPMM). In: Chang, K.C.-C., Wang, W., Chen, L., Ellis, C.A., Hsu, C.-H.,
     Tsoi, A.C., Wang, H. (eds.) APWeb/WAIM 2007. LNCS, vol. 4537, pp. 384–395.
     Springer, Heidelberg (2007)
[27] CMU/SEI, The Capability Maturity Model: Guides for Improving the Software Proc-
     ess. Addison Wesley, Reading (1994)

Extracting Database Interactions and Generating Test for Web Applications*

Bo Song, Huaikou Miao, and Bin Zhu

Abstract. Traditional testing techniques are not adequate for Web-based applications. Nowadays, Web applications depend more and more on the back-end database to provide much more functionalities. Additionally, database interactions make a great impact on the navigation of Web applications. In this paper, we propose an event-based dependence graph models (EDGMs) to model and extracting database interactions in Web applications. A FSM was used as a tool to model the presentation (.aspx) of Web page. And Control Flow Graph (CFG) and Database Interaction Flow Graph (DIFG) are employed to extracted and model its corresponding logical process (.cs) of each corresponding page. From the EDGMs, a FSM test-tree (FSM-TT) was constructed. Based on FSM-TT, the test sequences were generated. At last, by elaborating the test sequences with the FSMs of Web pages, we got the more detailed test sequences which can be easily instantiated and test executed.

1 Introduction

Nowadays, Web applications depend more and more on the back-end database to supply with correct and legal data. A large number of Web applications utilize a Database Management System (DBMS) and one or more databases. They are also referred to as Web database applications. Web applications consist of not only pure HTML, but also dynamic Web pages and much more interactions such as the interactions between users and Web browser, along with the interactions between users with a variety of Web forms, etc. Database interactions are seldom taken

* Supported by National High-Technology Research and Development Program (863 Program) of China under grant No. 2007AA01Z144, National Natural Science Foundation of China (NSFC) under grant No. 60673115 and 60433010, National Grand Basic Research Program (973 Program) of China under grant No. 2007CB310800, the Research Program of Shanghai Education Committee under grant No. 07ZZ06 and Shanghai Leading Academic Discipline Project, Project Number: J50103.

Bo Song, Huaikou Miao, and Bin Zhu
School of Computer Engineering and Science, Shanghai University, Shanghai, China
e-mail:{songbo, hkmiao}@shu.edu.cn, tozhubin@163.com

Bo Song
Shanghai Key Laboratory of Computer Software Evaluating and Testing, Shanghai, China

into account in modeling or testing of Web applications. Much work has been done which focuses on the extraction of database interactions from source code of Web applications [1, 2], but they did not take test generation to test Web applications with database interaction into account.

Traditional testing techniques are not adequate for Web-based applications, since they miss their additional features such as their multi-tier nature, hyperlink-based structure, and event-driven feature. Additionally, database interactions make a great impact on the navigations and functionalities of Web applications. Limited work has been done on taking database interactions into account in testing Web applications. The basic structure of Web applications consists of three tiers: the client, the server and the data store. The integration of HTML and other scripting languages has provided not only sophisticated Web applications but also new issues that have to be addressed. The recent Microsoft's .NET platform has lowered the barriers to Web software development [3]. Due to its widespread use, we focus on .NET Web applications in this paper.

ASP.NET supports event-driven programming. That is, objects on a Web page can expose events that can be processed by ASP.NET code. ASP.NET also provides a language-neutral execution framework for Web applications [3]. Traditional Web applications contain a mix of HTML and scripts, making the code difficult to read, test, and maintain. ASP.NET alleviates this problem by promoting the separation of code and content using the code-behind feature. Thus, a Web page consists mainly of a 'presentation file'(.aspx), that contains the HTML tags, and a 'code-behind' class (.cs) that contains methods and data items required for the class to perform specific actions.

A program in a Web application generally consists of many statements which execute the SQL data manipulation language (DML) operations such as *select*, *insert*, *update* or *delete*. We shall use the term *database interaction* to refer to an execution of such a statement.

In this paper, we pay special attention to database interactions in the source code of Web applications and an FSM is employed to model each page and Control Flow Graphs (CFG) and Database Interaction Flow Graph (DIFG) are used as analysis model to model .cs files. A FSM test-tree is constructed From FSM. Finally, an approach is proposed to generate test using test tree.

2 A Motivating Example

To the best of our knowledge, no approach to extracting functionalities from source code is capable of dealing with the dynamic nature of Web applications. This dynamic characteristic is caused by statements which are generated only at runtime and dependent on user inputs. To illustrate how to extract database interactions and generate tests from source code, consider Online *Student Information Management System* (SIMS) written in ASP.NET with C#. Fig. 1, 2, 3, 4, 5 are given in the end of this paper in order to facilitate type-setting. This system includes a *Login.aspx* page, which consists mainly of a presentation file. simplified source code is shown in Fig. 1, while a *Login.aspx.cs* file (see Fig. 2) is a class file for code-behind. This page displays a form for the student to input his/her name and password to enter *SIMS*. The user enters his/her username and password, and presses the *Login* button. Upon this pressing, the username and password are sent to the Web server for authentication. If the username and password are correct, a private page *UserView.aspx*(see Fig. 3) is loaded and displayed in Web browser to user. The content of *UserView* is something concerning the private information of a certain user, for example, user's self-information and some operations, such as modifying the user's password, etc. To simplify description, we give out only one link operation: *Modify User's Password*. When the student clicks this link, a *ModifyPwd.aspx* page (see Fig. 4) will emerge in the user's browser. Here, a *ModifyPwd.aspx.cs* (see Fig. 5) is a code behind class file for the page *ModifyPwd.aspx*. After inputting the old password and a new password, and pressing the *Modify* button, if the old password is correct, the user will see a message box "*Update success!*", or else, the user will see "*The old password is wrong!*".

3   Extraction of Database Interactions

ASP.NET separates the code and the content using code-behind feature, namely, the presentation file (.aspx) and the corresponding logical operation process (.cs) are set in different files. The actions on Web pages are realized by the relevant event implement in code-behind class. In order to extract database interactions and generate tests, an approach to extracting Web application models with database interactions based on events is proposed. FSM models are extracted from the presentation file (.aspx) of each page. Then, as for code-behind class file (.cs), we give out Control Flow Graphs (CFG) and Database Interactions Flow Graph (DIFG) triggered by the corresponding events for each .cs file as our analysis model. By identify database interaction node (i-node) in CFG, we propose an event-based dependence graph models (EDGMs) to model Web applications.

3.1   Statements of Database Interaction

To illustrate the problems in extracting of database interactions from source code, applying any of the current functional features extraction techniques [4], in general, we obtain the following statements for the database interaction features:

```
1. SqlCommand myCommand = new SqlCommand("select * from tbUserInfo
where UserID='" + userName + "' and UserPWD='" + pwd + "'", sqlConn);
   int row = myCommand.ExecuteNonQuery ();
2. SqlCommand comm = new SqlCommand("select * from tbUserInfo
where UserID='" + Session["user_name"] +"' and UserPWD='"+ oldPwd
+"'", sqlConn);
   SqlDataReader dr = comm.ExecuteReader();
3. SqlCommand myCommand = new SqlCommand("update tbUserInfo set
UserPWD='" + newPwd + "' where UserID='" + Session["user_name"] + "'
and UserPWD='" + oldPwd + "'", sqlConn);
   myCommand.ExecuteNonQuery();
```

Later, we will give the database interaction node (i-node) which is based on the statements of database interactions listed above.

3.2 Modeling Web Pages

Owing to ASP.NET Web application is an event-driven programming paradigm: all behaviors of Web pages are related to the corresponding events. Accordingly, we can describe Web pages via the sets of related inputs and actions. In addition, there may be some rules on the inputs. For example, some inputs may be required and others may be optional. A user may be allowed to enter inputs in any order, or a specific order may be required. Table 1 shows input constraints of both types; of course, other constraints may be defined in future work according to your needs.

Table 1 Constraints on Inputs

Input Choice	Order
Required (**R**) Optional (**O**)	Sequence (**S**) Any (**A**)

According to the input constraints as above mentioned, we can easily extract an FSM of each one page from its presentation file (.aspx). For example, as to Web page *Login.aspx*, the FSM of extraction results is as shown in Fig. 6. As Fig. 7 and Fig. 8 show the each FSM of page *UserView.aspx* and *ModifyPwd.aspx* respectively. Fig. 9 and Fig. 10 are FSMs for *Login.aspx* and *ModifyPwd.aspx* without input constrains.

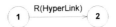

Fig. 6 FSM for *Login.aspx* with constrains on inputs

Fig. 7 FSM for UserView.aspx with constrains on inputs

Fig. 8 FSM for *ModifyPwd.aspx* with constrains on inputs

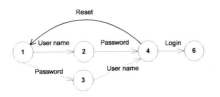

Fig. 9 FSM for *Login.aspx* without input constrains

Fig. 10 FSM for *ModifyPwd.aspx* without input constrains

3.3 Analysis Model

With regard to the corresponding logical operation process (.cs) of one page (.aspx), i.e, code-behind class, firstly, we give out Control Flow Graphs (CFG) triggered by the corresponding events for each .cs file. For example, when a user clicked the *Login* button in *Login.aspx* page, a Login_Click() event (see code in *Login.aspx.cs* file) was triggered. By this way, we can get its corresponding CFG as shown in Fig. 11. Secondly, a Database Interaction Flow Graph (DIFG) is identified from each control flow graph as the analysis model for our approach.

Definition 1. A Control Flow Graph (CFG) is a directed graph that consists of a set of *N* nodes and a set $E \subseteq N \times N$ of directed edges between nodes. Each node represents a program statement. Each edge (s, t) represents the flow of control from *s* to *t*. In a CFG, there is a "begin" node and an "end" node where computation starts and finishes, respectively.

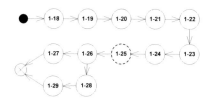

Fig. 11 CFG triggered by Login event in *Login.aspx*

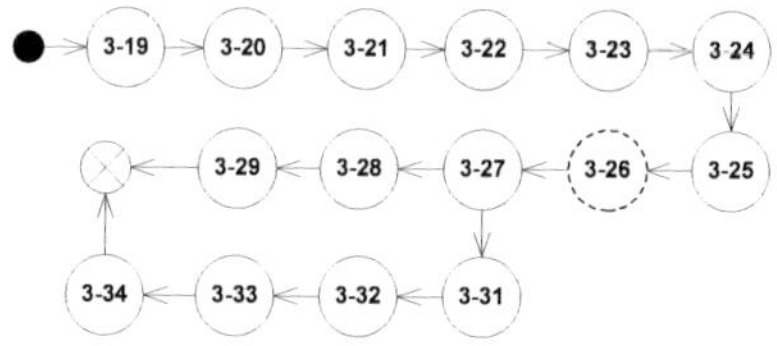

Fig. 13 CFG triggered by Modify event in *ModifyPwd.aspx*

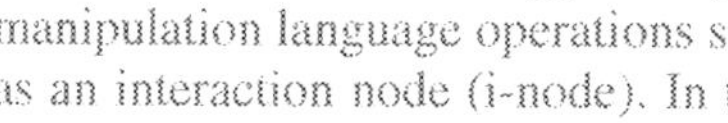

Fig. 12 CFG triggered by Reset event in *Login.aspx*

Fig. 14 CFG triggered by Reset event in *ModifyPwd.aspx*

A node in the CFG triggered by a event which executes one of the SQL data manipulation language operations such as select, insert, update or delete is defined as an interaction node (i-node). In the figures above, the nodes 1-25 and 3-26 are i-nodes. If there is at least one i-node in a CFG, we refer to this CFG as a DIFG.

As the definitions mentioned above, Fig. 11 and Fig. 13 are not only CFGs, but also DIFGs. And Fig. 12 and Fig. 14 are not DIFGs.

4 Generating Tests

This paper primarily concerns the extraction and test generation problems of database interactions. As a role, in virtue of Web applications face the short-time to market, database interactions were ignored for cost savings in modeling Web applications. The problems of database interactions were left to the developers without detailed specifications. The realization of database interactions is determined mainly by the developers' technical abilities, and it is unable to avoid the personal arbitrariness. Whether database interactions are successful or not is of vital importance to test Web applications.. In consequence, the database interactions should be taken into account at the time of later testing of Web applications. The existing literatures are almost interesting in abstraction of database interactions from source code of Web applications, test generation is not considered. This paper considers database interactions during the anaphase of testing Web applications. And our work can extract database interactions form source code to check whether they are consistent with the customer's requirement. Additionally, it can further been used to test Web applications by generating tests.

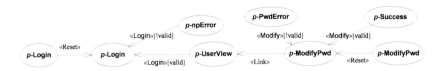

Fig. 15 EDGMs of *SIMS* with database interactions

In this section, we propose an event-based dependence graph models (EDGMs) of Web applications with database interactions to model the dependency between Web pages. When the FSMs of each page and the corresponding CFGs, especially DIFGs triggered by events are fell together, we can get the EDGMs of *SIMS* as shown in Fig. 15, where «Reset», «Login» and «Modify» are all action events, «Login»[valid] means that the user clicked the *Login* button on the condition that the username and the password he typed are all valid; on the contrary, it is «Login»[!valid]. «Modify»[valid] means that the user clicked the modify button on the condition that the old password he inputs is valid, and «Modify»[!valid] goes by contraries. According to the EDGMs of *SIMS* with database interactions, we can construct the state transition diagram of *SIMS* as shown in Fig. 16.

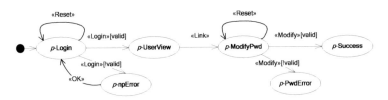

Fig. 16 FSM of *SIMS* with database interactions

The FSM test-tree represents an efficient set of tests, where each path from root to a leaf corresponds to a test sequence. So, from FSM of *SIMS* with database interactions (see Fig. 16), FSM test tree was construct using our construction algorithm [9]. By this means 5 test sequences are generated (shown in Fig. 17):

```
1. Login-«Reset»-Login;
2. Login-«Login»[!valid]-npError-«OK»-Login;
3. Login-«Login»[valid]-UserView-«Link»-ModifyPwd-«Modify»[!valid]-
   PwdError;
4. Login-«Login»[valid]-UserView-«Link»-ModifyPwd-«Modify»[valid]-
   Success;
5. Login-«Login»[valid]-UserView-«Link»-ModifyPwd-«Reset»-ModifyPwd.
```

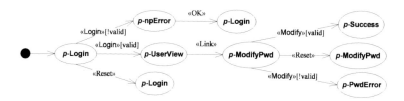

Fig. 17 A FSM-TT derived from the FSM of *SIMS*

From the FSMs of Web pages as shown in Fig. 6, 7, 8, 9 and 10, we can elaborate the test sequences above. For example, the test sequence 'Login-«Reset»-Login', according to Fig. 6, should be elaborated as: 'Login-*input username-input password*-«Reset»-Login' and 'Login- *input password-input username*-«Reset»-Login'. By this way, we can get the elaborated test sequences as follows:

```
1.Login-input username-input password-«Reset»-Login;
2.Login-input password-input username-«Reset»-Login;
3.Login-input illegal username- input password-«Login»[!valid]-
  npError-«OK»-Login;
4.Login- input illegal password-input username-«Login»[!valid]-
  npError-«OK»-Login;
5.Login- input legal username- input legal password «Login»[valid]-
  UserView-«Link»-ModifyPwd- input illegal old password-input new
  password- «Modify»[!valid]-PwdError;
6.Login- input legal username- input legal password «Login»[valid]-
  UserView-«Link»-ModifyPwd- input new password- input illegal old
  password- -«Modify»[!valid]-PwdError;
7.Login- input legal username- input legal password -«Login»[valid]-
  UserView- «Link»- ModifyPwd- input legal old password-input new
  password- «Modify»[valid]-Success;
8.Login- input legal username- input legal password -«Login»[valid]-
  UserView- «Link»- ModifyPwd-input new password - input legal old
  password - «Modify»[valid]-Success;
9.Login- input legal username- input legal password «Login»[valid]-
  UserView-«Link»-ModifyPwd- input new password - input old password
  -«Reset»-ModifyPwd;
10.Login- input legal username- input legal password «Login»[valid]-
   UserView-«Link»-ModifyPwd- input old password - input new password
   -«Reset»-ModifyPwd;
   ......
```

At last, by instantiating each test sequences above mentioned, we can get the instantiated test cases.

5   Related Work and Conclusions

A number of modeling and generating test techniques for Web applications have been already proposed, each of which has different origins and pursuing different goals for dealing with the unique characteristics of Web applications.

The literature [5, 6] proposed some methods for modeling and deriving tests from FSMs. But they did not care the Web database interactions.

There are also the interaction issues in modeling Web browser interactions have been addressed [7, 8, 9]. To the best of our knowledge, the issues on Web database interactions have seldom been addressed with the exception of [1, 2].

This paper primarily concerns the extraction and test generation problems of database interactions in Web applications. We used different approaches to modeling Web page and its corresponding code-behind class. At last, using the FSM test-tree, the test sequences are generated.

References

1. Ngo, M.N., Tan, H.B.: Applying static analysis for automated extraction of database interactions in web applications. Inf. Softw. Technol. 50(3), 160–175 (2008)
2. Ngo, M.N., Tan, H.B., Trinh, D.: Automated Extraction of Database Interactions in Web Applications. In: Proceedings of the 14th IEEE international Conference on Program Comprehension, pp. 117–126. IEEE Computer Society, Washington (2006)
3. Esposito, D.: Programming Microsoft ASP.NET 2.0 Core Reference. Microsoft Press (2005)
4. Lanubile, F., Visaggio, G.: Extracting Reusable Functions by Flow Graph-Based Program Slicing. IEEE Transactions on Software Engineering 23(4), 246–259 (1997)
5. Andrews, A., Offutt, J., Alexander, R.: Testing Web Applications by Modeling with FSMs. Software and Systems Modeling (2004)
6. Dargham, J., Al-Nasrawi, S.: FSM Behavioral Modeling Approach for Hypermedia Web Applications: FBM-HWA Approach. In: Proceedings of the Advanced International Conference on Telecommunications (AICT/ICIW 2006), French, pp. 199–204 (2006)
7. Lucca, G.A.D., Penta, M.D.: Considering Browser Interaction in Web Application Testing. In: Proceedings of the 5th IEEE International Workshop on Web Site Evolution, New York, USA, pp. 74–81. IEEE Press, Los Alamitos (2003)
8. Miao, H., et al.: Modeling Web Browser Interactions Using FSM. In: Proceedings of the 2nd IEEE Asia-Pacific Service Computing Conference, pp. 211–217 (2007)
9. Song, B., Miao, H., Chen, S.: Modeling Web Browser Interactions and Generating Tests. In: Proceding of the fourth International Conference on Computational Interlligence and Security (CIS 2008), vol. 2, pp. 399–404 (2008)

Appendix

```
1-1.  <%@ Page Language="C#" AutoEventWireup="true" CodeFile="Login.aspx.cs" Inherits="_Default"
%>
1-2.  <html xmlns="http://www.w3.org/1999/xhtml" >
1-3.  <head runat="server">
1-4      <title> Student Information Management System </title>
1-5.  </head>
1-6. <body>
1-7.     <form id="form1" runat="server">
1-8.       <div>
1-9         User Name:<asp:TextBox ID="TextBox1" runat="server"></asp:TextBox><br />
1-10. Password:<asp:TextBox ID="TextBox2" runat="server" TextMode="Password"></asp:TextBox><br />
1-11         <asp:Button ID="Login" runat="server" OnClick="Login_Click" Text="Login" />
1-12.        <asp:Button ID="Reset" runat="server" OnClick="Reset_Click" Text="Reset" /></div>
1-13.     </form>
1-14.  </body>
1-15. </html>
```

Fig. 1 Source code of *Login.aspx*

```
1-16. public partial class _Default : System.Web.UI.Page
{
1-17.    protected void Page_Load(object sender, EventArgs e) {
    }
1-18.    protected void Login_Click(object sender, EventArgs e)
    {
1-19.        string userName = this.TextBox1.Text.Trim();
1-20.        string pwd = this.TextBox2.Text.Trim();
1-21.        string strCon = "server=SONGBO;uid=sa;pwd=sa;database=USERDB";
1-22.        SqlConnection sqlConn = new SqlConnection(strCon);
1-23.        sqlConn.Open();
1-24.        SqlCommand myCommand = new SqlCommand("select * from tbUserInfo where UserID='" +
userName + "' and UserPWD='" + pwd + "'", sqlConn);
1-25.        int row = myCommand.ExecuteNonQuery ();
1-26.        if (row!=0) {
1-27.           Response.Redirect("UserView.aspx");}
       else{
1-28.           Response.Write("<script>alert('the username or password is wrong!')</script>");
1-29.           Response.Redirect("Login.aspx"); }
1-30.        sqlConn.Close();
    }
1-32.    protected void Reset_Click(object sender, EventArgs e) {
1-33.       TextBox1.Text = "";
1-34.       TextBox1.Text = "";
    }
}
```

Fig. 2 Source code of *Login.aspx.cs*

```
2-01     <%@     Page     Language="C#"     AutoEventWireup="true"     CodeFile="UserView.aspx.cs"
Inherits="UserView" %>
2-02 <html xmlns="http://www.w3.org/1999/xhtml" >
2-03 <head runat="server">
2-04     <title> Student Information Management System </title>
2-05 </head>
2-06 <body>
2-07     <form id="form1" runat="server">
2-08       <div>
2-09 <asp:HyperLink ID="ModifyPWD" runat="server" Height="13px" NavigateUrl="~/ModifyPwd.aspx"
2-10          Width="90px">Modify User's Password</asp:HyperLink></div>
2-11     </form>
2-12 </body>
2-13 </html>
```

Fig. 3 Source code of *UserView.aspx*

```
3-01 <%@      Page      Language="C#"      AutoEventWireup="true"      CodeFile="ModifyPwd.aspx.cs"
Inherits="ModifyPwd" %>
3-02 <html xmlns="http://www.w3.org/1999/xhtml" >
3-03 <head runat="server">
3-04     <title> Student Information Management System </title>
3-05 </head>
3-06 <body>
3-07     <form id="form1" runat="server">
3-08     <div>
3-09         Old Password:<asp:TextBox ID="TextBox1" runat="server"></asp:TextBox><br />
3-10         New Password:<asp:TextBox ID="TextBox2" runat="server"></asp:TextBox><br />
3-11         <br />
3-12         <asp:Button ID="ModifyButton" runat="server" OnClick="Modify_Click" Text="Modify" />
3-13         <asp:Button ID="Reset" runat="server" Text="Reset" OnClick="Reset_Click" /></div>
3-14     </form>
3-15 </body>
3-16 </html>
```

Fig. 4 Source code of *ModifyPwd.aspx*

```
3-17 public partial class ModifyPwd : System.Web.UI.Page
{
3-18     protected void Page_Load(object sender, EventArgs e) {
}
3-19     protected void Modify_Click(object sender, EventArgs e)
{
3-20         string oldPwd = Request.Form.Get("TextBox1");
3-21         string newPwd = Request.Form.Get("TextBox2");
3-22         string strCon = "server=SONGBO;uid=sa;pwd=sa;database=USERDB";
3-23         SqlConnection sqlConn = new SqlConnection(strCon);
3-24         sqlConn.Open();
3-25         SqlCommand comm = new SqlCommand("select * from tbUserInfo where UserID='" +
Session["user_name"] + "' and UserPWD='"+ oldPwd +"'", sqlConn);
3-26         SqlDataReader dr = comm.ExecuteReader();
3-27         if (!dr.Read()){
3-28             Response.Write("<script>alert('The old password is wrong!')</script>");
3-29             dr.Close();}
3-30         else{
3-31             SqlCommand myCommand = new SqlCommand("update tbUserInfo set UserPWD='" + newPwd
+ "' where UserID='" + Session["user_name"] + "' and UserPWD='" + oldPwd + "'", sqlConn);
3-32             myCommand.ExecuteNonQuery();
3-33             Response.Write("<script>alert('Update success!')</script>");
3-34             sqlConn.Close();
        }
    }
3-35     protected void Reset_Click(object sender, EventArgs e) {
3-36         TextBox1.Text = "";
3-37         TextBox2.Text = "";
    }
}
```

Fig. 5 Source code of *ModifyPwd.aspx.cs*

Middleware Frameworks for Ubiquitous Computing Environment

Eun Ju Park, Haeng-Kon Kim, and Roger Y. Lee

Abstract. For decades, software engineering continued to be an evolving discipline. Software developers are still after software development approaches that achieve reusability at more levels of abstraction. This is particularly true for developing applications for the enterprise; where business collaborations are a central concern. Software designers need a modeling framework that enables process engineering by assembling reusable services. Moreover, the development approach must enable parts of the enterprise to react quickly and reliably to changes in requirements and in available technologies.

In this paper, we are seeking to improve the workload efficiency and inference capability of context-aware process in real-time data processing middleware core of ubiquitous environment. We propose an architecture that apply a neural network model. Ubiquitous environment should cope with the requirements of context-aware by accepting a lot of information collected from USN(Ubiquitous Sensor Network) and inference. There exists a conventional context-aware method that extracts if-then rule and constructs a rule database. The conventional context-aware method can not recognize a situation when database is failed. Moreover, as rules get complicated, the performance of the method is reduced. This paper suggests the rule-based context-aware neural network model to solve these problems, and the empirical results are shown.

Keywords: Ubiquitous Middleware, Mobile Embedded Execution Model, Component Framework, Genetic Algorithm.

Eun Ju Park
Department of Computer information Communication Engineering,
Catholic Univ. of Daegu, Korea
e-mail: ejpark@cu.ac.kr

Haeng-Kon Kim
Department of Computer information Communication Engineering,
Catholic Univ. of Daegu, Korea
e-mail: hangkon@cu.ac.kr

Roger Y. Lee
Software Engineering Information Technology Institute,
Central Michigan University, USA
e-mail: lee1ry@cmich.edu

R. Lee, G. Hu, H. Miao (Eds.): Computer and Information Science 2009, SCI 208, pp. 213–228.
springerlink.com   © Springer-Verlag Berlin Heidelberg 2009

1 Introduction

The term middleware is used to describe separate products that serve as
the glue between two applications. It is, therefore, distinct from import and
export features that may be built into one of the applications. Middleware is
sometimes called plumbing because it connects two sides of an application and
passes data between them. Current Common middleware related environment
has been changed as following categories include:

- **Enterprise application integration**
  - need for integrating many applications/data sources within/across enterprises
  - large scale configuration, diverse interaction models
  - a chain of consecutive RPCs is too rigid
  - requirement for loosely-coupled interaction (autonomy of the various parties, spatial and temporal decoupling)

- **Internet applications**
  - number of users unpredictable, stateful session difficult to maintain, no QoS guarantees, Web-based and legacy applications must interoperate
  - requirement for autonomy, decentralized authority, intermittent connectivity, continual evolution, scalability

- **Quality of service**
  - need for response time, availability, data accuracy, consistency, security for commercial applications (charging for service) over a best-effort communication environment
  - research projects about extended CORBA-based platforms (real-time, replication, fault-tolerance)
  - integrating QoS management into middleware architectures is essential
  - ... but a procedure for doing so has yet to be agreed upon

- **Nomadic mobility**
  - need for accessing and processing information "anywhere and any time"
  - available resources vary widely and unpredictably
  - error rate, bandwidth, battery
  - location awareness
  - middleware must support the applications to explicitly accommodate these changes

- **Ubiquitous environment**
  - future computing environments will comprise diverse computing devices
  - from large computers to microscopic processing units

- addressing and naming a multitude of devices cannot be done with current technology cannot consume an IPv6 address for each item sold

The middleware for Ubiquitous integration control should meet a few conditions. Firstly, all kinds of sensors or devices should be independent and easy to use regardless of different types of hardware and software. Secondly, the technology that is able to interface with external environment easily without discontinued services and new technology or devices can be simply adapted to the middleware. Finally, it should have an adaptability, flexibility and recyclability in accordance with the demand of users and market.

Previous main ideas of researches of implementation for integration control were about system integration and synchronization on a basis of integrated databases by implementing EAI (Enterprise Application Integration). Even most integration controls are operated by EAI, however, the integration by EAI technology is not easy to agree on an interface with different systems and common core functions are overlapped in each system. Moreover, there is a little flexibility on the extension of systems and errors caused by system overload occur sometimes because systems stand on the basis of the integrated databases.

Although many researchers have been studying to find out answers for these problems, still a number of technologies are stick to ones on the basis of database.

The system on the basis of database collects information from sensors and store data at database and refers stored information repeatedly so that it is difficult to process data in real-time. The key point of integration control is not processing the past records but data from sensors happened at same time.

In this paper, We shows the system Middleware for integration control of Ubiquitous which ensures the independence of application services collecting different types of sensors data and supports flexible integration and also real-time processing collected data through shared memory and avoids duplicated development by providing common core functions. In addition, the middleware uses a frame structure in case the extension of new service. In order to implement the middleware not to be dependent of database, neural network is used instead of applying to the priority algorithm based on general rule-database.

The contents of the study are the followings. The brief explanation of neural network in chapter 2 and the middleware under Ubiquitous computing environment is shown in chapter 3. The whole structure of middleware for integration in Ubiquitous environment and neural network model for classifying events priority are explained in chapter 4. After that, the implementation of neural network for classifying events priority and the result of experiment are illustrated in chapter 5. Finally, in chapter 6, the result of this study and further study are suggested.

2   Related Works

2.1   CORBA [1, 2]

The following figure illustrates the primary components in the CORBA ORB
architecture. Descriptions of these components are shown in figure 1.

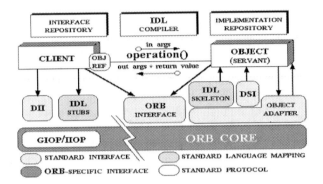

Fig. 1 CORBA components

- **Object** - This is a CORBA programming entity that consists of an identity, an interface, and an implementation, which is known as a Servant.

- **Servant** - This is an implementation programming language entity that
  defines the operations that support a CORBA IDL interface. Servants can
  be written in a variety of languages, including C, C++, Java, Smalltalk,
  and Ada.

- **Client** - This is the program entity that invokes an operation on an object implementation. Accessing the services of a remote object should be
  transparent to the caller.

- **Object Request Broker (ORB)** - The ORB provides a mechanism for
  transparently communicating client requests to target object implementations. The ORB simplifies distributed programming by decoupling the
  client from the details of the method invocations. This makes client requests appear to be local procedure calls.

- **ORB Interface** - An ORB is a logical entity that may be implemented
  in various ways (such as one or more processes or a set of libraries). To
  decouple applications from implementation details, the CORBA specification defines an abstract interface for an ORB.

- **CORBA IDL stubs and skeletons** - CORBA IDL stubs and skeletons serve as the "glue" between the client and server applications, respectively, and the ORB. The transformation between CORBA IDL definitions and the target programming language is automated by a CORBA IDL compiler.

- **Dynamic Invocation Interface (DII)** - This interface allows a client to directly access the underlying request mechanisms provided by an ORB. Applications use the DII to dynamically issue requests to objects without requiring IDL interface-specific stubs to be linked in.

- **Object Adapter** - This assists the ORB with delivering requests to the object and with activating the object. More importantly, an object adapter associates object implementations with the ORB.

2.2   Neural Network

Neural network learns experiences and approaches to a result from the learning. In the processing of the result, when inputs are given, General model outputs parameters and variables while neural network outputs proper results through the interaction between internal Units of Neural network and the strength of multiplicative Weights[3, 4].

Briefly, there are two major learning paradigms in Neural Network. The one is supervised learning with training sets and the other one is unsupervised learning with no training sets. In order to realize the model for classifying events priority under integration control situation, this study chooses supervised learning because the class values of each event can show the desired patterns of neural network. Furthermore, because of the problems of single-layer neural network, multilayer neural network that can learn desired patterns is needed.

Therefore, this study uses the Back-propagation Algorithm in order to meet condition that mentioned earlier. The back-propagation Algorithm was designed to decrease the error between output patterns and desired patterns by using theory of Gradient descent method.

2.3   Ubiquitous Middleware

The one of strengths of ubiquitous middleware is no fixed-form that can be dynamically realized by service conditions. Beside that, the middleware for Ubiquitous computing should have a difference with numerous middleware that has already developed[4, 5].

Various devices should be connected to sub-network and they should exchange and share information each other and this network should support

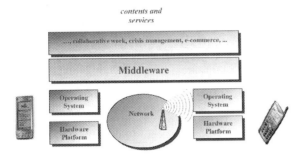

Fig. 2 Ubiquitous middleware

not only wire/wireless and but also wide-area mobile network and ad-hoc networking in the Ubiquitous computing environment as in figure 2.

As well as that, in order to supply reasonable services as recognizing and communicating between different devices, a technology of context-aware is also essential. Again, a context management skill, which is an effective management of resources, is needed because various devices are communicated and operated. These are all critical factors and roles in Ubiquitous middleware.

3   Middleware Design for Ubiquitous Computing Environment

3.1   An Architecture of Middleware

The proposed middleware consists of three layers, Application Connector Layer, External Layer and Internal Layer and Core is located in the center to handle with interfaces between different types of application devices and to process numerous data and events in real-time while controlling as in figure 3.

Application Connector Layer (ACL) supports environment to implement general application program. ACL can link to middleware with simply proceeding in accordance with application while excluding dependent with system, it can have implementations and applications.

External Layer (EL) roles as a Work Agent that receives external requests, processes them and display the stability of external communication and works. EL secures external interfaces through connectors and is also an External Agent to connect to internal layer through Listener.

Internal Layer (IL), where is around middleware core, is a core module. IL is in charge of the creation, destruction and processing of Agent. In addition, IL performs a key role for internal/ external interfaces and consists of HAS (Half System Agent) and SA (System Agent).

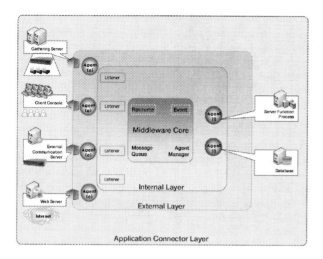

Fig. 3 Middleware Architecture

Middleware Core (MC) manages lifecycle of all Agents from creation to termination, as doing resource, events, messages, process and Agent management etc. MC also maintains and manages entire functions, in particular, all modules communication within system and distributing resources.

3.2 The Operation of Middleware

The operation of middleware in ubiquitous is shown as in figure 4.

This study shows that the middleware, which is in charge of server, was implemented in order to connect to multi-networks and link to several clients. figure 5 presents the implementation proceeding of server side.

According to figure 5, whenever clients request to connect through Listener and External Agents, at first, Listener receives a connection request and then checks if the connection has already done, or accessed connection. When the permission has made, the middleware informs Core to create an External Agent in charge of communication with clients and then each client creates each different External Agent in order to communicate independently between several clients and the middleware.

3.2.1 Middleware Connector Communication

Connector is an organization that can configure communication between application and middleware and transmits data of types of data-set from application to Middleware. As using SQL in application, Connector could transmit virtual DB that data are inputted to Middleware as in figure 5.

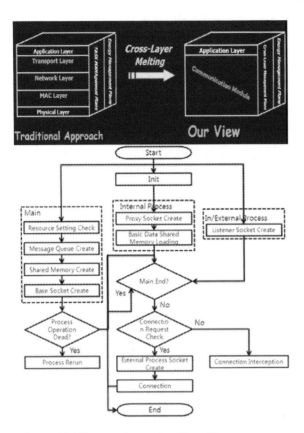

Fig. 4 Middleware Operation Flow Chart

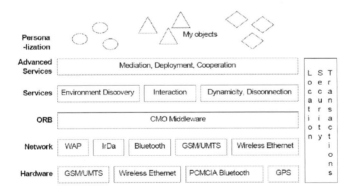

Fig. 5 Middleware connecting layers

3.2.2   Events Receive/Transmit between Middleware Reprocesses

An event handling between processes in the middleware repeatedly performs checking communication status, events test, and handling events. According

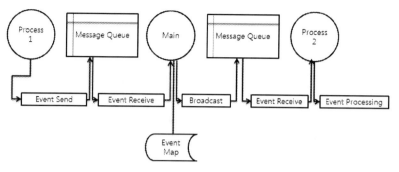

Fig. 6 Event Process Flow Chart

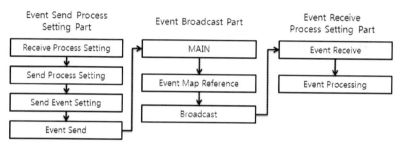

Fig. 7 Event Send Flow Chart

to figure 6, while process 1 sends event to Message Queue, Main receives the event of Message Queue and next Main broadcasts event to Message Queue as referring to event map. Finally process 5 handles event of Message Queue. Events are to be defined in header file in the same order indicated in event map. The figure 7 is shown about divided the event sending process; for example, event send, event broadcast and event receive.

Structures are defined in Middleware in order to write and read data in a shared memory. After defining structures in the shared memory, memory name is given to and memory map is configured. Authority, the type of structure members to read/write memory and the row number of memory should be assigned when the shared memory is configured.

3.3 Neural Network for Event Conflict Processing

Generally, conflict processing system in integration control already collects and stores priority information in accordance with the national disaster management act. When conflicts occur in an information gathering interface, the stored information can use comparison data. In addition, traffic and general information are stored and the priority should decide to send which messages go first.

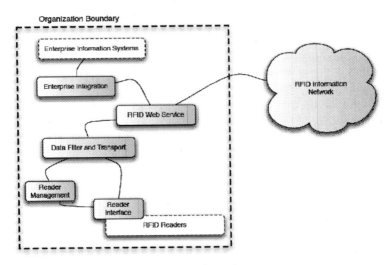

Fig. 8 Neural networks organization boundary

The priority has relatively importance according to the national disaster management act and the guide of Ministry of land, Transport and Maritime Affairs. This guide includes degree of safety, necessity of users, the frequency of accident, and importance of damage. The priority has the following order-fatality, disaster, emergency information, traffic information and general information- and has already stored in database and becomes a standard data for conflict processing.

The Priority Order Algorithm - when more than two processes are in conflict, each process is given a priority and higher priority process has a processing authority - is a universal way to handle with conflict processing; However, if there is continuous overload in a system, the Priority Order Algorithm is not able to handle with low level of priority and also exceptional events or data behind database, this Algorithm can not function properly. Even though Aging method- when there is massive time delay in processing low level of priority, as increasing one step of the priority order in every a certain time period, this low priority could have the highest priority then-could be used, there are still problems about handling exceptional events not to be stored in Database.

Therefore, this study uses Back-Propagation Algorithm, one of neural network learning, to handle with conflict processing. Neural network classifies many cases with one algorithm and then after learning, it can processes conflict processing without referring to Database as in figure 8.

In order to define the priority order forward situation information in Integration Control Environment, this study refers to "the national disaster management act[5]", "the guide of Ministry of land, Transport and Maritime Affairs[6]" and the materials of "u-City association[7]" and makes "Situation code Pattern" for Neural Network Learning. table 1 shows example of the priority order.

Table 1 Situation Code System Construction Example

P	SC	Explanation
1	1-6-1	In prevention of crimes/disasters service a conflagration(large department store, multistory building, forest fire) occurrence
2	4-1-4	In public facility service large inundation(over radius 10 m) occurrence
3	3-3-12	In environment watch service industrial-strength yellow dust (minuteness dust density $800\mu g/m^3$) occurrence
4	3-3-15	In environment watch service industrial-strength ozone(over ozonic density 0.5ppm/h) occurrence
5	2-6-20	In road/traffic service big traffic accident(more than 2 seriously, more than car damage 3, death) occurrence
6	4-4-21	In public facility service building collapse
7	3-3-22	In environment watch service poisonous gas leak
8	1-6-2	In prevention of crimes/disasters service a moderate fire(office, factory, education, sanitary facilities) occurrence
9	4-1-5	In public facility service of scale inundation(radius 5 10 m) occurrence
10	3-3-11	In environment watch service strong yellow dust (minuteness dust density 400 $800\mu g/m^3$) occurrence
11	3-3-14	In environment watch service strong ozone(over ozonic density 0.3ppm/h) occurrence
12	2-6-19	In road/traffic service medium size traffic accident(more than 2 seriously, more than car damage 3) occurrence
13	1-6-3	In prevention of crimes/disasters service a small scale fire(residing equipment) occurrence
14	4-1-6	In public facility service small scale inundation(radius 1m) occurrence
15	3-3-10	In environment watch service very bad minuteness dust(density $201\mu g/m^3$ $400\mu g/m^3$) occurrence
16	3-3-13	In environment watch service weak ozone(over ozonic density 0.12ppm/h) occurrence
17	2-6-18	In road/traffic service small size traffic accident(more than 1 seriously) occurrence
18	1-5-23	In prevention of crimes/ disasters service threat(urgency request) occurrence
19	3-3-19	In environment watch service worse minuteness dust(density 151 200 $\mu g/m^3$) occurrence
20	2-6-17	In road/traffic service small size traffic accident(more than 2 slightly wounded person) occurrence
21	1-5-24	In prevention of crimes/disasters service help request occurrence by invasion
22	3-3-18	In environment watch service bad minuteness dust(density 101 150 $\mu g/m^3$) occurrence
23	2-4-16	In road/traffic service small size traffic accident(minor collision, less than 1 slightly wounded person) occurrence
24	3-3-7	In environment watch service worse minuteness dust(density $100\mu g/m^3$ low) occurrence
25	2-4-25	In road/traffic service assembly and demonstration occurrence

Table 2 Learning Parameters

Input Unit	Hidden Layer	Hidden Unit	α	β	Learning Number
3	1	60	1.2	0.003	68.431

This research focuses on developing "Priority Classification Algorithm" in Integration Control Environment and "Supervised learning" that handles with values of the priority order in accordance with events as training patterns of neural network is proper for the study. In addition, multi-layer neural network model is needed rather than single-layer because multi-layer neural network is able to learn effectively training patterns. After considering all these conditions, Back-propagation Algorithm, one of neural network learning methods is selected on this study. In table 1, service code, event code and context code are used as input vales for neural network learning and the priority order is used as training patterns.

Multi-layer neural network could be revised in accordance with how network is configured against inputs, such as the number of hidden layers, the number of units of hidden layers, types of activation functions, learning iteration, initial weights, learning rate and etc.

Generally, if the number of hidden layers or units of hidden layers is large, the convergence time for neural network optimization increases.

Therefore, the proposed neural network has simply three units (service, event, and contents code) of input layer and fixed one hidden layer and also adjusts learning pattern while increasing the number of units of hidden layer every 2 times.

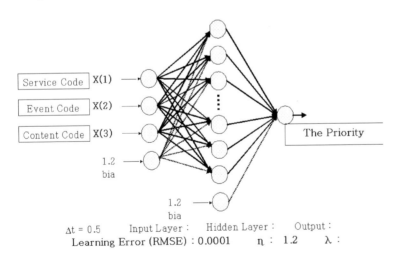

Fig. 9 Neural Network Structure for Event Priority Classification

Furthermore, momentum term, which is related to neural network settling, value of learning constant and complementing local minimum problem, is not fixed so that it is usually selected between 0.5 and 0.9. Learning rate is selected between 0.1 and 2. Even the more iteration of learning in neural network, the less learning error rate, the proposed model has a maximum 10, 000,000 iteration of learning until a learning error is less than 0.0001.

Thus, the case that the error and iteration is the smallest value among attempts was selected the optimized learning alternative.

The table 2 shows parameters (the number of Layers, units of each layer, learning constant, momentum term) and the structure of neural network is shown in figure 9.

4 Examples and Evaluation

4.1 Evaluation Result

Before performing the priority order classification, at first, the proposed network was learnt by 500 situation codes (Training Set) and then the test was done by using 100 situation codes, which was not learned by the neural network, and learned 500 situation codes (events data). The result of test is given in the table 3. In order to evaluate the result, evaluation indexes that are a recognition rate and a processing speed were given in this study. The execution environment structure is shown as in figure 10.

The recognition rate defines the numbers of the event which were precisely classified among to be classified, except for unclassified events of the total test events.

500 out of total 600 situation codes of the priority order were precisely classified and also 78 of total unlearning 100 situation codes of the priority order were classified. Therefore, as 578 out of 600 situation codes were classified, the recognition rate was 96.3% and the processing speed was 0.01 seconds.

According to whether learning exists or not, the recognition rate showed 100% (500 out of 500) for events being learnt the recognition rate 78% (78 out of 100) for events not being learnt, even though 22 events were not classified, similar outputs came out.

Table 3 Experiment Result

Experiment Result	LearnedEvent (500)	NotLarned Event (100)	Total(600)
	500 classified	78 classified	578 classified
Recognition Rate	100%	78%	96.3%

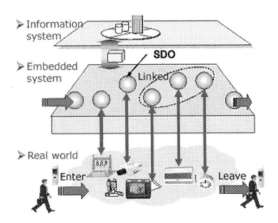

Fig. 10 Execution environment

4.2   The Evaluation of the Event Priority Algorithm

The table 4 shows the comparison between the event priority algorithm and the rule-database Algorithm. The rule-database classifies the priority order of events by referring to rule table which had already stored in database while the event priority algorithm classifies the priority order by using neural network regardless of database.

The conventional event priority algorithm puts rule-data in accordance with events into database and when events occur, this algorithm refers to priority information of database and then classifies priority. Thus it can not classify priority if there are events which are stored or errors on database. However, the proposed the event priority algorithm by using neural network carries out neural network learning as situation codes in accordance with events, when events occur, learnt neural network automatically classify priority, even if there are no learning events, the proposed algorithm can classify regardless of database errors. In case of the even priority algorithm on basis of rule-database, its the recognition rate was 83.3%.

Among 600 events, only 500 events that have already stored in database were classified and 100 events that were not stored were not classified.

Table 4 Event Priority Algorithm Comparison

	Rule-Database	Neural Network
Saved Data or Learned Event	500 classified of 500	500 classified of 500
Not Saved Data or Not Learned Event	0 classified of 100	78 classified of 100
Process Velocity	Average 0.1	Average 0.01
Recognition Number	500 classified of 600	578 classified of 600
Recognition Rate	83.3%	96.3%

The proposed event priority algorithm classified 500 events is the same as the conventional event priority algorithm but the difference was that 78 out of 100 events were classified and its the recognition rate was 96.3% (578 out of 600). These 100 events were not learnt or store in database. Therefore the proposed event priority algorithm has more efficient data processing than the conventional event priority algorithm in integration control environment.

5   Conclusion

This study shows implementation of neural network model to classify priority of events in accordance with situation in integration control environment of Ubiquitous computing environment and defines functions of the middleware minutely.

The proposed method has more excellent recognition rate (96.3%) and processing speed (0.01 seconds) compared with the conventional event priority algorithm on the basis of rule-database. In other words, in classifying event priority, new approach reduces processing time and performs effective classification.

Although this study covered four types of u-service, six events and 25 contents that might occurs in u-City, extra researches are needed about more various and complicated situations and also different types of algorithms except for Back-propagation Algorithm, such as RBF(Radialbasis Function), Grossberg-Network, Fuzzy and Genetic Algorithm, will be considered in near future.

References

1. Sameshima, S., Arbanowski, S.J.: OMG Super Distributed Objects. White Paper, Super Distributed Objects Domain SIG, Object Management Group (2001)
2. Conan, D., Taconet, C., Ayed, D., et al.: A Pro-Active Middleware Platform for Mobile Environments. In: Proc. Mobile Computing Systems in Dynamic Environments Workshop, special session of the IASTED International Multi-Conference on Software Engineering SE 2004, Innsbruck, Austria, pp. 701–706 (2004)
3. Nasser, N., Al-Manthari, B., Hassanein, H.: A Performance Comparison of Class-based Scheduling Algorithms in Future UMTS Access. In: Proc. of the IEEE IPCCC, Phoenix, Arizona, pp. 437–441 (2005)
4. Nasser, N., Hasswa, A., Hassanein, H.: Handoffs in Fourth Generation Heterogeneous Networks. IEEE Communications Magazine 44(10), 96–103 (2006)
5. Hasswa, A., Nasser, N., Hassanein, H.: Tramcar: A Context-Aware, Cross-layer Architecture for Next Generation Heterogeneous Wireless Networks. In: IEEE International Conference on Communications (ICC), Istanbul, Turkey, pp. 240–245 (2006)

6. Zhu, F.. McNair, J.: Optimizations for Vertical Handoff Decision Algorithms. In: IEEE Wireless Communications and Networking Conference (WCNC). pp. 867–872 (2004)
7. Ylianttila, M.. Makela. J., Mahonen. P.: Optimization scheme for mobile users performing vertical handoffs between IEEE 802.11 and GPRS/EDGE networks. In: IEEE PIMRC. vol. 1, pp. 15–18 (2002)
8. Zurada, J.M.: Introduction to Artificial Neural System. West Publishing Company (1992)
9. Lee, J.S.: A Study on the Fingerprint Recogni-tion Method Directional Feature Detection using Neural Networks. Korea Maritime University Masters thesis (2001)

Structural Relationship Preservation in Volume Rendering

Jianlong Zhou and Masahiro Takatsuka

Abstract. This paper presents structural relationship preservation as a technique for improving efficiency of volume rendering-based 3D data analysis. In the presented approach, a mapping between data and renderings is defined and volumetric object appearances are determined by structural relationships of interest in the data space. Two typical relationships of inclusion and neighboring are defined, extracted, represented, and revealed in volume rendering in the pipeline. The structural relationship representation, which is controlled by a contour tree, allows users to perceive relationships, and to control how and what structural relationships are revealed in the pipeline. The experimental results show that structural relationship preservation in volume rendering provides more intuitive and physically meaningful renderings.

1 Introduction

Traditional computer graphics is a unidirectional projection from a 3D objective scene to a 2D image. In the context of medical volume visualization, it is usual to think in rational that how effectively the visualization depicts anatomical structures and how well features of anatomical structures can be discerned in order to make decisions. Most of conventional volume rendering methods place emphasis in conveying details of desired features or structures by exposing them clearly to viewers in results. However, relations between structures in a volume are also of interest to viewers [3]. For example, in surgical planning, radiologists are interested in not only shape of structures of interest, but also their neighboring information (e.g., how

Jianlong Zhou
School of Information Technologies, The University of Sydney, Australia;
National ICT Australia
e-mail: zhou@it.usyd.edu.au

Masahiro Takatsuka
School of Information Technologies, The University of Sydney, Australia
e-mail: masa@vislab.usyd.edu.au

R. Lee, G. Hu, H. Miao (Eds.): Computer and Information Science 2009, SCI 208, pp. 229–238.
springerlink.com   © Springer-Verlag Berlin Heidelberg 2009

close they are). Such relation information of structures is crucial for visual analysis and understanding of volumetric data in various applications [3].

As Bertin [1] pointed out, information is about relationships between things, and similarly a picture is about relationships between graphical marks. We expect a good visualization to be a structural relationship preserved mapping between an information domain and a perceptual domain, such that mental models and understanding of the data set are improved during data analysis. Traditional visualization approaches do not take into account the structure of the underlying data. Therefore, the structural relationship analysis part is often taken as a post-processing step in image space [12] and depends on manual inspection to reveal various relations [3].

This paper presents an approach on depicting structural relationships between objects in volume rendering through a concept of structural relationship preserved mapping. Two typical relationships of inclusion and neighboring are defined and depicted in volume rendering respectively. The contour tree controlled structural relationship depiction allows users to perceive structural relationship in a more direct way. The objective of this work is to deliver a new visualization paradigm for better understanding, presentation, and visualization of volumetric data based on structural relationships. The advantage of the structural relationship preservation approach is that it allows analysis of volumetric data to focus on revealing high-level topological relations instead of low-level rendering parameter modulation, and thus improves understanding of volumetric data.

The paper is organized as follows. We first introduce previous work in Section 2. In section 3, the concept of structural relationship preserved mapping is presented, and approaches of depicting relationships in volume rendering are discussed. Section 4 reports experimental results and discussions of the proposed approach. Finally, Section 5 concludes this paper and refers to future work.

2   Related Work

Relationship and its uses have been studied in several research fields. In computer graphics, scene graph [8] is a commonly used tree data structure for representing the hierarchical relation of geometric objects. The major purpose of scene graph is to model and organize objects in a scene.

In 3D visualization, Viola et al. [10] used a focus+context approach to depict importance relationship of objects. Chan et al. [3] presented a relation-aware volume exploration pipeline. However, all of these operations highly depend on segmented data set. If the segmentation information is not available, the relation definition lacks object boundary information and has difficulty to set up. Image segmentation is acknowledged as a very difficult and complex task. It is often desirable to be avoided or at least postponed for after a first inspection of 3D data with volume rendering. This work motivates us to introduce topological relationship into volume rendering to analyze volume data sets.

The investigated approaches try to represent different relationships in a data set based on pure rendering techniques, but they do not touch the basic theory behind the problem — topology. Representing structures with different relationships belongs to the problem of revealing topological information between them. Kniss et al. [6] proposed a framework to encode topology information of a data set in raster-based representation. The topological attributes in a volume were also considered in a framework to define transfer functions and depict inclusion relationships in volume rendering [9]. The contour tree has been used to explore the relation between iso-surfaces and their evolution [2]. It was also used to index subregions in transfer function specifications [11].

In contrast, our work focuses on data analysis. Thus, more relations other than hierarchy as used in computer graphics are considered. Because object relations in a data set are related to topology of the data set, our work also focuses on using the contour tree to reveal various relationships in volumetric data. Instead of depending on segmented data set as used in [3] to define spatial relations, our approach creates the contour tree directly from the data set and then simplify it to represent topology.

3 From Perception to Structural Relationship Preservation

Volume visualization aims at making the mapping from data to renderings as effective as possible. The effectiveness of a visualization can be measured in terms of easiness and directness of acquiring its intended interpretation [5]. In the context of volume visualization, the structural correspondence between volume data and perceptual visual elements in the rendering should be set up in order to get the effective volume visualization. The volume visualization process generates rendering images and the interpretation process uses the rendered image from the visualization process to get properties of the original data. Because the interpretation of a rendered image depends on human visual perception, Dastani [5] pointed out that a visualization is effective if the intended structure of the data and the perceptual structure of the visualization coincide. Based on this view, we combine perceptual structure into the data analysis pipeline as shown in Fig. 1. The figure shows how a volume data set is rendered and explained during the data analysis pipeline.

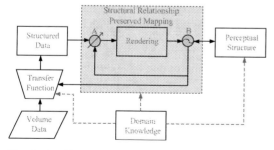

Fig. 1 Combine perceptual structure into the volume rendering based data analysis pipeline

In Fig. 1, the volume rendering process begins with an input data set. The first step is to determine the structure of the data. This step is a part of classification in volume rendering. It transforms the input data to structured data. The second step is to render the structured data in a rendering display. Usually, it is not possible to display all classified objects in the rendering at the same time. It is necessary to determine which structures will be displayed and how they will be displayed. This is determined at the point A in Fig. 1. Then the structured data set is rendered as the rendering display. The rendering display is then interpreted at the point B in Fig. 1, such that the perceptual structure is obtained from the interpretation. On the other way round, through interpretation, the user compares perceptual structure and rendering display in order to see whether the perceptual structure from the domain knowledge and the rendering display coincide. If this is not true, the user needs to return to the point A to modulate rendering options. The user may also return to the transfer function module to classify the data and get new structured data. During this pipeline, the domain knowledge is used in transfer function modulation, rendering interpretation, and perceptual structure understanding.

3.1   Structural Relationship Preserved Mapping

Based on the analysis pipeline as shown in Fig. 1, volume visualization is defined as a mapping between structured data elements and visual perceptual elements. We assume that there are certain relations existed between structured data elements, e.g. topological relations, and similar relations existed between visual perceptual elements. An effective volume visualization should define a mapping between data and renderings such that every relationship between objects in the data space is mapped onto relationships of objects' appearances in the rendering space. The structural relationships between structured data are preserved during this mapping. The shaded region in Fig. 1 shows how structural relationships are mapped and preserved during the data analysis pipeline.

3.2   Topological Relationships in Data Set

This paper uses the contour tree to represent the topological structure of the volume data set. In this paper, we confine our interest on relationships that are used to enhance the understanding and perception of structures in the rendering. As shown in Fig. 1, volume rendering acts as a bridge to map structural relationships from structured data to perceptual structure. The relationships are preserved during this mapping. So it is necessary to develop appropriate volume rendering methods in order to reveal various relationships. We believe that appropriate depiction of various structural relationships will produce better understanding of a 3D data set. Usually, not all relationships between objects can be revealed completely in a 2D image view during data analysis. To provide a formal basis, we give definitions of two typical

relationships of inclusion and neighboring, and show how to depict these relationships based on the contour tree in volume rendering as follows.

3.2.1 Inclusion Relationship

In 3D space, structures of interest are often included by other structures because of
their spatial positions. The inclusion relationship refers to the situation of objects
included by others in 3D space, which creates inner objects and outer objects respectively. The inclusion relationship is commonly seen in a volume data set and
heavily affects understanding of it.

In this paper, we assume that the contour tree of a data set is represented as a
graph (see Fig. 2(b)). As shown in Fig. 2(b), given a contour c as the "outside"
isosurface — i.e. a point in the tree, and choose two points p and q in the tree. In
order to determine whether p is "inside" q, we find the path P (dashed line in blue
in Fig. 2(b)) from c to p in the contour tree. There is only one such path, because
it is a tree. Similarly, we find the path Q (dashed line in red in Fig. 2) from c to q.
p is "inside" q if Q is a prefix of P (i.e. we can only get to p by passing through q
first). The points c, p, and q belong to various arcs. The subregions that these arcs
correspond to are displayed in Fig. 2(a). From this figure, we see that the inclusion
relationship of c, p, and q represented in the contour tree graph is clearly revealed
between subregions in the data set.

We use the branch decomposition [7] to analyze the inclusion relationship in this
paper. The branch decomposition provides a direct way to track inclusion relationship based on branches. Because a branch is usually a concatenation of a sequence
of arcs in the contour tree, the subregion corresponding to a branch is a combination of subregions corresponding to arcs. We assume that there exists an outermost

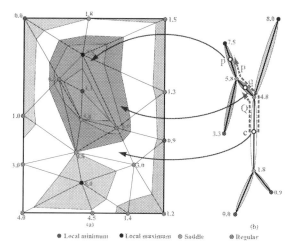

● Local minimum   ● Local maximum   ◉ Saddle   ● Regular

Fig. 2 A 2D mesh and its contour tree. The 2D mesh is segmented into subregions and
indexed with the contour tree.

surface which includes objects of interest in a data set. So we track the inclusion relationship in the branch decomposition from the branch corresponding to the outside surface. According to this basic theory, this branch includes all of its children. The parent branch includes its child branch, and there is no inclusion relationship between child branches which have the same parent.

In order to depict inclusion relationship, we introduce *inclusion opacity* based on the nesting depth of branches. The inclusion opacity is used to show the visibility of subregions corresponding to different branches. In order to emphasize inner structures and deemphasize outside structures, branches with lower depth value are applied with lower opacity.

At this step, there are two opacities: one is the inclusion opacity α_d, and another is the opacity α_s specified by the scalar transfer function. Two opacities need to be combined in order to get the final opacity α applied to objects as follows:

$$\alpha = f(\alpha_d, \alpha_s), \qquad (1)$$

where f is the function used to combine two opacities. A typical combination is:

$$\alpha = \alpha_d \cdot \alpha_s. \qquad (2)$$

Given two branches a and b, we assume branch a is included by branch b and represented as $a \in b$. The final opacity for a and b is α_a and α_b respectively. α_a and α_b should meet following inequality in order to preserve inclusion relationship in the rendering:

$$\alpha_a > \alpha_b. \qquad (3)$$

Compared with approaches in [9], our method analyze inclusion relationships based on the branch decomposition. It overcomes the disadvantage of "oversegmentation" by the general contour tree, and thus avoids the unnecessary inclusion analysis. Furthermore, our method applies transfer functions of different subregions corresponding to branches locally but not globally to depict inclusion relationship.

3.2.2 Neighboring Relationship

Usually, a volume data set can be divided into various subregions which represent different objects. Neighboring relationship refers to the situation that objects are extremely connected or there is a gap interval which can be transparent or opaque between them. The neighboring information has the role of controlling perceptual information of structures, for example, it is used to modulate optical properties of structures to improve perception of difference between structures. It can also be used in surgical planning to reveal closeness of structures in order to make decisions.

In this subsection, the neighboring relationship between subregions in a data set is depicted in the volume rendering pipeline based on the branch decomposition. The branch decomposition encodes the neighboring relationship between subregions of the data set directly: two subregions have neighboring relationship if their corresponding branches are connected; otherwise, the neighboring relationship does

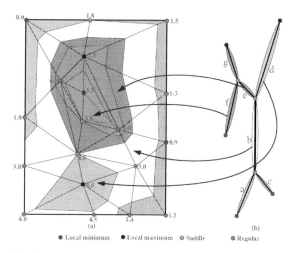

● Local minimum   ● Local maximum   ● Saddle   ● Regular

Fig. 3 Neighboring relationship analysis in the contour tree

not exist. In order to reveal this relationship, we check branch connection information and use this information to emphasize structures in volume rendering pipeline.

In the contour tree, neighbors of a given branch includes its parent, children, and sibling branches. For example, as shown in Fig. 3(b), d's neighbors include its parent b and sibling e. f is not connected with d, so it is not the neighbor of d. Fig. 3(a) shows these corresponding neighboring relationships in the data domain. From this figure, we see that the contour tree can represent neighboring relationship of objects faithfully.

In order to reveal neighboring relationship in volume rendered images, a color range based approach is proposed to encode neighboring relationship presented in the contour tree. We use different colors to render different regions to reveal neighboring relationship in this paper. As shown in Fig. 4, a hue wheel in the HSV color space is divided into four ranges: current (one point), children, parent (one point), and siblings. We use a harmonic color template X-Type [4] as the neighboring relationship preservation template in this paper. In Fig. 4, the size of gray regions for children and sibling branches is $93.6°$ respectively, and the position of the current branch and parent branch is at the center of white regions respectively. Based on

Fig. 4 The neighboring relation color wheel: different color ranges are used to encode neighboring relationship of branches in the contour tree

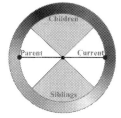

this approach, neighboring relationships between branches in the contour tree are preserved in the rendering with various colors. Users may perceive neighboring relationships of various subregions using the color wheel as a legend describing the neighboring relationships.

4 Experimental Results and Discussions

We conducted experiments to demonstrate the effectiveness and utility of our approach. Our system was run on Ubuntu on a Dell machine (Intel Core2Duo CPU E4400, 3G RAM) equipped with an NVIDIA GeForce 8300GS graphics card.

Fig. 5 shows renderings of "nucleon" (the first three images) and "neghip" (the last two images) data sets (see http://www.volvis.org/) to depict inclusion relationship. In these data sets, small inner structures are included by outer surfaces. Users need to depict this inclusion relationship in order to reveal inner structures. The first and forth images in Fig. 5 are rendered using direct volume rendering. It shows that the general direct volume rendering cannot depict inclusion relationship effectively. The third and last images are rendered using the approach proposed in this paper. The corresponding contour tree used for the third image is shown in the second image. Through the comparison of two results, we see that the proposed approach can depict inner structures clearly and this provides richer information for the user to analyze the data set. This depiction allows users to easily understand topological relationships of structures inside the data set.

Fig. 6 shows the rendering result of depicting neighboring relationship in the "fuel" data set (see http://www.volvis.org/). The contour tree and relation color wheel are presented at the same time. In the left image, the current branch and

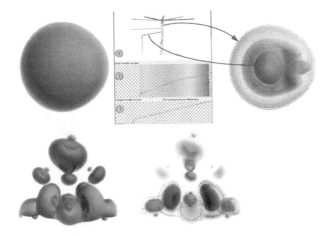

Fig. 5 Inclusion relationship analysis based on the contour tree: 1) Contour tree. 2) Scalar transfer function. 3) Inclusion transfer function

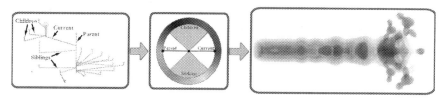

Fig. 6 Depiction of neighboring relationship of "fuel" data set

its neighboring branches are pointed out in order to show their neighboring relationships. In the right image, the subregion corresponding to the user selected current branch and its neighboring structures are emphasized and rendered in various colors based on the relation color wheel presented at the middle of the figure. In this way, users can perceive neighboring structures of a selected subregion clearly. It helps users to understand which structures are affected during interaction with a specific subregion and improve the perception of neighboring relationship.

From the experiments, we see that the proposed approach can effectively extract and visualize relations between objects in volumetric data. The proposed approach provides users a framework to extract, display, and query relations. The concept of relationship preservation mapping helps users to improve understanding of volumetric data. For general volume rendering, our approach can help detect whether any relations are unexpectedly revealed in the final rendering image through interacting with the contour tree. For the exploration of unknown data sets, our approach helps users easily understand structures and their relations through interacting with the rendering image and the contour tree, and thus improve understanding of data sets.

Compared with the work in [3], we analyze two types of relations: neighboring and inclusion. Actually, neighboring relation proposed in our approach include both separate and touch relations in [3]. Furthermore, our approach is not viewpoint dependent as used in [3]. So our approach does not consider overlap relation as shown in [3]. The contour tree is a data structure to depict topological relations more directly than the relation graph presented in [3]. These properties allows the proposed approach to reveal and analyze relations in volumetric data more effectively.

5 Conclusions and Future Work

In this paper, we investigated structural relationship preserved mapping as a technique to enhance and improve volume rendering based 3D data set analysis. We defined and depicted two typical relationships of inclusion and neighboring in volume rendering respectively. The contour tree controlled structural relationship depiction allows users to perceive structural relationship more effectively. An open issue for future work is that how to do the structural relationship selection and depiction automatically. Various features of a 3D data set can be detected to define structural relationships, and integrated into the architecture to depict structural relationships.

Acknowledgements. The authors wish to thank Scott Dillard for some basic contour tree codes, and National ICT Australia (NICTA) and China Scholarship Council (CSC) for their support for this research.

References

1. Bertin, J.: Graphics and graphic information processing. Walter de Gruter, Berlin (1981)
2. Carr, H., Snoeyink, J.: Path seeds and flexible isosurfaces using topology for exploratory visualization. In: Proceedings of the symposium on Data visualisation 2003, pp. 49–58 (2003)
3. Chan, M.Y., Qu, H., Chung, K.K., Mak, W.H., Wu, Y.: Relation-aware volume exploration pipeline. IEEE Transactions on Visualization and Computer Graphics 14(6), 1683–1690 (2008)
4. Cohen-Or, D., Sorkine, O., Gal, R., Leyvand, T., Xu, Y.-Q.: Color harmonization. ACM Transactions on Graphics 25(3), 624–630 (2006)
5. Dastani, M.: The role of visual perception in data visualization. Journal of Visual Languages and Computing 13, 601–622 (2002)
6. Kniss, J., Hunt, W., Potter, K., Sen, P.: Istar: A raster representation for scalable image and volume data. IEEE Transactions on Visualization and Computer Graphics 13(6), 1424–1431 (2007)
7. Pascucci, V., Cole-McLaughlin, K., Scorzelli, G.: Multi-resolution computation and presentation of contour trees. In: Proceedings of the IASTED conference on Visualization, Imaging, and Image Processing, pp. 290–452 (2004)
8. Sowizral, H.: Scene graphs in the new millennium. IEEE Computer Graphics Applications 20(1), 56–57 (2000)
9. Takahashi, S., Takeshima, Y., Fujishiro, I., Nielson, G.M.: Emphasizing Isosurface Embeddings in Direct Volume Rendering. In: Scientific Visualization: The Visual Extraction of Knowledge from Data, pp. 185–206. Springer, Heidelberg (2005)
10. Viola, I., Kanitsar, A., Gröller, M.E.: Importance-driven feature enhancement in volume visualization. IEEE Transactions on Visualization and Computer Graphics 11(4), 408–418 (2005)
11. Weber, G.H., Dillard, S.E., Carr, H., Pascucci, V., Hamann, B.: Topology-controlled volume rendering. IEEE Transactions on Visualization and Computer Graphics 13(2), 330–341 (2007)
12. Westermann, R., Ertl, T.: A multiscale approach to integrated volume segmentation and rendering. In: Computer Graphics Forum (Proceedings of Eurographics 1997), vol. 16(3), pp. 117–129 (1997)

Performance of Quasi-synchronous Frequency-Hopping Multiple-Access System with OFDM Scheme and Application of the No-Hit-Zone Codes

Qi Zeng, Daiyuan Peng, and Dao Chen

Abstract. In this paper, a quasi-synchronous frequency-hopping multiple-access communications system with OFDM-BPSK scheme, i.e., QS-OFDM-BPSK-FHMA system, is introduced. The bit error rate (BER) performance of the proposed system is first analyzed in additive white Gaussian noise and Rayleigh fading channels. A novel frequency-hopping code, i.e., No-Hit-Zone (NHZ) code is applied in the proposed system. By the numerical and simulation results, the BER performance of the system with NHZ code can be improved substantially and the frequency resource of system is saved notably, in compared with other kinds of codes.

1 Introduction

Orthogonal frequency division multiplexing (OFDM) system is a kind of the multi-carrier modulation technology, which can transmit data with high-rate in wireless fading channel. In OFDM system, the data is actually transmitted in the narrowband of parallel subchannels, so the characteristics of subchannels are nearly flat. Up to now, the OFDM technology has been widely applied in the wide-band digital communications systems, and also has been proposed in several wireless standards, such as IEEE 802.11a, IEEE 802.16a and so on.

Frequency-hopping (FH) multiple-access (MA) is robust to the hostile jamming and is able to accommodate more active users. The *Hamming-correlation* properties can exactly describe the performance of FH sequence. Thus, the construction of the FH sequence with perfect Hamming-correlation properties is important to improve the FH system performance [1, 2]. *No-hit-zone* (NHZ) FH code is a kind of novel

Qi Zeng and Daiyuan Peng
Provincial Key Lab of Information Coding and Transmission, Institute of Mobile
Communications, Southwest Jiaotong University, Chengdu, Sichuan 610031, P. R. C.
e-mail: zeng_qi@yahoo.com.cn, dypeng@home.swjtu.edu.cn

Dao Chen
School of Information Science and Technology, Southwest Jiaotong University,
Chengdu, Sichuan 610031, P.R.C.
e-mail: cd_2005@162.com

R. Lee, G. Hu, H. Miao (Eds.): Computer and Information Science 2009, SCI 208, pp. 239–248.
springerlink.com   © Springer-Verlag Berlin Heidelberg 2009

sequence, which has *ideal* Hamming-correlation properties in the vicinity of zero delay. If the delay of the user is restricted within no-hit-zone, the multiple-access interference (MAI) will be reduced efficiently [3].

A combination of OFDM and FHMA system has the advantage of the both technologies. In [4], the performance of frequency-hopping multiple-access with non-coherent OFDM-MASK system (OFDM-MASK-FHMA) is analyzed. But the ability of anti-interference of MASK is poor in digital communications. In [5], a spectrally efficient system, i.e., synchronous OFDM-FHMA system with phase-shift keying modulation is proposed, where the FH frequency is collision-free. But in the quasi-synchronous system, the system is not necessarily frequency collision-free.

In recent year, the main study of the OFDM-FHMA system is a synchronous system with random frequency-hopping codes [4-6]. But in the practical communication system, it is more significant to analyze the performance of quasi- synchronous system. In this paper, we will extend the research of Ref. [6] and analyze a *quasi-synchronous* system which is proposed in [6]. The research highlights are that, the theoretical BER performance is derived and the performance of the system with NHZ codes is simulated. With the same parameters, the BER performance of QS-OFDM-BPSK-FHMA system with NHZ code is substantially improved and the band-width is smaller than other kinds of FH codes.

The outline of this paper is as follows. In Section 2, the NHZ codes and the description of the system are presented briefly. The multiuser performance analysis of the quasi-synchronous system in AWGN and Rayleigh fading channels are developed in Section 3. The numerical and simulation results are presented in Section 4. Finally, conclusions are drawn in Section 5.

2 NHZ FH Codes and System Description

2.1 NHZ FH Codes

Firstly, the definition of NHZ code is presented. Let $S=\{S^{(k)}, k=1,2,...,M\}$ denotes a family of M frequency-hopping sequences, where $S^{(k)}=\{c_i^{(k)} | c_i^{(k)} \in GF(q-1), i=0,1,2,..., L-1\}$ that is the frequency-hopping sequence for user k with length L and the size of frequency slots q. The terms Z_{ANH} and Z_{CNH} denote the NHZ width of the Hamming auto-correlation function (HAF) and the Hamming cross-correlation function (HCF) respectively, which can be defined as

$$Z_{ANH}=\max\{T | H_{uu}(\tau)=0, \forall S^{(u)} \in S, 0<\tau \leq T\}$$

$$Z_{CNH}=\max\{T | H_{uv}(\tau)=0, \forall S^{(u)}, S^{(v)} \in S, u \neq v, 0 \leq \tau \leq T\}$$

where the Hamming-correlation function can be defined as

$$H_{uv}(\tau)=\sum_{i=0}^{L-1} h[c_i^{(u)}, c_{i+\tau}^{(v)}], \qquad (0 \leq \tau < L)$$

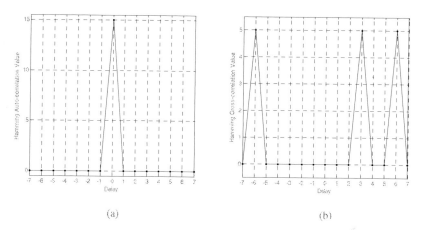

(a) (b)

Fig. 1 Hamming auto-correlation function (a) and Hamming cross-correlation function (b) of NHZ codes

where the subscript addition $i+\tau$ is performed modulo L. If x is equal to y, then $h[x,y]=1$. Otherwise, $h[x,y]=0$. When $u=v$, $H_{uv}(\tau)$ denotes the HAF. Otherwise, $H_{uv}(\tau)$ denotes the HCF. Finally, the no-hit-zone width of the codes is defined as

$$Z_{NH} = \min\{Z_{ANH}, Z_{CNH}\}.$$

If the family of sequences S satisfies the definitions mentioned above, S is namely a family of NHZ code, which is denoted as $S(q, L, M, Z_{NH})$. NHZ codes have the ideal Hamming correlation properties within the delay which is restricted in Z_{NH}. The theoretical bound of NHZ code is established in [7].

In [2], the NHZ code set is constructed by the matrix transform method, which is denoted by $S(q, L, M, Z_{NH}) = S(M(Z_{NH}+1), N(Z_{NH}+1), M, Z_{NH})$, where N is a positive integer. The typical figures of the HAF and the HCF of these NHZ codes can be shown in Fig. 1. From Fig. 1(a), if the delay is not equal to zero, the HCF value will be constant equal to zero. In Fig. 1(b), the HCF value is constantly equal to zero as the delay is restricted within no-hit-zone ($Z_{NH} = 2$).

2.2 System Description

Assuming the frequency-hopping rate of the system is equal to 1 bit per hop, which belongs to slow FH system and the phase coherent between the transmitter and receiver can be maintained easily, so it is feasible to apply coherent BPSK to the proposed system [8, 9]. The transmitter block diagram of OFDM-BPSK-FHMA system is shown in Fig. 2.

The transmitter of the proposed system consists of the OFDM system and the FH system. OFDM modulation is implemented easily by using inverse fast *Fourier transform* (IFFT) circuit. The working frequency of each user is hopped according

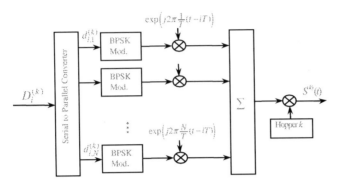

Fig. 2 Transmitter block diagram of OFDM-BPSK-FHMA for user k

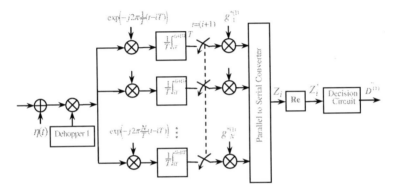

Fig. 3 Receiver block diagram for user 1

to the assigned FH pattern. After the OFDM modulation system and FH modulation system, the transmitted signal of user k is generated, which can be written as

$$S^{(k)}(t)=\sum_i\left[\sum_{l=1}^{N}\sqrt{2P}d_{i,l}^{(k)}\cdot\exp\left(j2\pi\frac{l}{T}(t-iT)\right)\exp\left(j2\pi c_i^{(k)}f_dN(t-iT)\right)\cdot P_l(t-iT)\right] \qquad (1)$$

where $d_{i,l}^{(k)}$ is a bit information in lth subchannel of the user k, P is the average energy of the subcarrier, T denotes the bit time interval in subchannel $(f_d=1/T)$, $P_l(t)$ is a unit rectangular pulse over interval $[1, T]$, $c_i^{(k)}$ is an element of the sequence and f_dN is the minimum hopping interval.

The communication channel that we consider is a frequency-selective Rayleigh fading channel with AWGN. Let $g_l^{(k)}=\alpha_l^{(k)}e^{j\theta_l^{(k)}}$ be the channel complex gain observed by the lth subchannel of user k, where factors $\alpha_l^{(k)}$ and $\theta_l^{(k)}$ respectively denote the *amplitude fading coefficient* and the *phase fading coefficient*. In case of

AWGN channel, we assume that $g_i^{(k)}$ are constantly equal to 1. Considering the above channel model, the total received signal can be written as

$$r(t)=\sum_{k=1}^{N_u}\sum_{i}\sum_{j=1}^{N}\sqrt{2P}d_{ij}^{(k)}g_i^{(k)}\cdot\exp\!\left(j2\pi\frac{j}{T}(t-iT-\tau_k)\right)\!\exp\!\left(j2\pi c_i^{(k)}f_dN(t-iT-\tau_k)\right)\cdot P_i(t-iT-\tau_k)\right]+n(t) \quad (2)$$

where $n(t)$ is AWGN with a two-sided power spectral density of $N_0/2$ and τ_k is the access delay of user k.

In the receiver, we assume that, the FH pattern is the same as that of the transmitter and the desired user is user 1. Thus, the delay τ_1 is assumed to be zero. The receiver block diagram for user 1 is presented in Fig. 3. The received signal r(t) is fed into dehopper firstly, then is processed by OFDM demodulator. The factor $g_m^{*(1)}$ is the conjugate of the channel complex gain. Z_i is the serial-data after the demodulation, namely, $Z_i=[Z_{i,1}, Z_{i,2}, \ldots, Z_{i,N}]$, where the demodulated signal observed in mth subchannel can be expressed as

$$Z_{i,m}=\frac{1}{T}\int_{iT}^{(i+1)T} r(t)\cdot g_m^{*(1)}\cdot\exp\!\left(-j2\pi c_i^{(1)}f_dN(t-iT)\right)\cdot\exp\!\left(-j2\pi\frac{m}{T}(t-iT)\right)\!dt. \quad (3)$$

The real part of $Z_{i,m}$ (i.e. $Z'_{i,m}$) is the decision variable, which is fed into the maximum likelihood detector.

3 BER Performance Analysis

In order to analyze conveniently, we assume that the FH frequency slots are randomly chosen from the set with size q [4, 7]. Since all users in transmitter send the binary data with the equivalent probability, it is reasonable to assume that the delay τ_k has a uniform distribution in $[0,T)$.

3.1 BER Analysis in AWGN Channel

The receiver block diagram in AWGN channel can be shown in Fig. 3. Especially, $g_m^{(1)}$ is constantly equal to 1. The decision variable $Z_{i,1}$, which is observed in 1st subchannel, can be expressed as

$$Z'_{i,1}=\frac{1}{T}\int_{iT}^{(i+1)T}\sum_{i}\sum_{k=1}^{N_u}\sum_{j}\sqrt{2P}d_{ij}^{(k)}\cos\!\left[2\pi\!\left(\frac{j}{T}(t-\tau_k)-\frac{1}{T}t\right)+2\pi\!\left(c_i^{(k)}(t-\tau_k)-c_i^{(1)}t\right)f_dN\right]\!dt+v' . \quad (4)$$

We only consider the serial-data in $i=0$. So, the $Z'_{0,1}$ can be simply rewritten as

$$Z'_{0,1}=\sqrt{2P}d_{0,1}^{(1)}+I+v' . \quad (5)$$

When $k\neq1$, the total interference I is equal to

$$I=\sum_{k\neq1}^{N_u}\sum_{j=1}^{N}\underbrace{\frac{1}{T}\!\left[\sqrt{2P}d_{-1,j}^{(k)}\!\left(\gamma_{j,1}(l,\tau_k)+\mu_{j,1}(l,\tau_k)\right)+\sqrt{2P}d_{0,j}^{(k)}\!\left(\hat{\gamma}_{j,1}(l,\tau_k)+\hat{\mu}_{j,1}(l,\tau_k)\right)\right]}_{I_j^{(k)}} \quad (6)$$

where $d_{-1,l}^{(k)}$ denotes a bit in lth subchannel previous to $d_{0,l}^{(k)}$ and the other factors are denoted as the following equation

$$\gamma_{k,1}(l,\tau_k) = \int_0^{T_b} \cos\left[2\pi(t-\tau_k)\left(\frac{1}{T}+c_{-1}^{(k)}f_dN\right)\right]\cos\left[2\pi t\left(\frac{1}{T}+c_0^{(1)}f_dN\right)\right]dt \ ,$$

$$\mu_{k,1}(l,\tau_k) = \int_0^{T_b} \sin\left[2\pi(t-\tau_k)\left(\frac{1}{T}+c_{-1}^{(k)}f_dN\right)\right]\sin\left[2\pi t\left(\frac{1}{T}+c_0^{(1)}f_dN\right)\right]dt \ ,$$

$$\hat{\gamma}_{k,1}(l,\tau_k) = \int_{\tau_k} \cos\left[2\pi(t-\tau_k)\left(\frac{1}{T}+c_0^{(k)}f_dN\right)\right]\cos\left[2\pi t\left(\frac{1}{T}+c_0^{(1)}f_dN\right)\right]dt \ ,$$

$$\hat{\mu}_{k,1}(l,\tau_k) = \int_{\tau_k} \sin\left[2\pi(t-\tau_k)\left(\frac{1}{T}+c_0^{(k)}f_dN\right)\right]\sin\left[2\pi t\left(\frac{1}{T}+c_0^{(1)}f_dN\right)\right]dt \ .$$

The term $I_l^{(k)}$ in (6) denotes the interference due to the user k in lth subchannel. For different k and l, $I_l^{(k)}$ is independent. Thus the variance of I is computed as

$$D(I) = \frac{2P}{T^2}\sum_{k=1}^{N_u}\sum_{l=1}^{N}\underbrace{\left[\frac{1}{q^2T}\int_0^T \sum_{(c_{-1}^{(k)},c_0^{(1)},c_0^{(k)})} (\gamma_{k,1}(l,\tau_k)+\mu_{k,1}(l,\tau_k))^2+(\hat{\gamma}_{k,1}(l,\tau_k)+\hat{\mu}_{k,1}(l,\tau_k))^2 d\tau_k\right]}_{\psi_{k,1}(l)}$$

$$=\frac{2P}{T^3q^2}\sum_{k=1}^{N_u}\sum_{l=1}^{N}\psi_{k,1}(l) \ . \tag{7}$$

Based on the central limit theorem, when the number of the user accommodated in the system is large enough, $Z'_{0,1}$ has the Gaussian distribution with the following mean $E(Z'_{0,1})$ and variance $D(Z'_{0,1})$

$$E(Z'_{0,1}) = \sqrt{2P}d_{0,1}^{(1)} \quad , \quad D(Z'_{0,1}) = \sum_{k=1}^{N_u}\sum_{l=1}^{N}\frac{2P}{q^2T^3}\psi_{k,1}(l) + \frac{N_0}{T} \ . \tag{8}$$

Assuming the transmitted bit of user 1 in the 1st subchannel is assumed to be 1, i.e., $d_{0,1}^{(1)}=1$, and the zero-level is considered as the decision threshold. Thus the BER performance of quasi-synchronous OFDM-BPSK-FHMA system in AWGN channel is expressed as

$$P_b = P(Z'_{0,1}<0) = Q\left(\sqrt{2P\left/\left(\frac{2P}{T^3q^2}\sum_{k=1}^{N_u}\sum_{l=1}^{N}\psi_{k,1}(l)+\frac{N_0}{T}\right)\right.}\right) \ . \tag{9}$$

In particular, when the delay τ_k is equal to nT, where n is a integer, Eq. (9) can be simplified as

$$P_b = Q\left(1\left/\sqrt{\alpha(N_u-1)+\frac{1}{2(E_b/N_0)}}\right.\right) \tag{10}$$

where $E_b=PT$, which is the energy per bit, and $\alpha=1/q$, which is the probability of one frequency hit. Thus, when the number of active users is fixed, the BER can be improved by the increase of the size of the frequency slots set q.

3.2   BER Analysis in Rayleigh Fading Channel

In Rayleigh channel, the channel complex gain $g_l^{(k)}$ is equal to $\alpha_l^{(k)} e^{j\theta_l^{(k)}}$, where factors $\alpha_l^{(k)}$ and $\theta_l^{(k)}$ respectively denote the amplitude fading coefficient which has *Rayleigh distribution* with square mean value $E\left[\left(\alpha_l^{(k)}\right)^2\right]=\Omega$ and the phase fading coefficient which has *uniform distribution* over the interval $[0, 2\pi)$ in the lth subchannel of user k. For different k and l, factors $g_l^{(k)}$, $\alpha_l^{(k)}$ and $\theta_l^{(k)}$ are independent. Thus, the demodulated data of user 1 in the 1st subchannel can be expressed as

$$Z_{0,1}^{'}=\sqrt{2P}d_{0,1}^{(1)}\alpha_1^{(1)}+I+v^{'} . \tag{11}$$

In (11), the total interference I and AWGN interference $v^{'}$ can be shown as

$$I=\frac{1}{T}\sum_{k\neq 1}^{N}\sum_{l=1}^{N}\alpha_l^{(k)}\cos\left(\theta_l^{(k)}\right)\left[\sqrt{2P}d_{-1,l}^{(k)}\left(\gamma_{k,1}(l,\tau_k)+\mu_{k,1}(l,\tau_k)\right)+\sqrt{2P}d_{0,l}^{(k)}\left(\hat{\gamma}_{k,1}(l,\tau_k)+\hat{\mu}_{k,1}(l,\tau_k)\right)\right] \tag{12}$$

$$v^{'}=\frac{1}{T}\int_0^T\cos\left[2\pi\left(c_0^{(1)}f_dN+\frac{1}{T}\right)t\right]n(t)dt \tag{13}$$

The expressions $\gamma_{k,1}(\cdot)$, $\mu_{k,1}(\cdot)$, $\hat{\gamma}_{k,1}(\cdot)$ and $\hat{\mu}_{k,1}(\cdot)$ in (12) are as defined in (6). Conditioned on the amplitude fading coefficient $\alpha_1^{(1)}$, the mean and the variance of the random variable $(I+v^{'})$ can be computed as

$$E\left(I+v^{'}\right)=0, \quad D\left(I+v^{'}\right)=\frac{1}{T}\left[\sum_{k=1}^{N_k}\sum_{l=1}^{N}\frac{P\Omega}{q^2T^2}\psi_{k,1}(l)+N_0\right] \tag{14}$$

where $\psi_{k,1}(l)$ is as defined in (7). Assuming the transmitted data $d_{0,1}^{(1)}$ is "1", then the BER performance conditioned on the amplitude fading coefficient $\alpha_1^{(1)}$ can be obtained as

$$P\left(e\middle|\alpha_1^{(1)}\right)=P\left(I+v^{'}<-\sqrt{2P}\alpha_1^{(1)}\right)=Q\left(\frac{\sqrt{2P}\alpha_1^{(1)}}{\sqrt{D\left(I+v^{'}\right)}}\right) . \tag{15}$$

We define $\gamma=\left(\alpha_1^{(1)}\right)$. Since γ has a Rayleigh distribution of which probability density function is denoted as $p_\gamma(\gamma)$, the average BER of quasi-synchronous OFDM-BPSK-FHMA system can be computed as

$$\overline{P}_b=\int_0^\infty P\left(e\middle|\gamma\right)\cdot p_\gamma(\gamma)d\gamma . \tag{16}$$

In particular, when the delay τ_k is equal to nT, which n is a integer, Eq. (16) can be simplified as [10]

$$\overline{P}_b = \frac{1}{2} - \frac{1}{2}\sqrt{\frac{\overline{E}_b/N_0}{1+(\beta+\alpha N_u)\cdot(\overline{E}_b/N_0)}}$$ (17)

where the average energy per bit is denoted by $\overline{E}_b$, which is equal to $PT\Omega$. Factors α and β denote the probability of a hit and no hit respectively, please see (10).

4 Simulation Results and Analysis

In order to reduce the MAI of quasi-synchronous OFDM-BPSK-FHMA system, we propose to apply NHZ code, which is presented in Ref. [2]. In the simulation results, the BER performance of NHZ code system is compared with the Reed-Solomon (RS) code system. The BER curve of RS code system is shown by solid line in Fig. 4 and Fig. 5.

Fig. 4 presents the plots of BER versus the signal-to-noise rate (E_b/N_0) in AWGN channel. The BER of single-user system is the low-bounds of the system. When the delay is restricted within no-hit-zone ($\tau_k \leq Z_{NHZ}T$), the BER of the NHZ code system is lower than that of the RS code system. This is because that, the Hamming cross-correlation properties of NHZ code are ideal in this case, so that the MAI can be eliminated efficiently. Otherwise, the performance will get worse due to the fact that the frequency hit will appear when $\tau_k > Z_{NHZ}T$. Under the desired level of BER and the same number of the FH slots, the band-width of the system with NHZ code is saved in comparison with the system with RS code.

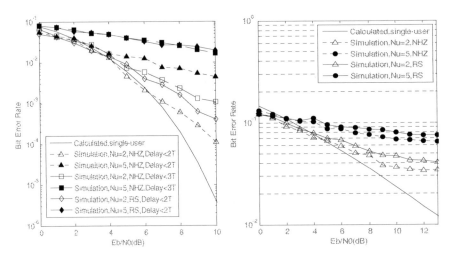

Fig. 4 BER of the OFDM-BPSK-FHMA system in AWGN channel

Fig. 5 BER of the OFDM-BPSK-FHMA system in fading channel

Fig. 5 presents the plots of BER versus E_b/N_0 in Rayleigh fading channel. The delay is restricted within no-hit-zone ($\tau_k \leq Z_{NH}T$). In Fig.5, compared with RS code, the NHZ code system performs significantly better. For example, at the same BER level and $N_u=5$, the E_b/N_0 of the NHZ code system has approximately 3dB gains (namely the difference of E_b/N_0 between the 3rd curve and the 5th curve in legend). We expect that the BER of the multiuser system with NHZ code approaches to the single user system. However this ideal performance isn't achieved in generally because each interference user's delay is a random value, which it will lead to the result of integral is not equal to zero (please see Eq.(4)). In particular, when the delay is equal to nT, the BER of NHZ code system is close to single user system and in this case the theoretical BER results is obtained as Eq.(10) and Eq.(17), where $N_u=1$.

5 Conclusions

In this paper, the BER performance of quasi-synchronous OFDM-BPSK-FHMA system is first analyzed in AWGN and Rayleigh fading channels. According to the characteristics of NHZ hopping codes, the performance of proposed system with NHZ code notably perform better than that of system with other kinds of hopping codes such as RS code and random codes, when the delay is restricted within no-hit-zone. Meanwhile the NHZ code system band-width is saved efficiently. Thus, it is more significant to apply the NHZ code in the quasi-synchronous FH systems. Especially, utilizing the diversity receiver in Rayleigh fading channel, the performance will be improved by the increase of the number of subchannel.

Acknowledgments. This work is supported by the National Natural Science Foundation of China (No. 60572142) and the Application and Basic Research Foundation of Sichuan Province, China (No. 2006J13-112).

References

1. Ding, C., Moisio, M.J., Yuan, J.: Algebraic constructions of optimal frequency-hopping sequences. IEEE Trans. Inf. Theory 53, 2606–2610 (2007)
2. Ye, W., Pingzhi, F., Gabidulin, E.M.: Consrtuction of non-repeating frequency-hopping sequences with no-hit zone. Electronics letters (2006)
3. Lifang, F., Xiaoning, W., Wenxia, Y., et al.: Quasi-synchronous scheme for frequency hopping system based on no-hit-zone hopping code. Journal of Southwest of Jiaotong University China 39, 776–779 (2004)
4. Al-Dweik, A., Xiong, F.: Frequency-hopped multiplex-access communication with noncoherent M-ary OFDM-ASK. IEEE Trans. Commun. 51, 33–36 (2003)
5. Li, T.T., Ling, Q., Ren, J.: A spectrally efficient frequency hopping system. In: IEEE GLOBECOM 2007 proceedings, pp. 2997–3001 (2007)
6. Qi, Z., Daiyuan, P., Hongbin, L.: Performance analysis of frequency-hopping multiple-access communications with OFDM-BPSK scheme. In: Third International Conference on Communications and Networking in China (IEEE ChinaCom 2008), pp. 25–27 (2008)

7. Daiyuan, P., Pingzhi, F., Lee, M.H.: Lower bounds on the periodic Hamming correlations of frequency hopping sequences with low hit zone. Science in China 49, 1–11 (2006)
8. Berens, F., Rüegg, A., Scholand, T., et al.: Fast frequency hopping diversity scheme for OFDM-based UWB systems. Electronic Letters 43(1) (2007)
9. Wenhua, M., Subo, W.: Frequency Hopping Communications. Press of Defense Industry, China Bei Jing (2005)
10. Molisch, A.F.: Wireless Communications. Wiley, New York (2005)

Efficient DAG Scheduling with Resource-Aware Clustering for Heterogeneous Systems

Behrouz Jedari and Mahdi Dehghan

Abstract. Task scheduling on *Heterogeneous Distributed Computing Systems* (HeDCSs) with the purpose of efficiency and reduction of execution time is of paramount importance. In this paper a novel task scheduling algorithm, called *Resource-Aware Clustering* (RAC) for *Directed Acyclic Graphs* (DAGs) is proposed. The objective of this algorithm is to keep the relative load balancing and efficiency increase between processors with different processor capabilities. To aim this fact, RAC by giving a dynamic score function to each task, performs task clustering and allocation according to processing capability of cooperative processors. In execution phase, *Modified Bottom-Level* (MBL) quantity of each task with a complexity of $O(v+e)$ would be calculated and applied for tasks priority identification. Comparing simulation results of RAC algorithm with famous scheduling approaches such as MCP, MD and DSC shows that RAC substantially increases efficiency while in most cases reduces completion time.

1 Introduction

The Heterogeneous Distributed Computing System (HeDCS) is composed of a group of systems having different resources or processors connected via a high speed network. To execute programs in environments that processors have different capabilities, application will be partitioned by a code partitioning algorithm into tasks with different instructions, so among the tasks, there will be minimum dependency in the most optimal state. Abstract model of partitioned application is usually indicated with a *Directed Acyclic Graph* (DAG) that optimal scheduling of the DAG in multiprocessor systems is known as NP-complete [1]. It becomes more complex on heterogeneous computing systems.

An optimal scheduling algorithm should consist of specific characteristics such as short response time, high efficiency and acceptable search speed during scheduling process. So far, several Static algorithms for graph scheduling are presented.

Behrouz Jedari
Islamic Azad University of Tabriz, Young Researchers Club of Tabriz, Iran
e-mail: behrouz_jedari@yahoo.com

Mahdi Dehghan
Computer Engineering Department, Amirkabir University of Technology, Tehran, Iran
e-mail: dehghan@aut.ac.ir

R. Lee, G. Hu, H. Miao (Eds.): Computer and Information Science 2009, SCI 208, pp. 249–261.
springerlink.com   © Springer-Verlag Berlin Heidelberg 2009

These algorithms can be broadly classified into two main categories: heuristic algorithms and guided random algorithms [2]. The heuristic algorithms in each step of their operation prevent unknown errors and converge purposely toward optimal response of problem. On the contrary, algorithms based on evolutionary computing as genetic algorithms [3] will be efficient when, by choosing suitable policy, they converge quickly to sub-optimal schedules.

Heuristic algorithms for DAG scheduling consist of list-based algorithms, clustering algorithms, and duplication-based algorithms. In the list scheduling algorithms, a well arranged list of tasks establish with priority allocation to each task and the task with high priority will be allocated to idle processor. HLFET [4], MCP [5], DSC [6] and MD algorithm [7] are examples of above mentioned algorithms and some of them will be compared and examined in this paper. Clustering algorithms try to schedule the tasks which communicate heavily onto the same processors, even if other processors are idle. There are also known three phase scheduling as: Grouping or clustering, Mapping and de-clustering. TDS [7] and CHP [8] are some well known task scheduling algorithms based on clustering approach. In the duplication algorithms, tasks are duplicated on more than one processor to reduce the waiting time of the dependent tasks so that it can be executed in one of the processors promptly. In these types of algorithms, choosing an appropriate processor for performing task and informing other processor about task performance enhances the communication cost and network traffic. CPFD [9] and HCNF [10] are two well known algorithms proposed in the literature using task duplication.

In this paper, overall framework of task scheduling on HeDCSs is presented and a cluster-based algorithm called Resource-Aware Clustering (RAC) for DAG scheduling on HeDCSs is proposed.

The objective of our proposed algorithm is to:

- Increase the efficiency of processors with preserving the completion time.
- Load balancing with awareness of available resources.

The rest of the paper is organized as follows. In section 2, three most well-known scheduling algorithms are introduced. In section 3, Framework of scheduling in heterogeneous computing systems is presented. In section 4, we present RAC scheduling algorithm. We present experimental evaluation of our proposed algorithm in section 5. Section 6 concludes the paper.

2 Related Work

In recent years, numerous static scheduling algorithms for different environments are proposed. The most important features of these approaches are allocating tasks to processors at the start of scheduling operation which decreases network traffic. In overall examination of proposed algorithm in this domain, list-based methods have less time-complexity compared to others, and are taken into consideration more. The major idea of these methods is based on establishing a sorted list of tasks with allocation of priority to each of the tasks and then continuous execution of selected tasks from list in idle processors. However, the most significant difference of these algorithms is in assigning the task execution priority. Likewise, the most important shortcoming of list-based algorithms in scheduling is the lack of their capabilities in exact priority allocation of static tasks according to importance and relative priority. In the following, three reported scheduling algorithms MCP, MD and DSC are illustrated.

The Modified Critical Path (MCP) algorithm [5] is designed based on an attribute called the latest possible start time of a node. The node's latest possible start time is determined through the As-Late-As-Possible (ALAP) binding. This is done by traversing the task graph upward from the exit nodes to the entry nodes and by pulling the nodes start times downwards as much as possible. The latest possible start time of the node itself is followed by a decreasing order of the latest possible start time of its successor nodes. The MCP algorithm then constructs a list of nodes in an increasing lexicographical order of the latest possible start times' lists. At each scheduling step the first node is removed from the list and scheduled to a processor that allows for the earliest start time.

The Mobility Directed (MD) algorithm [5] selects a node at each step for scheduling based on an attribute called relative mobility. Mobility of the node is defined as the difference between the node's earliest start time and latest start time. Similar to the ALAP binding mentioned in MCP algorithm, the earliest possible start time is assigned to each node through the As-Soon-As-Possible (ASAP) binding, which is done by traversing the task graph downward from the entry nodes to the exit nodes and by pulling the nodes upward as much as possible. The MD schedules the node with the smallest mobility to the first processor which has a large enough time slot to accommodate the node without considering the minimization of the node's start time. After the node is scheduled, all the relative mobility is updated.

The Dominant Sequence Clustering (DSC) algorithm [6] is based on an attribute called the dominant sequence, which is essentially the critical path of the partially scheduled task graph at each step. At each step, the DSC algorithm checks whether the highest Critical Path (CP) node is a ready node. If so, the DSC algorithm schedules it to a processor that allows the minimum start time. On the other hand, if not, the DSC algorithm does not select it for scheduling. Instead, the DSC algorithm selects the highest node that lies on a path reaching the CP for scheduling. The DSC algorithm schedules it to the processor that allows the minimum start time of the node, provided that such a processor selection will not delay the start time of an unscheduled CP node.

3 Task Scheduling on HeDCSs

To represent efficient algorithms on heterogeneous environments, determining overall framework of these environments is of great importance. To facilitate and utilize all available resources on distributed environments, establishing middleware technology between operating system and application is essential. Figure 1 shows overall framework of the scheduling on Heterogeneous Computing Systems (HCSs). In the following, formal definitions of task scheduling problem are presented.

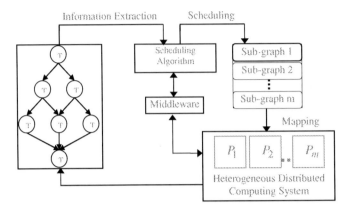

Fig. 1 Framework of Task Scheduling on Heterogeneous Distributed Computing Systems

3.1 Problem Definition

DAG scheduling problem consists of allocating T task to P processor where $T>>P$. In general, the distributed computing model can be represented as a DAG, $G=(T,<.E)$. to which T is a suite of tasks in a way that each of which is constituted from a number of successive instructions and E includes a set of communication edges that $\tau_{(i,k)}$ is Time taken by data transfer from T_i to T_k . Also $<$ shows processing priorities, where $\tau_{(i,k)}$ represents the partial order between T_i and T_k nodes, defining the node T_i as a *parent node* and the node T_k as a *child node*. Partial order dictates that task T_k cannot be executed unless all its parents have been executed. Note that the term node denotes task, and the cost is similar to time. If two tasks are scheduled to the same processor. the communication cost between them is assumed to be negligible. Each task T_i is assumed to have an estimated execution time $P(T_i)$. A node without parent node is *entry node* and a node without child node is *exit node.* DAG usually started with a node and ends with one node. A path from T_i to T_k is defined as including all nodes and all edges between T_i and T_k. The length of a path is the sum of all the processing cost of nodes and edge weights along the path. The longest path in the DAG is defined as Critical Path (CP). pred(T_i) is the set of immediate predecessors of T_i and succ(T_i) is the set of immediate successors of T_i. The problem with optimal scheduling of a DAG is the assignment of the tasks to processors in such a way that precedence relations are maintained as the following equation:

$$t_k - p(T_k) - \tau_{(i,k)} \geq t_i , \ T_i > T_k , \tag{1}$$

$$\forall i, k \ , t_i \geq 0 \ \forall i$$

Where t_i is equal to completion time of T_i. Also in optimal way all tasks are finished in the shortest possible time as presented in the equation (2):

$$\min f = \max_{i} = \{t_i\} \qquad (2)$$

In this equation, f is Finish Time (FT) or Makespan. Top-Level or TL(T_i) is equal the longest path of entry node to T_i and Bottom-Level or BL(T_i) is equal the longest path of T_i to exit node. It should be noted that processing time of T_i is added in BL(T_i) calculation, where in TL(T_i) processing time of T_i is not added to this value. To compute *Critical Path* (CP), equation (3) is also used as:

$$CP = BL(T_i) + TL(T_i) \text{ with max length for all } i \qquad (3)$$

With regard to explicit illustration of other parameters, TL and BL of a task are considered as significant parameters that are exploited for determining task execution priority of DAG scheduling using proposed algorithm with some alternations in this paper. The main objective of current scheduling algorithms is to minimize the scheduling length (makespan), while satisfying the tasks' dependencies.

4 RAC Algorithm

RAC algorithm like MCP can be classified in scheduling algorithm for bounded number of processors. Most of the existing algorithms don't consider the processors capabilities for optimizing load balancing and increasing efficiency. The most important property of RAC algorithm is to distribute load according to processing power of processors, and also to determine priority of tasks, considering Modified B-Level (MBL) parameter. RAC performs scheduling of DAG such as clustering, task allocation and determining execution priority between tasks in 2 steps. In first stage, after adapting the capacities of each cluster with DAG processing capacity, clustering operation is done. Each cluster includes a coordinator as a node with maximum score which is selected at the beginning of the clustering process and is changed dynamically during task allocation. The operation of the task membership is also continued in its cluster by determining clustering direction in DAG. For assigning priority in clustering, a sort of scoring policy is used for every task and these scores or task allocation priority vary, while during clustering process. At the end of the clustering, calculated MBL for each task are utilized independently. The independence among processors for assigning task execution priority reduces communication cost during execution. In the following, clustering and DAG scheduling are stated with RAC algorithm.

4.1 Clustering and Task Allocation

In first stage, the DAG provides DAG clustering and task allocation with awareness of processing power of cooperative processors. It is assumed that middleware

Procedure RAC (DAG.Resources)
1:    Initialize capacity of all clusters (C) to zero;
2:    For each in processor set P **DO**
3:       **EndIf**
4:       Select P_m with minimum $Power_m$;
5:       For each C_e **DO**
6:           If C_e before allocated **Then**
7:                continue;
8:           **EndIf**
9:       Set Coordinator of the C_e to node with max {Score (T_i)}
10:        $C_e = C_e + P(T_{coor})$ //Add processig time of Coordinator
11:       **While** $C_e < P_m$ **DO**
12:            For each $succ(T_{coor})$ and $pred(T_{coor})$ **DO**
13:                Find neighbor node (NN) with maximum score:
                                $(\tau_{(succ(T_i),i)}/P(T_{NN}))$
14:                If $Ce <= P_m$ Then
15:                    Set Coordinator of the Cj to NN
16:                    $C_e = C_e + P(T_{coor})$
17:                **EndIf**
18:                For each $succ(T_{coor})$ and $pred(T_{coor})$ **DO**
19:                    Reduce Score of node according to equation 6
20:                **EndFor**
21:            **EndFor**
22:       **End while**
23:       Allocate C_e to P_m
24:    **EndFor**
End Procedure

Fig. 2 RAC Algorithm for Task Scheduling

of distributed environment informs a-priori free resources in computational environment. So the owner machine of DAG is able to perform clustering operation
fairly. In this algorithm, processing power of the processor m has been shown by
$Power_m$ where m is an index of mth processor. Before starting clustering process,
clusters the same as the numbers of existing processors, are initialized in a way
that the capacity of each cluster is proportional to processing power of processors
and processing value of each DAG. The capacity of each cluster is calculated as
equation (4):

$$Ce = \frac{Power_m}{\sum_{k=1}^{P} Power_k} \times \sum_{i=1}^{T} P(T_i) \qquad (4)$$

Where C_e is the maximum processing load can be allocated to cluster e, $P(T_i)$ Required time to process T_i and $Power_m$ is Processing Power of P_m. For determining

each cluster coordinator, task with maximum $Score(T_j)$ is selected according to moving direction, where scoring function are defined in equation (5):

$$Score\ (T_j) = \begin{cases} \dfrac{\sum \tau_{(pre(T_j),i)}}{P(Ti)} & \text{Upward} & \text{(a)} \\ \\ \dfrac{\sum \tau_{(i,succ(T_j))}}{P(T_j)} & \text{Downward} & \text{(b)} \end{cases} \qquad (5)$$

When clustering direction is upward, the ratio of communication cost of entry node to the processing time of node 5(a) is calculated and in case it is downward, equation 5(b) is used. After assigning cluster centroid, the membership operation of neighboring node is performed according to the clustering direction and the maximum scores of neighboring node with cluster centroid. The score of centroid and selected node is also reduced in accordance with communication edge between them as equation (6):

$$Score(T_j) = Score(T_j) - (\tau_{(i,succ}(T_j)/P(T_j)) \qquad (6)$$

In continue, latest allocated node ($succ(T_j)$ or $pred(T_j)$) as a cluster centroid continues clustering process. Clustering operation continues as long as the processing capacity of cluster during membership is not more than the processing capacity of the smallest cooperative resource. This process carries on until a cluster with the highest processing capacity is allocated to a processor with the lowest processing resource and it will continue so that each of clusters is allocated to the machines with suitable processing capabilities. RAC algorithm is presented in Figure 2.

4.2 Scheduling and Task Execution

After clustering and allocating each group of tasks to the related processor, in this stage, information of DAG space has been computed for each task so that the information are transmitted to a processor along with per task cluster.

MTL and MBL are of those essential parameters in processing priority of each task that is equal the longest path of per task to the beginning and the end of the graph. These values are computed by eliminating communication cost inside per sub-graph or interconnected tasks that are allocated to the common processor. MBL computational pseudo-code is indicated in Figure 3. MTL computation resembles MBL algorithm except that the longest path of the DAG entry node to node i is calculated with omission of communication cost in which includes $O(v+e)$ time complexity. After computing these values, a task with the most MBL amount or the longest path to exit node will have higher priority for execution and will be selected for execution too.

```
Procedure MBL(DAG)
1:  For all nodes as T_j in DAG Do
2:      If (Succ (T_j)) Then
3:          For each Succ(T_j) Do
4:              If ( ClusterID( T_j ) =ClusterID(Succ(T_j)) ) Then
5:                  level(T_j) = MBL (Succ(T_j)) + P(T_j)
6:              Else
7:                  level(T_j) = MBL(Succ(T_j)) +P(T_j) + τ(i.succ(Ti))
8:              EndIf
9:              If (mbl(T_j)< level(T_j)) Then
10:                 mbl(T_j)= level(T_j))
11:             EndIf
12:         EndFor
13:     Else
14:         mbl(T_j)= P(T_j)
15:     EndIf
End Procedure
```

Fig. 3 MBL Algorithm for each task.

5 Comparison of Results

In this section, evaluation results of RAC algorithm as compared with three current list-based algorithms MCP, DSC and MD are presented. Two DAG [11] with regard to a number of different tasks and specific communication properties in Figures 4 and 6 are chosen for efficient evaluation of RAC algorithm.

Fig. 4 Example DAG 1 with 9 tasks [11]

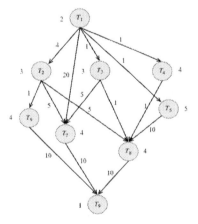

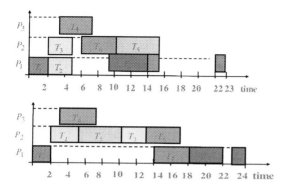

Fig. 5 Gantt chart for DAG 1 generated by RAC algorithm and 3 processors with RAC-1 resources

Fig. 6 Example DAG 2 with 18 tasks [11]

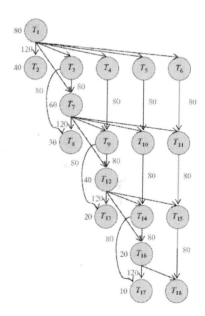

5.1 Comparison Metrics

The comparison of different algorithms is based on the following three metrics:

Finish Time: The scheduling length is called makespan; the shorter the makespan, the better.

Speedup: Speedup is the ratio of the sequential execution time to the parallel execution time.

Efficiency: Efficiency is the ratio of the speedup to the number of processor used to scheduling.

5.2 Simulation Results

In the first experiment for scheduling of DAG 1 two sets of processors with homogeneous (RAC-I) and heterogeneous (RAC-II) processing powers are utilized. In RAC-I, processors with homogeneous processing capability and in RAC-II, processors with processing capability in proportion to 6, 10 and 12 are used.

Table 1 Comparing RAC algorithm results, for example DAG 1 with 3 current scheduling algorithms

Algorithm	MCP	MD	DSC	RAC-I	RAC-II
No. processor	3	2	4	3	3
Finish Time	29	32	27	23	24
Speedup	1.03	0.93	1.11	1.3	1.25
Efficiency	34.4	46.9	27.8	43	41.6

Table 2 Calculating MTL, MBL and P_{id} for all tasks in DAG 2 with RAC-II resources

$Node_{id}$	$P(T_i)$	MTL	MBL	P_{id}
1	80	0	500	1
2	40	80	40	1
3	40	80	420	1
4	40	80	360	1
5	40	80	320	1
6	40	80	100	1
7	60	200	300	2
8	30	380	30	1
9	30	260	240	2
10	30	260	200	2
11	30	380	60	1
12	40	290	210	2
13	20	330	20	2
14	20	330	170	2
15	20	450	30	1
16	20	350	150	2
17	10	370	10	2
18	10	490	10	1

At the outset of scheduling operation and for processors with the same processing capabilities, processing capability of the 3 clusters are computed using equation (4) and regarding total sum of DAG process time is 30, and this value is equal 12. At first, T_9 nodes as a cluster coordinator with score 30 starts clustering operation by assigning clustering direction to the upward of graph. This operation will continue as long as cluster capacity is equal or less than 12. This process is performed for all clusters and in every step; the score of each task is reduced as the ratio of communication costs between tasks.

With regard to the results of Table 1, Finish Time and processor efficiency by utilizing RAC algorithm and same resources (RAC-I) are dramatically reduced as compared with other list scheduling algorithm such as MCP, DSC and MD. With existence of heterogeneous resources of processors in RAC-II as well, the results of proposed scheduling are resistance, despite many bounded restrictions, and are comparable with the clustering results.

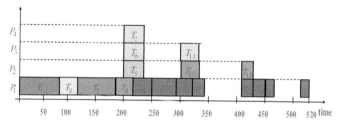

(a).Gantt chart: The Schedule Generated by MCP algorithm.

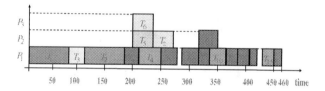

(b).Gantt chart: The Schedule Generated by MD algorithm.

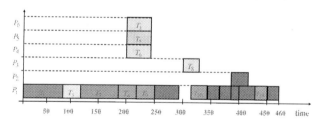

(c).Gantt chart: The Schedule Generated by DSC algorithm.

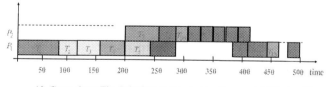

(d).Gantt chart: The Schedule Generated by Proposed RAC algorithm.

Fig. 7 Gantt chart for checking the scheduling results for example DAG 2. Schedules generated with (a) MCP algorithm, (b) MD algorithm, (c) DSC algorithm and (d) proposed RAC algorithm

In next experiment, Example DAG 2 is used for scheduling and comparison of RAC algorithm with the other algorithms. This DAG is selected by the purpose of evaluation and measuring given algorithm with the other algorithms for application with high capacity. For scheduling of the graph by means of RAC algorithm, two processors in proportion to 200 and 400 processing capabilities are chosen.

The results of DAG 2 clustering operation for a system with heterogeneous processing resources are shown in Table 2. In this tables, $P(T_i)$ is processing MTL and MBL columns are computed for each task (T_i) after clustering operation and at last P_{id} shows the processor id that their related work has allocated. The values of MTL and MBL for each task are calculated by eliminating communication cost between nearby tasks inter-cluster. Figure 7 is also indicates comparison of RAC scheduling Gantt chart with 3 algorithms MCP, MD and DSC for DAG 2.

Acquired results display that the algorithm has preserved finish time, despite restrictions in resources and the number of processors, while efficiency of processors have increasingly improved. Figure 8 shows the average efficiency of processors in different time intervals for DAG 2.

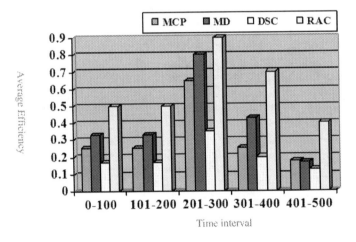

Fig. 8 Efficiency comparison of RAC with 3 current algorithms for DAG 2

6 Conclusion

In this paper, a Resource-Aware Clustering (RAC) algorithm has exploited for the task graph scheduling. Most of the existing algorithms don't take into account the processors capabilities for optimizing load balancing and increasing efficiency. The most important property of RAC algorithm is to distribute load according to processing power of processors, and also to schedule priority of tasks, considering Modified Bottom Level (MBL) parameter. Comparing simulation results of RAC algorithm with well-known scheduling approaches such as MCP, MD and DSC shows that RAC substantially it increases speedup and efficiency while in most cases reduces finish time. Optimal selection of the available processors for increasing efficiency is of great importance. So, this problem has been less taken into consideration. In future work this problem will be put under close study.

References

1. Graham, R.L., Lawler, E.L., Lenstra, J.K., Rinnooy Kan, A.H.G.: Optimization and Approximation in Deterministic Sequencing and Scheduling: a Survey. Annual Discrete Mathematics 5, 287–326 (1979)
2. Topcuoglu, H., Hariri, S., Wu, M.Y.: Performance- Effective and Low-Complexity Task Scheduling for Heterogeneous Computing. IEEE Transactions Parallel and Distributed Systems 13(3), 260–274 (2002)
3. Davidovic, T., Crainic, T.G.: Benchmark-Problem Instances for Static Scheduling of Task Graphs with Communication Delays on Homogeneous Multiprocessor Systems. C.R.T.'s Publications (2004)
4. Adam, T.L., Chandy, K.M., Dicksoni, J.R.: A Comparison of List Schedules for Parallel Processing Systems. Communications of the ACM, 685–690 (1974)
5. Wu, M.Y., Gajski, D.D.: Hypertool: a Programming Aid for Message-Passing Systems. IEEE Transactions on Parallel and Distributed Systems 1(3), 330–343 (1990)
6. Yang, T., Gerasoulis, A.: DSC: Scheduling Parallel Tasks on an Unbounded Number of Processors. IEEE Transactions on Parallel and Distributed Systems 5(9), 951–967 (1994)
7. Boeres, C., Viterbo Filho, J., Rebello, E.F.: A Cluster-Based Strategy for Scheduling Task on Heterogeneous Processors. In: Proceedings of 16th Symposium on Computer Architecture and High Performance Computing, SBAC-PAD (2004)
8. Haupt, R.L., Haupt, S.E.: Parallel genetic algorithms. John Wiley & Sons, Chichester (2004)
9. Ahmad, I., Kwok, Y.K.: On Exploiting Task Duplication in Parallel Program Scheduling. IEEE Transactions on Parallel and Distributed Systems 9, 872–892 (1998)
10. Bajaj, R., Agrawal, D.P.: Improving Scheduling of Tasks in a Heterogeneous Environment. IEEE Transactions on Parallel and Distributed Systems 15, 107–118 (2004)
11. Hwang, R., Gen, M., Katayama, H.: A Comparison of Multiprocessor Task Scheduling Algorithms with Communication Costs. Computers & Operations Research 35, 976–993 (2008)

Research on Multilingual Indexing and Query Processing in Uyghur, Kazak, and Kyrgyz Multilingual Information Retrieval System

Dilmurat Tursun, Turdi Tohti, and Askar Hamdulla

Abstract. The Uyghur, Kazak, and Kyrgyz languages no language ID and the some letters in this languages are sharing code points in Unicode area, so it is difficult to distinguish between Uyghur, Kazak, and Kyrgyz letters in information exchange, automatic word segmentation and retrieval applications, existing linguistic ambiguity. In addition, in the region in alphabetical order with the Arabic alphabet, Uyghur, Kazak, and Kyrgyz letter is the order of chaos, this will led to great difficulties for Uyghur, Kazak, and Kyrgyz multilingual data indexing, query processing and sorting process. In this paper, studied and proposed the most effective solutions and ideas for above actual problems: in view of the problem of linguistic ambiguity, proposed a Relocated Unicode Format (short for RuniForm) Encoding Method; For multilingual indexing, proposed a multilingual indexing technology based on MD5 encryption and related query processing approach in Uyghur, Kazak, and Kyrgyz information retrieval system (UKKIRS). The experimental results indicated that, the proposed algorithms solved well the problems mentioned above, and are very dedicated to this UKKIRS.

1 Introduction

International Unicode association released the Unicode 4.0 edition in august 27, 2003, and assigned corresponding Unicode code points for all Uyghur, Kazak, and Kyrgyz characters, make up a blank on Uyghur, Kazak, and Kyrgyz information processing. (Before this they share code points with other languages or coding by themselves).It brings enormous breakthrough to the computer processing of Uyghur, Kazak, and Kyrgyz information [1]. Uyghur, Kazak, and Kyrgyz belong to

Dilmurat Tursun, Turdi Tohti, and Askar Hamdulla
School of Information Science and Engineering, Xinjiang University
Urumqi, Xinjiang, China, 830046

Askar Hamdulla
Corresponding author: askar@zju.edu.cn

R. Lee, G. Hu, H. Miao (Eds.): Computer and Information Science 2009, SCI 208, pp. 263–271.
springerlink.com   © Springer-Verlag Berlin Heidelberg 2009

the identical language family (Turkish Language group of west Huns of Altaic Language family). The alphabet (majority of the letters are the same) and the grammar structure (is one stickiness language) are very similar. So in the Unicode, Uyghur, Kazak, and Kyrgyz letters are arranged in the Arabic area (0600~06FF).In this region Uyghur, Kazak, and Kyrgyz share some letters, moreover they does not have the language marking information. Therefore, it is very difficult to distinguish Uyghur, Kazak, and Kyrgyz letters in the exchange of information or the automatic word segmentation and the recognition application. Moreover, in this region, the letter order conforms to the Arabic alphabet, the order of Uyghur, Kazak, and Kyrgyz letter is chaotic. And the majority database systems do not support the user defined character collating sequence.

In this case, the traditional method is manual classification saving, management and indexing of Uyghur, Kazak, and Kyrgyz information resource (increases language ID) [1]. The realizing of this method is not conveniently and the efficiency is very low. Therefore, we study one kind of multilingual data interior storage format which can be facilitate transformed to Unicode and propose one multilingual indexing technology based on MD5 encryption and related query processing approaches based on root [3].

The rest of the paper is structured as follows. Section 2 gives a brief Description about Uyghur, Kazak, and Kyrgyz language features, Unicode and the main reasons for the above problems caused by, also introduced the most effective solutions: RuniForm Encoding Method. Section 3 details Uyghur, Kazak, and Kyrgyz multilingual indexing approach, problems and gives solutions: multilingual indexing approach based on RuniForm Encoding Method and MD5 encryption. In section 4, analyzed the problems on query processing in UKKIRS and then has given the unique processing method and experimental results. Section 5 concludes the paper.

2 Uyghur, Kazak, and Kyrgyz Language Characteristics and UNICODE

Uighur, Kazak and Kyrgyz language belongs to Turkish Language group of west Huns of Altaic Language family, and also belongs to the agglutinating language on structure grammar. The Uyghur alphabet consists of 32 letters, the Kazak alphabet consists of 33 letters, and the Kyrgyz alphabet consists of 30 letters, and the majority of letters in their alphabet are completely identical. Moreover they are also similar in the grammar structure, very closely to Arabic letters. Because of the above reasons, the international Unicode association has appointed the Unicode code-point for Uyghur, Kazak, and Kyrgyz letter in the Arabic Unicode area to differentiate with other language character code completely. Uyghur (Uy), Kazak (Ka), and Kyrgyz (Ky) alphabets and corresponding Unicode codes as shown in Table 1.

Table 1 Uyghur, Kazak, and Kyrgyz alphabets and Unicode

No.	Uy	code	Ka	code	Ky	code
1	ا	0627	ا	0627	ا	0627
2	ە	06D5	ٵ	0675	ب	0628
3	ب	0628	ب	0628	پ	067E
4	پ	067E	ۆ	06C6	ت	0646
5	ت	062A	گ	06AF	ت	062A
6	ج	062C	ع	0639	ج	062C
7	چ	0686	د	062F	چ	0686
8	خ	062E	ە	06D5	ح	062D
9	د	062F	ج	062C	ف	0641
10	ر	0631	ز	0632	ق	0642
11	ز	0632	ي	064A	ع	0639
12	ژ	0698	ك	0643	ك	0643
13	س	0633	ق	0642	گ	06AF
14	ش	0634	ل	0644	ڭ	06AD
15	غ	063A	م	0645	ل	0644
16	ف	0641	ن	0646	م	0645
17	ق	0642	ڭ	06AD	و	0648
18	ك	0643	و	0648	ۅ	06C5
19	گ	06AF	ۇ	0676	ۇ	06C7
20	ڭ	06AD	پ	067E	ۉ	06C9
21	ل	0644	ر	0631	ۋ	06CB
22	م	0645	س	0633	س	0633
23	ن	0646	ت	062A	ش	0634
24	ھ	06BE	ۋ	06CB	د	062F
25	و	0648	ۆ	06C7	ر	0631
26	ۇ	06C7	ۇ	0677	ز	0632
27	ۆ	06C6	ف	0641	ە	06D5
28	ۈ	06C8	ح	062D	ى	0649
29	ۉ	06CB	ھ	06BE	ى	0626
30	ى	06D0	چ	0686	ي	064A
31	ی	0649	ش	0634		
32	ي	064A	ى	0649		
33			ی	0678		
	ء	0626	ء	0621	ء	0626

Although Unicode has solved the chaotic coding problem in Uyghur, Kazak, and Kyrgyz characters, there are new questions caused by coding for us to solve. The questions are as follows:

1. Majority of Uyghur, Kazak, and Kyrgyz letters share the same UNICODE code-point (its Unicode is completely same), so unable to realize multilingual indexing because of the ambiguity in the language.
2. The distribution order of Unicode code points does not conform to the Uyghur, Kazak, and Kyrgyz alphabets order. Thus appears confusion during sorting and query processing in multilingual information.
3. The character attachment part "ء"is regarded as an independent character to appoint the Unicode code-point. Thus it causes the character operation mistake.

2.1 Uniforms Encoding Method

The above questions in Uyghur, Kazak, and Kyrgyz Unicode encoding method has brought major difficulties in Uyghur, Kazak, and Kyrgyz multilingual indexing, sorting and query processing, also decrease the efficiency. Therefore, we study one kind of multilingual data interior storage format which is facilitating to transform into Unicode and effectively solved a series of questions caused by character coding in multilingual indexing, multilingual sorting and the multilingual query processing.

Our method is to use Unicode coding schemes based on the relocated technology (Relocated Unicode Format, short for RuniForm). In this code scheme, Uyghur, Kazak, and Kyrgyz letters are arranged separately in three different regions (still in 0600~06FF) according to their respective alphabet order. They obtained the respective language marking information automatically. The experimental results indicated that, this method has effectively eliminated the ambiguity in the language and it is easy to establish multilingual index, to do multilingual sorting and the multilingual query processing. The interior Unicode code-point in Uyghur, Kazak, and Kyrgyz as shown in Table 2.

Table 2 RuniForm code point assignment for Uyghur, Kazak, and Kyrgyz

Uyghur letters(32)	Kazak letters(33)	Kyrgyz letters(30)	Other symbols
0600 (ﺍ) – 0631 (ﻯ)	0632 (ﺍ) – 0664 (ﻯ)	0665 (ﺍ) – 0694 (ﻯ)	0695 – –

Because of the relocated Unicode code point and the respectively language marking information (based on Unicode code-point region) for Uyghur, Kazak, and Kyrgyz, has completely solved the problems that effectively integration, sorting and query processing of Uyghur, Kazak, and Kyrgyz multilingual information resource.

3 Multilingual Indexing Based on RuniForm

In Uyghur, Kazak, and Kyrgyz multilingual search engine, we have to collect massive Uyghur, Kazak, and Kyrgyz web information and establish indexing list. We used the most popular technology at present that inverted index [4], and uses the Hash table as its data structure directly established in the server memory, in the node in Hash table saves the indexing information of each key word, including the key word, page ID that contains this key word, the starting position, appearing times, length and so on. In the process of establishing inverted index list, first we extract the root of every index item (words separated by blank space), then we obtain 128bit hash values from the MD5 encryption algorithm to determine corresponding Hash link and insert it to this link. So, the root of the entire index item (keywords) on the same Hash link is identical (no matter it is correctly spelt or miss-spelt). The organization structure of inverted index based on root is shown in Figure 1.

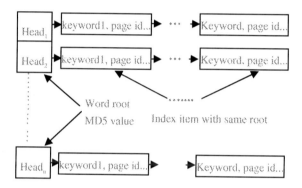

Fig. 1 The organization structure of inverted index based on root MD5 encryption

In normal condition, because of the ambiguity in the language (the expressions and meanings of many words in these three languages are identical), it is unable to establish an index list for these three languages. For instance, the expression of the word "schools "in Uyghur, Kazak, and Kyrgyz is completely same, namely" مەكتەپ"i.e. we are unable to distinguish whether it is Uyghur, Kazak, or Kyrgyz word. So, we need to establish three index lists separately for text materials of these three languages.

To solve the above problem, we encoding the Uyghur, Kazak, and Kyrgyz text information to RuniForm and then establish an index list to provide services for the three languages query. Because after the RuniForm encoding, the character codes of a single word in three languages are different and the hash value obtaining through the MD5 encryption are different, too. In this way, we could be identifying the three languages. The comparison relations of Unicode, RuniForm and MD5 value is shown in Table 3.

Besides, there is not need to indicate the query language in Web Server, and the search results could be shown in three languages on the same user interface. There is obviously improvement on the index establishment and query processing. Uyghur, Kazak, and Kyrgyz multi-language index structure Based on the RuniForm as shown in Figure 2.

Table 3 Comparison relations of Uyghur, Kazak, and Kyrgyz Unicode, RuniForm and D5 value

word (snow)	قار (Uyghur)	قار (Kazak)	قار (Kyrgyz)
Unicode	064206270631	064206270631	064206270631
MD5 value (128bit)	CA-8A-DC-67-A1-91-99-44-53-EB-05-CB-7A-34-35-F8	CA-8A-DC-67-A1-91-99-44-53-EB-05-CB-7A-34-35-F8	CA-8A-DC-67-A1-91-99-44-53-EB-05-CB-7A-34-35-F8
RuniForm	061606000609	064406320652	067406650689
MD5 value (128bit)	70-13-17-B6-7A-FB-BC-91-53-93-4E-DF-E0-B3-BD-C1	84-FB-34-02-7C-F5-53-4A-4F-B7-9E-A2-E9-BB-59-41	E8-45-DF-45-80-05-94-16-54-46-06-3F-BF-11-67-AA

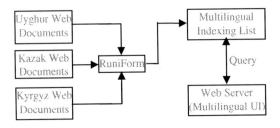

Fig. 2 Uyghur, Kazak, and Kyrgyz multilingual index based on RuniForm

4   Multilingual Query Processing

4.1   Preprocessing the Query Phases

The preprocessing of the query phrases is first step of query processing in this search engine and it is also a key step that immediately influences the query efficiency [5]. The Uyghur, Kazak, and Kyrgyz languages are all stickiness languages. In this kind of language, word is the smallest independent linguistic unit for utilization, and the query phrases the users submit are word sequences which separated by blank spaces. Therefore, effectively organizing these word sequences, deleting words or spatial characters without any query significance, correcting spelling mistakes and extracting roots and reconstructing query phrase based on original query words as well as the word roots are premises to realize effective query and the documents match .

4.1.1 Filtering Spaces and Stop Words

The query phrase correctly input should be the word sequences which separated by a single blank space (e.g. يەرشارى), but irregular input which contains several blank spaces between words in query phrases (e.g. يەر□□□□شارى). So, when doing word segmentation on query phrase, the insignificant blank spaces is also regarded as query words to be inquired and matched. It will seriously influence query efficiency as well as cause the wrong query results. In addition, there are a large number of stop words in Uyghur, Kazak, and Kyrgyz languages. If we don't effectively filter these words that without any retrieval values, the search engine is unable to guarantee it gives the correct related search results to help reducing the searching range, will be reduce the search efficiency. Therefore, we have counted and collected the majority of stop words in Uyghur, Kazak, and Kyrgyz languages. So, the system automatically filters the stop words, deletes unnecessary blank spaces and reorganizes the query word sequence before searching and matching.

4.1.2 Query Phrase Re-construction

Words in Uyghur, Kazak, and Kyrgyz are composed by root and additional parts (Suffixes).for example: the user input query phrase as "جۇڭگونىڭ نوپۇسى"(population of china). in this phrase.نوپۇس = ى + نوپۇس، جۇڭگونىڭ= نىڭ +جۇڭگو. We need to obtain the query word's root"جۇڭگو" (china) and"نوپۇس" (population) to submit to the indexer, and the indexer will query and matching according to the root, and the recall will be improved much more. Also in addition, we must consider that the WebPages which contains the original query words (جۇڭگونىڭ and نوپۇسى) should be presented at the front row when sorting the query results. So the query phrases which provide for the indexer should also contain the original query words.

But, if the original query words presents the spelling mistakes and this will be leads to the problem that are unable to extract the correct root[5]. For example, the correct spelling word" پولات " (steel), it is the root of its own, but wrongly spells as"پۇلات" generally, and the root segmentation model will be returns the"پۇل" (money) as its root. Therefore, the search results includes the information that related to the query word " پولات " . also includes the useless information that about"پۇل". reduced the accuracy of UKKIRS.

In order to solve the above problem, we counted and have collected about more than 10,000 words (general spelling error ratio above 95%) misspellings and the correct spellings, and have established the comparative list. during the process, carries on the spelling correction for each original query words (kw$_i$) in query phrase $q= (kw_1 kw_2..... kw_i)$ and extract its root (kr$_i$), then re-construction the query phrase as $q= (kw_1, kr_1 \ kw_2, kr_2 .... \ Kw_i, kr_i)$. For example. original query phrase from user is "جۇڭگونىڭ   ئىقتىسادى   ۋە   نوپۇسى", the query phrase is preprocessed as follows:

step1: filtering unnecessary blank spaces:"جۇڭگونىڭ ئىقتىسادى ۋە نوپۇسى"

step2: filtering stop words:" ۋە " : "جۇڭگونىڭ ئىقتىسادى نوپۇسى"

step3: reconstruct query phrase : "جۇڭگونىڭ،جۇڭگو   ئىقتىساد،ئىقتىسادى   نوپۇس،نوپۇسى"

Before searching, we make above processing on the original query phrase that user input, on one hand it reduces the search range and enhances the relevance, on the other hand it realizes query expansion based on the root and raises the recall. Query phrase preprocessing as shown in Figure 3.

Fig. 3 The algorithm flow of query phrase preprocessing

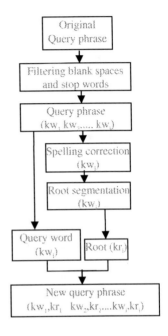

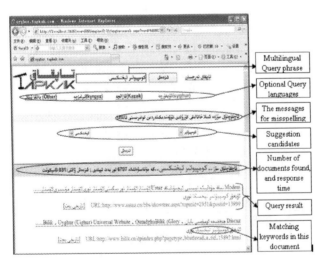

Fig. 4 The snapshot of retrieved information

After above treatment for query phrase, obtain the related query results immediately from multilingual indexing table, and showing the search results to users in special type. The search results as shown in Figure 4.

5 Conclusions

In this article, taking the Uyghur, Kazak, and Kyrgyz multilingual search engines for example introduced a multilingual indexing technology and multilingual query processing approach. In view of the effective integration of the non-language ID Uyghur, Kazak, and Kyrgyz multilingual text information, we propose one relocated Unicode Format (RuniForm) coding method and has solved the language recognition problem. Based on this, we propose a multilingual inverted index technology based on the MD5 encryption and the unique query processing approach. The experimental results indicate that, this method effectively solved the problem that storing Uyghur, Kazak, and Kyrgyz multilingual data information .At the same time, it simplifies the multilingual index structures greatly, raises the query efficiency obviously.

Acknowledgements. This work has been supported by the Project of Natural Science Fund of Xinjiang Uyghur autonomous region (No. 200612115), Scientific Research Program of the Higher Education Institution of Xinjiang (XJEDU2006113).

References

1. Tohti, T., Musajan, M., Hamdulla, A.: Character Code Conversion and Misspelled Word Processing in Uyghur, Kazak, Kyrgyz Multilingual Information Retrieval System [J]. In: 7th International Conference on Advanced Language Processing and web information technology (ALPIT 2008), Dalian, China, pp. 139–144 (2008)
2. Tohti, T., Musajan, M., Hamdulla, A.: Design the Uyghur, Kazak, Kyrgyz Full-text Search Engine Indexer and Its Implementation. Journal of Information (10), 49–51 (2008)
3. Yi, H.G., She, M.G.: MD5 Arithmetic and Digital Signature. Computer & Digital Engineering, China 34(5), 44–46 (2006)
4. Scholer, F., Williams, H.E., Yiannis, J., Zobel, J.: Compression of Inverted Indexes for Fast Query Evaluation. In: Proceedings of 25th ACM-SIGIR, Finland, pp. 222–229 (2002)
5. Tohti, T., Musajan, M., Hamdulla, A.: Research on Query Processing and Implementation in Uyghur, Kazak, and Kyrgyz Full-text Search Engine [J]. In: 4th National Conference on Information Retrieval and Content Security, Beijing, China, pp. 217–223 (2008)

A Proof System for Projection Temporal Logic*

Zhenhua Duan and Xinfeng Shu

Abstract. This paper presents a proof system for projection temporal logic (PTL) over finite domains. To this end, the syntax and semantics of PTL are briefly presented; a set of axioms and inference rules is formalized; also some theorems are summarized and proved. Further, an example is given to illustrate how the axioms and rules work.

1 Introduction

Projection Temporal Logic (PTL) [3, 4, 5] is an interval based first order temporal logic. It extends interval temporal logic (ITL) [11] by introducing a new projection construct, $(P_1, \ldots, P_m)\ prj\ Q$, and supporting both finite and infinite time. PTL is a useful formalism for specification and verification for concurrent and reactive systems [6, 13].

To verify the properties of concurrent systems, basically, two verification approaches, model checking [2] and theorem proving [1], are widely used in practice. Model checking is an automatic verification approach based on model theory. The advantage of model checking is that the verification can be done automatically. However, the checking is limited to finite state space, and suffers from the state explosion problem. Thus, it is less suitable for data intensive applications since the

Zhenhua Duan

Institute of Computing Theory and Technology, Xidian University, Xi'an 710071, P.R.China
e-mail: zhenhua_duan@126.com

Xinfeng Shu

Xi'an Institute of Posts and Telecommunications, Xi'an 710121, P.R.China
State Key Laboratory of Software Engineering, Wuhan University, Wuhan 430072, P.R.China
e-mail: shuxinfeng@gmail.com

* This research is supported by the NSFC Grant No. 60433010 and No. 60873018, DPRP No. 51315050105, SRFDP 200807010012, and SKLSE 20080713.

treatment of the data usually produces infinite state spaces [10]. With theorem proving, to verify whether or not a system S satisfies a property P is to prove whether or not $\vdash S \rightarrow P$ is a theorem within the proof system. The advantage is that theorem proving avoids the state explosion problem and can verify both finite and infinite systems, and the verification can be done semi-automatically. Therefore, it is suitable for data intensive applications. However, within the verification process, lots of assertions need to be inserted in the context of the program modeling the system, and the use of theorem prover requires considerable expertise to guide and assist the verification process. As a complement of model checking, the theorem proving approach is still useful in practice for verifying properties of programs. With PTL, a decision procedure for checking the satisfiability of formulas and a model checking approach based on SPIN have been given in [4, 6]. However, to reason about concurrent systems based on a deductive approach, a proof system of PTL is required. Therefore, in this paper, we are motivated to formalize a proof system for PTL.

In the past two decades, a number of axiomatizations for temporal logics have been proposed to verify properties of concurrent and reactive systems with success. Within ITL community, Dutertre [7] gives two complete proof systems for first order ITL with finite time on possible worlds and abstract interval semantics respectively. The restriction on his logic is that it contains no temporal operators other than chop and admits only constant domains and uses of rigid assignments to variables. Guelev [8] extends the Dutertre's work for abstract time ITL to infinite time and supporting more temporal operators as chop-star and projection. Moszkowski [12] presents a complete axiomatization of ITL with a fixed finite domain based on the work of Kesten and Pnueli [9]. The version of ITL contains no functions and predicates, which restricts the application of the proof system.

Similar to the axiomatic system for QPTL and ITL, we take a finite domain to achieve completeness. However, the domain can be any finite domains, e.g., integers, lists, sets, etc., and not be limited within just boolean domain or a subset of integers. Further, PTL allows temporal terms(terms with temporal operators), functions and predicates, which to the best of our knowledge is the first time for an axiomatization of a first order temporal logic equipped with such items.

The paper is organized as follows. The next section briefly presents the syntax and semantics of the underlying logic. Section 3 gives the axioms, inference rules of the proof system, and some useful theorems. In Section 4, an example is given to show how our proof system works. Finally, conclusions are drawn in Section 5.

2 Projection Temporal Logic

2.1 Syntax

Let *Prop* be a countable set of atomic propositions and V a countable set of typed variables. $B = \{true, false\}$ represents the boolean domain. D denotes the finite data domain of the underlying logic, which can be a finite subset of integers, a finite list, a finite set, etc. The terms e and formulas P of PTL are inductively defined as follows:

$$e ::= d \mid a \mid x \mid \bigcirc e \mid f(e_1, \ldots, e_m)$$
$$P ::= p \mid e_1 = e_2 \mid \rho(e_1, \ldots, e_m) \mid \neg P \mid P_1 \wedge P_2 \mid \exists v P \mid \bigcirc P \mid (P_1, \ldots, P_m) \, prj \, P$$

where $d \in D$ is a constant, $a \in V$ a static variable, $x \in V$ a dynamic variable, $v \in V$ either a static variable or a dynamic one; $p \in Prop$ is an atomic proposition; f is a function and ρ a predicate both defined over D.

Note that some primitive functions used in underlying logic are partial functions which may return an undefined value. To solve the problem, we extend the partial functions to total functions by explicitly introducing the undefined value nil. Thus, the total data domain is denoted by $D' = D \cup \{nil\}$. Further, we extend definition of all the primitive functions $f(e_1, \ldots, e_m)$ $(m \geq 1)$ from over D to over D' by letting the functions' value be nil whenever one of the parameters e_i $(1 \leq i \leq m)$ is assigned with nil. Moreover, the definitions of primitive predicates over D' are kept unchanged from over D.

The conventional constructs $true$, $false$, $\wedge$, $\rightarrow$ as well as $\leftrightarrow$ are defined as usual. Furthermore, we use the following abbreviations:

$$\varepsilon \overset{def}{=} \neg \bigcirc true \qquad\qquad \overline{\varepsilon} \overset{def}{=} \neg \varepsilon$$
$$\odot P \overset{def}{=} \neg \bigcirc \neg P \qquad\qquad P;Q \overset{def}{=} (P,Q) \, prj \, \varepsilon$$
$$\bigcirc^0 P \overset{def}{=} P \qquad\qquad \bigcirc^n P \overset{def}{=} \bigcirc \bigcirc^{n-1} P, \ (n > 0)$$
$$\Diamond P \overset{def}{=} true ; P \qquad\qquad len(n) \overset{def}{=} \bigcirc^n \varepsilon$$
$$\Box P \overset{def}{=} \neg \Diamond \neg P \qquad\qquad keep(P) \overset{def}{=} \Box(\overline{\varepsilon} \rightarrow P)$$
$$skip \overset{def}{=} \bigcirc \varepsilon \qquad\qquad halt(P) \overset{def}{=} \Box(\varepsilon \leftrightarrow P)$$
$$\forall v P \overset{def}{=} \neg \exists v \neg P \qquad\qquad fin(P) \overset{def}{=} \Box(\varepsilon \rightarrow P)$$
$$P \| Q \overset{def}{=} ((P;true) \wedge Q) \vee (P \wedge (Q;true)) \vee (P \wedge Q)$$

To avoid an excessive number of parentheses, the following precedence rules are used (1 =highest and 8 =lowest):

1. $\neg$	2. $\bigcirc, \odot, \Diamond, \Box$	3. $\exists, \forall$	4. $=$
5. $\wedge$	6. $\vee, \|$	7. $\rightarrow, \leftrightarrow$	8. prj , ;

A formula (term) is called *static* if it does not refer to any dynamic variables. A formula (term) is called a *state* formula (term) if it does not contain any temporal operators; otherwise it is called a *temporal* formula (term). For convenience, we let lowercase letters $p, q, \ldots$ with subscripts (e.g., p_s, q_i) denote any state formulas.

Let τ be a formula or term. If t is a term and v is a variable used in τ, then $\tau[t/v]$ denotes the result of simultaneously replacement of all free occurrences of v by t in τ. The replacement is called *compatible* if either both of v and t are static or v dynamic. The replacement $\tau[t/v]$ is called *admissible* for $\tau(x)$ if it is compatible and none of the variables appearing in t is quantified in τ. We also say that t is admissible for v in $\tau(v)$ and write $\tau(t)$ to denote $\tau[t/v]$.

2.2 Semantics

A state s is a pair of assignments (I_p, I_v), which I_p assigns each atomic proposition $p \in Prop$ a truth value in B, whereas I_v assigns each variable $v \in V$ a value in D.

An interval (i.e., model) σ is a non-empty sequence of states, which can be finite or infinite. The length of σ, denoted by $|\sigma|$, is ω if σ is infinite, or the number of states minus one if σ is finite. To have a uniform notation for both finite and infinite intervals, we will use extended integers as indices. That is, we consider the set N_0 of non-negative integers and ω, $N_\omega = N_0 \cup \{\omega\}$, and extend the comparison operators, $=, <, \leq$, to N_ω by considering $\omega = \omega$, and for all $i \in N_0$, $i < \omega$. Moreover we define $\preceq$ as $\leq - \{(\omega, \omega)\}$. To simplify definition, we will denote σ as $<s_0, \ldots, s_{|\sigma|}>$, where $s_{|\sigma|}$ is undefined if σ is infinite. We use notation $\sigma_{(i..j)}$ to mean that a subinterval $<s_i, \ldots, s_j>$ of σ with $0 \leq i \preceq j \leq |\sigma|$. The *concatenation* of a finite interval $\sigma = <s_0, \ldots, s_{|\sigma|}>$ with another interval $\sigma' = <s'_0, \ldots, s'_{|\sigma'|}>$ (may be infinite) is denoted by $\sigma \bullet \sigma'$ and $\sigma \bullet \sigma' = <s_0, \ldots, s_{|\sigma|}, s'_0, \ldots, s'_{|\sigma'|}>$.

To define the semantics of the projection operator we need an auxiliary operator. Let $\sigma = <s_0, s_1, \ldots, s_{|\sigma|}>$ be an interval and $r_1, \ldots, r_h$ be integers ($h \geq 1$) such that $0 \leq r_1 \leq r_2 \leq \ldots \leq r_h \preceq |\sigma|$. The projection of σ onto $r_1, \ldots, r_h$ is the interval (called projected interval)

$$\sigma \downarrow (r_1, \ldots, r_h) = <s_{t_1}, \ldots, s_{t_l}>, \quad (t_1 < t_2 < \ldots < t_l).$$

where $t_1, \ldots, t_l$ is obtained from $r_1, \ldots, r_h$ by deleting all duplicates. In other words, $t_1, \ldots, t_l$ is the longest strictly increasing subsequence of $r_1, \ldots, r_h$. For example,

$$<s_0, s_1, s_2, s_3, s_4, s_5> \downarrow (0, 2, 2, 2, 4, 4, 5) = <s_0, s_2, s_4, s_5>.$$

An interpretation, as for PTL, is a triple $\mathscr{I} = (\sigma, i, j)$, where σ is an interval, $i \in N_0$ and $j \in N_\omega$, and $0 \leq i \preceq j \leq |\sigma|$. We use notation (σ, i, j) to mean that a term or a formula is interpreted over a subinterval $<s_i, \ldots, s_j>$ of σ with the current state being s_i. Then, for every term e, the evaluation of e relative to $\mathscr{I}$, denoted by $\mathscr{I}[e]$, is defined by induction on the structure of the term as follows:

$$
\begin{aligned}
\mathscr{I}[d] &= d, \text{ if } d \in D' \text{ is a constant value.} \\
\mathscr{I}[a] &= I_v^i[a] = I_v^0[a], \text{ if } a \text{ is typed static variable.} \\
\mathscr{I}[x] &= I_v^i[x], \text{ if } x \text{ is typed dynamic variable.} \\
\mathscr{I}[\bigcirc e] &= \begin{cases} (\sigma, i+1, j)[e], & \text{if } i < j \\ nil, & \text{otherwise} \end{cases} \\
\mathscr{I}[f(e_1, \ldots, e_m)] &= \begin{cases} nil, \text{ if } \mathscr{I}[e_h] = nil \text{ for some } h(1 \leq h \leq m) \\ f(\mathscr{I}[e_1], \ldots, \mathscr{I}[e_m]), \text{ otherwise} \end{cases}
\end{aligned}
$$

For a variable v (static or dynamic), two intervals σ and σ' are v-equivalent, denoted by $\sigma \overset{v}{=} \sigma'$, whenever σ' is the same as σ except that different values can be assigned to v. It is assumed that *static* variable remains the same over an interval whereas a *dynamic* variable can have different values at different states. The satisfaction relation ($\models$) for PTL formulas is inductively defined as follows:

$\mathscr{I} \models p$ iff $I_p^i[p] = true$, for any given atomic proposition p.

$\mathscr{I} \models \rho(e_1,...,e_m)$ iff ρ is a primitive predicate other than $=$ and, for all $h(1 \leq h \leq m)$, $\mathscr{I}[e_h] \neq nil$ and $\rho(\mathscr{I}[e_1],...,\mathscr{I}[e_m]) = true$.

$\mathscr{I} \models e_1 = e_2$ iff $\mathscr{I}[e_1] = \mathscr{I}[e_2]$.

$\mathscr{I} \models \neg P$ iff $\mathscr{I} \not\models P$.

$\mathscr{I} \models P \wedge Q$ iff $\mathscr{I} \models P$ and $\mathscr{I} \models Q$.

$\mathscr{I} \models \exists vP$ iff $(\sigma',i,j) \models P$ for some interval σ', $\sigma'_{(i..j)} \overset{v}{=} \sigma'_{(i..j)}$.

$\mathscr{I} \models \bigcirc P$ iff $i < j$ and $(\sigma,i+1,j) \models P$.

$\mathscr{I} \models (P_1,...,P_m) \, prj \, Q$ iff there exist integers $i = r_0 \leq ... \leq r_{m-1} \leq r_m \preceq j$ such that $(\sigma,r_{l-1},r_l) \models P_l$ for all $1 \leq l \leq m$, and $(\sigma',0,|\sigma'|) \models Q$ for one of the following σ':

    (1) $r_m < j$ and $\sigma' = \sigma \downarrow (r_0,...,r_m) \bullet \sigma_{(r_m+1..j)}$.

    (2) $r_m = j$ and $\sigma' = \sigma \downarrow (r_0,...,r_h)$ for some $0 \leq h \leq m$.

A formula P is satisfied by an interval σ, denoted by $\sigma \models P$, if $(\sigma,0,|\sigma|) \models P$. A formula P is called *satisfiable* if $\sigma \models P$ for some σ. A formula P is *valid*, denoted by $\models P$, if $\sigma \models P$ for all σ. Sometimes, we denote $\models (P \leftrightarrow Q)$ by $P \equiv Q$ and $\models (P \to Q)$ by $P \supset Q$.

3   An Axiomatization for PTL

The axioms and inference rules of our axiomatic system are given in Table 1. Let P and Q be any PTL formulas, for convenience of deduction, sometimes we denote $\vdash (P \leftrightarrow Q)$ by $P \cong Q$ and $\vdash (P \to Q)$ by $P \sqsupset Q$.

A formula P deduced from the axiom system is called a PTL *theorem*, denoted by $\vdash P$. A set of selected theorems and derived inference rules is given in Table 2 and we only choose one of them to prove. The others can be proved in a similar way.

PROOF OF TCSA

$$
\begin{aligned}
(p_s \wedge P;Q) &\cong (p_s \wedge P.Q) \, prj \, \varepsilon & \text{APEB2} \\
&\cong p_s \wedge (P.Q) \, prj \, \varepsilon & \text{APSF} \\
&\cong p_s \wedge (P;Q) & \text{APEB2, IRS}
\end{aligned}
$$

Theorem 1 (Soundness). The axiomatic system is sound, i.e., for any PTL formula P, $\vdash P \Rightarrow \models P$.

Proof. It is readily to prove all the axioms are valid and all the inference rules preserve validity in model theory. The detail is omitted here.

4   Example

In this section, we given an example to show how the axiomatic system works in the correctness verification of a process scheduler in a simple computer system. As shown in Fig. 1, the hardware of the computer system consists of a single-core CPU,

Table 1 Axioms and Inference rules

Axioms:			

AXA	$\bigcirc(P\wedge Q) \cong \bigcirc P \wedge \bigcirc Q$	AXC	$\bigcirc(P;Q) \cong \bigcirc P;Q$
AXN	$\neg\bigcirc P \cong \varepsilon \vee \bigcirc\neg P$	APEF	$\varepsilon\ prj\ Q \cong Q$

APEB1 $P\ prj\ \varepsilon \cong P$

APEB2 $(P_1,\ldots,P_m)\ prj\ \varepsilon \cong (P_1;\ldots;P_m)$, where $m>1$

APFT $(P_1,\ldots,P_i,\ldots,P_m)\ prj\ Q \cong (P_1,\ldots,P_i\wedge\diamondsuit\varepsilon,\ldots,P_m)\ prj\ Q$, where $1\le i<m$

APOF $(P_1,\ldots,(P_i\vee P_i'),\ldots,P_m)\ prj\ Q$
        $\cong ((P_1,\ldots,P_i,\ldots,P_m)\ prj\ Q)\vee((P_1,\ldots,P_i',\ldots,P_m)\ prj\ Q)$

APOB $(P_1,\ldots,P_m)\ prj\ (Q_1\vee Q_2) \cong ((P_1,\ldots,P_m)\ prj\ Q_1)\vee((P_1,\ldots,P_m)\ prj\ Q_2)$

APSF $(p_s\wedge P_1,P_2,\ldots,P_m)\ prj\ Q \cong p_s\wedge(P_1,\ldots,P_m)\ prj\ Q$

APSB $(P_1,\ldots,P_m)\ prj\ (q_s\wedge Q) \cong q_s\wedge(P_1,\ldots,P_m)\ prj\ Q$

APSEF $(P_1,\ldots,p_s\wedge\varepsilon,P_i,\ldots,P_m)\ prj\ Q \cong (P_1,\ldots,p_s\wedge P_i,\ldots,P_m)\ prj\ Q$

APSEB $(P_1,\ldots,P_i,p_s\wedge\varepsilon,\ldots,P_m)\ prj\ Q \cong (P_1,\ldots,P_i\wedge\bigcirc(p_s\wedge\varepsilon),\ldots,P_m)\ prj\ Q$

APX1 $P\wedge\overline{\varepsilon}\ prj\ \bigcirc Q \cong P_1\wedge\overline{\varepsilon};Q$

APX2 $(P_1\wedge\overline{\varepsilon},\ldots,P_m)\ prj\ \bigcirc Q \cong P_1\wedge\overline{\varepsilon};(P_2,\ldots,P_m)\ prj\ Q$

ATSX $\overline{\varepsilon} \rightarrow (e_s = \bigcirc e_s)$, where e_s is a static state term.

ATSR $e_s = e_s'\wedge P(e_s) \cong e_s = e_s'\wedge P(e_s')$, where $P(v)$ is a formula-expression, e_s and e_s' are
        static state terms which are admissible for v in $P(v)$.

ATXE $e_1 = e_2(\bigcirc t) \cong \exists a(e_1 = e_2(a)\wedge\bigcirc(t = a))$, where $e_2(v)$ is a term-expression, e_1 and
        $\bigcirc t$ are terms, a is a fresh static variable which does not appear in $e_1 = e_2(\bigcirc t)$

ATXP $\rho(t_1,\ldots,\bigcirc t_i,\ldots,t_m) \cong \exists a(\rho(t_1,\ldots,a,\ldots,t_m)\wedge\bigcirc(t_i = a))$, where $t_1,\ldots,\bigcirc t_i,\ldots,t_m$
        are terms, a is a fresh static variable which does not appear in $\rho(e_1,\ldots,\bigcirc e_i,\ldots,e_m)$

AEX $\exists x(p_s\wedge\bigcirc P) \cong \exists xp_s\wedge\bigcirc\exists xP$

AEPF $\exists v(P_1,\ldots,P_i,\ldots,P_m)\ prj\ Q \cong (P_1,\ldots,\exists vP_i,\ldots,P_m)\ prj\ Q$, where v does not occur
        freely in sub-formulas $P_1,\ldots,P_m$(except forP_i) and Q

AEPB $\exists v(P_1,\ldots,P_m)\ prj\ Q \cong (P_1,\ldots,P_m)\ prj\ \exists vQ$,
        where v does not occur freely in sub-formulas $P_1,\ldots,P_m$.

AEI $P(e) \sqsupset \exists vP(v)$, where e is a state term which is admissible for v in $P(v)$
        and v does not quantified in $P(v)$

AUD $\forall vP(v) \sqsupset P(t)$, where t is admissible for v in $P(v)$.

ADFT $f(d_1,\ldots,d_m) = d$, where $d_1,\ldots,d_m,d \in D'$ and the input of the function $d_1,\ldots,d_m$
        and d having mapping relation f over D'.

ADPT $\rho(d_1,\ldots,d_m) \cong true$, where $d_1,\ldots,d_m \in D'$, and $d_1,\ldots,d_m$ having the m-place
        relation ρ over D'.

ADPF $\rho(d_1,\ldots,d_m) \cong false$, where $d_1,\ldots,d_m \in D'$, and $d_1,\ldots,d_m$ not having the m-place
        relation ρ over D'.

AT $\vdash P$, where P is a substitution instance of classical first order tautology.

Inference Rules	

IRS $P_1 \cong P_2 \implies Q \cong Q[P_1/P_2]$, where $Q[P_1/P_2]$ denotes the formula given by replacing
        some occurrences of P_2 in Q by P_1.

IRPG $P_1 \sqsupset P_1',\ \ldots,\ P_m \sqsupset P_m',\ Q \sqsupset Q' \implies (P_1,\ldots,P_m)\ prj\ Q \sqsupset (P_1',\ldots,P_m')\ prj\ Q'$.

IRMP $\vdash P \rightarrow Q, \vdash P \implies \vdash Q$

IRAG $\vdash P \implies \vdash \Box P$

IRXR $P \sqsupset \bigcirc P\vee Q \implies P \sqsupset \Box\bigcirc P\vee\diamondsuit Q$

IRUG $\vdash P \implies \vdash \forall vP$, for any variable v

IRCF $\exists_1 v_1\ldots\exists_m v_m((P_1\wedge len(n);P_2)\wedge Q) \sqsupset R$ for all $n \in N_0$
        $\implies \exists_1 v_1\ldots\exists_m v_m((P_1\wedge\diamondsuit\varepsilon;P_2)\wedge Q) \sqsupset R$

Table 2 Theorems and Derived Inference rules

TXM	$\bigcirc P$	$\cong$	$\bar{\varepsilon}\wedge\bigcirc P$	TXC	$\bigcirc P$	$\cong$	$len(1);Q$
TXF	$\bigcirc false$	$\cong$	$false$	TXO	$\bigcirc(P\vee Q)$	$\cong$	$\bigcirc P\vee\bigcirc Q$
TSR	$\Diamond P$	$\cong$	$P\vee\bigcirc\Diamond P$	TCSA	$(p_s\wedge P;Q)$	$\cong$	$p_s\wedge(P;Q)$
TAR	$\Box P$	$\cong$	$P\wedge\bigcirc\Box P$	TAA	$\Box(P\wedge Q)$	$\cong$	$\Box P\wedge\Box Q$
TCEF	$\varepsilon:P$	$\cong$	P	TCEB	$P:\varepsilon$	$\cong$	$P\wedge\Diamond\varepsilon$
TCF	$false:P$	$\cong$	$P;false$	$\cong$	$false$		
TRSA	$(p_s\wedge P)\|Q$	$\cong$	$p_s\wedge(P\|Q)$	$\cong$	$P\|(p_s\wedge Q)$		
TECF	$\exists v(P;Q)$	$\cong$	$\exists vP;Q$, where v does not occur freely in Q				
TECB	$\exists v(P;Q)$	$\cong$	$P;\exists vQ$, where v does not occur freely in P				
TRX	$(p_s\wedge\bigcirc P)\|(q_s\wedge\bigcirc Q)$	$\cong$	$p_s\wedge q_s\wedge\bigcirc(P\|Q)$				
DIRXG	$P\sqsupset P'$	$\Rightarrow$	$\bigcirc P\sqsupset\bigcirc P'$				
DIRCG	$P_1\sqsupset P'_1, P_2\sqsupset P'_2$	$\Rightarrow$	$(P_1:P_2)\sqsupset(P'_1:P'_2)$				
DIRRG	$P_1\sqsupset P'_1, P_2\sqsupset P'_2$	$\Rightarrow$	$(P_1\|P_2)\sqsupset(P'_1\|P'_2)$				

and two different I/O devices named by Device1 and Device2 which only support exclusive access mode; the Process Queue manages all the process applying for the computer resources, i.e., CPU, Device1 or Device2; whenever a resource is free, the Process Scheduler will select an appropriate process with a standing request for the resource and put it to execute. For simplicity, we use an algorithm of first-come, first-serviced (FIFS), a non-preemptive scheduling algorithm, to select a process. The computer system allows many processes to execute concurrently, however, at any time one resource can only be occupied by one process.

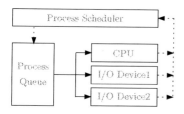

Fig. 1 A simple computer system

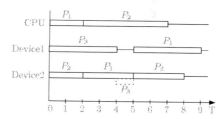

Fig. 2 Scheduling Graph of P_1, P_2 and P_3

Suppose three processes P_1, P_2, P_3 are running now in the computer and their ongoing requirements for the resources are as follows:

$$P_1 : \text{CPU}(2), \text{Device2}(3), \text{Device1}(4)$$
$$P_2 : \text{Device2}(2), \text{CPU}(5)$$
$$P_3 : \text{Device1}(4), \text{Device2}(3)$$

where notation of the form CPU(2) means the process needing CPU for 2 units of system time.

According to FIFS scheduling algorithm, the scheduling graph of processes P_1, P_2 and P_3 is shown in Fig. 2. Although P_1, P_2, P_3 begin to execute at the same time, after a period of concurrently execution in the computer, P_1, P_2 and P_3 terminate in the sequence of P_2, P_3, P_1. In the following, we employ PTL formulas to describe the process scheduler, and then prove the correctness of the scheduling algorithm.

4.1   Formal Specification

For ease of system specification, we assign each resource a unique identifier, and let identifers of CPU, Device1 and Device2 be 1,2 and 3 respectively. Let set of resource identifers be Γ, $\Gamma = \{1, 2, 3\}$. Further, let set of the subscribes of the processes' names running in the computer be Δ and $\Delta = \{1, 2, 3\}$.

We use atomic formula $W_t^i = r$ $(r \in \Gamma, i \in \Delta)$ to denote that process P_i is now waiting for resource r; atomic formula $U_s^i = r$ $(r \in \Gamma, i \in \Delta)$ to denote that process P_i is now using the resource r. Particularly, $W_t^i = 0$ denotes process P_i is running now and currently does not wait for other resource. Further, we employ atomic proposition R^r $(r \in \Gamma)$ to denote resource r is free now and hence can be allocated to a waiting process.

For a process $P_i (i \in \Delta)$, the waiting function for a resource $r(r \in \Gamma)$ can be defined as:

$$ W(i,r) \stackrel{\text{def}}{=} T_w^i = 0 \wedge keep(\bigcirc T_w^i = T_w^i + 1) \wedge \Box (W_t^i = r) \wedge halt(U_s^i = r) $$

where, variable T_w^i is used to recorded the waiting time for resource r, which helps the scheduler to select a process according to the policy of FIFS; sub-formula $\Box(W_t^i = r) \wedge halt(U_s^i = r)$ denotes process P_i keeps on waiting for resource r until it acquires the resource. For process P_i, the using function of resource r for T units of system time is defined as :

$$ U(i,r,T) \stackrel{\text{def}}{=} len(T) \wedge keep(U_s^i = r) \wedge (T > 0 \rightarrow \bigcirc keep(W_t^i = 0)) $$

where, sub-formula $keep(U_s^i = r)$ denotes P_i keeps on using resource r; sub-formula $(T > 0 \rightarrow \bigcirc keep(W_t^i = 0))$ denotes that P_i does not wait for other resource within the period of using r.

Intuitively, the executing of a process in the computer corresponds to a sequence of standing for resources and using the resources. Thus, the processes P_1, P_2 and P_3 can be depicted as follows:

$$ P_1 \stackrel{\text{def}}{=} (W(1,1), U(1,1,2), W(1,3), U(1,3,3), W(1,2), U(1,2,4), E_1 \wedge \varepsilon) $$
$$ P_2 \stackrel{\text{def}}{=} (W(2,3), U(2,3,2), W(2,1), U(2,1,5), E_2 \wedge \varepsilon) $$
$$ P_3 \stackrel{\text{def}}{=} (W(3,2), U(3,2,4), W(3,3), U(3,3,3), E_3 \wedge \varepsilon) $$

where atomic proposition $E_i (i \in \Delta)$ denotes P_i finishes successfully.

A process P_i can use resource r only if it has a standing request, i.e., $C_1 \stackrel{\text{def}}{=} U_s^i = r \rightarrow W_t^i = r$. Further, at most one process can use r at any time, i.e., $C_2 \stackrel{\text{def}}{=} U_s^i = r \rightarrow \bigwedge_{j \neq i} \neg (U_s^j = r)$. Moreover, whenever there some standing request for r, then some process must acquire r to run, i.e., $C_3 \stackrel{\text{def}}{=} W_t^i = r \rightarrow \bigvee_{j \in \Delta} U_s^j = r$. In addition, once a process finishes successfully, it does not apply for any resource from then on. i.e., $C_4 \stackrel{\text{def}}{=} E_i \rightarrow \Box(W_t^i = 0)$. The above scheduling conditions hold for any process and any resource at any time, that is,

$$Con \stackrel{\text{def}}{=} \bigwedge_{r \in \Gamma} \bigwedge_{i \in \Delta} \Box(C_1 \wedge C_2 \wedge C_3 \wedge C_4).$$

A process P_i has a higher priority over process P_j $(i, j \in \Delta)$ to acquire resource $r(r \in \Gamma)$, only if either P_i and P_j are both waiting for r and the former has a longer waiting time, or P_i but not P_j is waiting for r, i.e.,

$$Prior(i, j, r) \stackrel{\text{def}}{=} (W_t^i = r \wedge W_t^j = r \wedge T_w^i > T_w^j) \vee (W_t^i = r \wedge \neg(W_t^j = r)).$$

For any resource, the scheduler must guarantee that one among the highest priority processes will acquire the resources at any time. It is equivalent to a lower priority process cannot acquire the resource when a higher priority process has a standing request,

$$Pol \stackrel{\text{def}}{=} \bigwedge_{r \in \Gamma} \bigwedge_{i \neq j} \Box \neg(Prior(i, j, r) \wedge U_s^j = r \wedge W_t^i = r).$$

The formulas Con and Pol together specifies scheduling strategy of the Process Scheduler:

$$Sch \stackrel{\text{def}}{=} Con \wedge Pol \wedge \Box \, \overline{\varepsilon}.$$

The scheduling of process P_i can be depicted by a projection formula $P_i \, prj \, Sch$. Since processes P_1, P_2 and P_3 execute concurrently in the computer system and the scheduler is fair to them, the scheduling of all the three processes can be described by the following PTL formula:

$$Sys \stackrel{\text{def}}{=} (P_1 \, prj \, Sch) \| (P_2 \, prj \, Sch) \| (P_3 \, prj \, Sch).$$

4.2 System Verification

As mentioned above, the execution of processes P_1, P_2 and P_3 must be finished in the sequence P_2, P_3, P_1. In the following, we employ the axiomatic system to prove the correctness of the scheduler, i.e.,

$$Sys \sqsupset \Diamond(E_2; \bigcirc E_3; \bigcirc E_1).$$

PROOF. Let $sch \cong \bigwedge_{r \in \Gamma} \bigwedge_{i \neq j} \neg(Prior(i, j, r) \wedge U_s^j = r \wedge W_t^i = r) \wedge \bigwedge_{r \in \Gamma} \bigwedge_{i \in \Delta} (C_1 \wedge C_2 \wedge C_3 \wedge C_4)$, according to axiom AXA and theorem TAR, we have $Sch \cong sch \wedge \bigcirc Sch$. Further, by axioms AT, APSF, APSB, inference rule IRS, and theorems TAR, TRSA,

$$Sys \cong (W_t^1 = 1 \wedge W_t^2 = 3 \wedge W_t^3 = 2) \wedge \bigwedge_{i \in \Delta} (T_w^i = 0) \wedge sch \wedge Sys \qquad (1)$$

Moreover, by classical predicate calculus, it is not hard to prove

$$(W_t^1 = 1 \wedge W_t^2 = 3 \wedge W_t^3 = 2) \wedge \bigwedge_{i \in \Delta} (T_w^i = 0) \wedge sch$$
$$\sqsupset (U_s^1 = 1 \wedge U_s^2 = 3 \wedge U_s^3 = 2) \tag{2}$$

In addition, by axioms AXC, APSA and APX2, we get

$$U_s^1 = 1 \wedge (P_1 \, prj \, Sch) \quad \sqsupset \quad (U_s^1 = 1 \wedge \varepsilon, U(1,1,2), \ldots, E_1 \wedge \varepsilon) \, prj \, Sch$$
$$\sqsupset \quad U(1,1,2); ((W(1,3), \ldots, E_1 \wedge \varepsilon) \, prj \, Sch) \tag{3}$$
$$\sqsupset \quad \bigcirc^2 ((W(1,3), \ldots, E_1 \wedge \varepsilon) \, prj \, Sch)$$

$$U_s^2 = 3 \wedge (P_2 \, prj \, Sch) \quad \sqsupset \quad \bigcirc^2 ((W(2,1), \ldots, E_2 \wedge \varepsilon) \, prj \, Sch) \tag{4}$$

$$U_s^3 = 2 \wedge (P_3 \, prj \, Sch) \quad \sqsupset \quad \bigcirc^2 (\, W_t^3 = 0 \wedge U_s^3 = 2$$
$$\wedge \bigcirc^2 ((W(3,3), \ldots, E_3 \wedge \varepsilon) \, prj \, Sch) \,) \tag{5}$$

Thus, by (1)–(5) and theorems TRX, DIRRG,

$$Sys \quad \sqsupset \quad \bigcirc^2 (\, ((W(1,3), \ldots, E_1 \wedge \varepsilon) \, prj \, Sch)$$
$$\| \, ((W(2,1), \ldots, E_2 \wedge \varepsilon) \, prj \, Sch) \tag{6}$$
$$\| \, (W_t^3 = 0 \wedge U_s^3 = 2 \wedge \bigcirc^2 ((W(3,3), \ldots, E_3 \wedge \varepsilon) \, prj \, Sch)) \,)$$

Similarly to the above analysis, we can prove

$$W_t^3 = 0 \wedge (\, ((W(1,3), U(1,3,3), \ldots, E_1 \wedge \varepsilon) \, prj \, Sch)$$
$$\| \, ((W(2,1), U(2,1,5), E_2 \wedge \varepsilon) \, prj \, Sch) \,)$$
$$\sqsupset \bigcirc^2 (U_s^1 = 3 \wedge \bigcirc ((W(1,2), \ldots, E_1 \wedge \varepsilon) \, prj \, Sch)) \tag{7}$$
$$\| \bigcirc^2 (\, U_s^2 = 1 \wedge \bigwedge_{1 \leq k \leq 2} \bigcirc^k (U_s^2 = 1 \wedge W_t^2 = 0) \wedge \bigcirc^3 (E_2 \wedge W_t^2 = 0) \,)$$

From (6), (7) and theorems TRX, DIRXG, DIRRG,

$$Sys \quad \sqsupset \quad \bigcirc^4 (\, U_s^1 = 3 \wedge U_s^2 = 1 \wedge (\bigcirc ((W(1,2), U(1,2,4), E_1 \wedge \varepsilon) \, prj \, Sch)$$
$$\| \, (\bigwedge_{1 \leq k \leq 2} \bigcirc^k (U_s^2 = 1 \wedge W_t^2 = 0) \wedge \bigcirc^3 (E_2 \wedge W_t^2 = 0) \,) \tag{8}$$
$$\| \, ((W(3,3), U(3,3,3), E_3 \wedge \varepsilon) \, prj \, Sch) \,) \,)$$

Since $U_s^1 = 3 \to \neg (U_s^3 = 3)$, by axioms APX, APSF, ATXE, ATSR, AT and theorems DIRXG and DIRCG,

$$U_s^1 = 3 \wedge ((W(3,3), U(3,3,3), E_3 \wedge \varepsilon) \, prj \, Sch)$$
$$\sqsupset \bigcirc (T_w^3 = 1 \wedge W_t^3 = 3 \wedge halt(U_s^3 = 3)); (U(3,3,3), E_3 \wedge \varepsilon) \, prj \, Sch \tag{9}$$

Further, by classical predicate calculus,

$$(W_t^1 = 2 \wedge W_t^2 = 0 \wedge W_t^3 = 3 \wedge T_w^3 = 1 \wedge U_s^2 = 1) \wedge sch$$
$$\sqsupset U_s^1 = 2 \wedge U_s^2 = 1 \wedge U_s^3 = 3 \tag{10}$$

Thus, by (8)–(10), we obtain

$$Sys \quad \sqsupset \quad \bigcirc^5(\bigcirc^4 E_1 \| \bigcirc^2 E_2 \| \bigcirc^3 E_3)$$
$$\sqsupset \quad \Diamond(E_2 : \bigcirc E_3 : \bigcirc E_1)$$

5 Conclusion

In this paper, we present a proof system for the projection temporal logic supporting both finite and infinite time under the condition of finite data domain. However, due to the space limitation, we have not investigated the completeness of the proof system. It will be studied elsewhere. Further, in the future, as case studies, we will further apply our proof system to verify properties of protocols, software and hardware systems. In particular, we are interested in modeling and verifying of composite web-services using projection temporal logic. To do so, we also need develop a verification environment with a group of supporting tools.

References

1. Bledsoe, W., Loveland, D.: Automating Theorem Proving: After 25 Years. Amer Mathematical Society, USA (1984)
2. Clarke, E., Emerson, E.: Design and Synthesis of Synchronization Skeletons Using Branching-Time Temporal Logic. Logic of Programs, 52–71 (1981)
3. Duan, Z.: An Extended Interval Temporal Logic and A Framing Technique for Interval Temporal Logic Programming. Ph.D Thesis, University of Newcastle Upon Tyne (May 1996)
4. Duan, Z., Tian, C., Zhang, L.: A decision procedure for propositional projection temporal logic with infinite models. Acta Inf. 45(1), 43–78 (2008)
5. Duan, Z., Yang, X., Koutny, M.: Framed temporal logic programming. Sci. Comput. Program. 70(1), 31–61 (2008)
6. Duan, Z., Tian, C.: A Unified Model Checking Approach with Projection Temporal Logic. In: Liu, S., Maibaum, T., Araki, K. (eds.) ICFEM 2008. LNCS, vol. 5256, pp. 167–186. Springer, Heidelberg (2008)
7. Dutertre, B.: Complete proof systems for first order interval temporal logic. In: Proc. 10th LICS, pp. 36–43. IEEE Computer Society, Los Alamitos (1995)
8. Guelev, D.P.: A Complete Proof System for First-order Interval Temporal Logic with Projection. Journal of Logic and Computation 14(2), 215–249 (2004)
9. Kesten, Y., Pnueli, A.: A Complete Proof Systems for QPTL. In: LICS 1995, pp. 2–12 (1995)
10. McMillan, K.: Symbolic Model Checking: An Approach to the State Explosion Problem. Kluwer Academic Publisher, Dordrecht (1993)
11. Moszkowski, B.: Executing temporal logic programs. Cambridge University Press, Cambridge (1986)
12. Moszkowski, B.: A complete axiomatization of interval temporal logic with infinite time. In: LICS 2000, pp. 241–252 (2000)
13. Tian, C., Duan, Z.: Model Checking Propositional Projection Temporal Logic Based on SPIN. In: Butler, M., Hinchey, M.G., Larrondo-Petrie, M.M. (eds.) ICFEM 2007. LNCS, vol. 4789, pp. 246–265. Springer, Heidelberg (2007)

Investigation into TCP Congestion Control Performance for a Wireless ISP

J. Lillis, L. Guan, X.G. Wang, A. Grigg, and W. Dargie

Abstract. Wireless technology and devices are becoming ever more pervasive and embedded in our lives. In particular, wireless access to the internet is defining the way that we learn, work and socialize. The primary motivation for undertaking this investigation was the author's involvement in a community broadband project. The contributions of the paper have real impacts for industry and the results of this investigation have been used to provide recommendations to a community wireless internet provider and highlight the wider impact on the UK broadband network. Five major contributions have been achieved: (1) Results from both wired and wireless scenarios are critically evaluated and then concluded; (2) Accurate simulations of 802.11b network in NS2 are validated with live network tests; (3) Evidence is given through both simulation and real network tests to show that current TCP standards are inefficient on wireless networks; (4) A new TCP congestion control algorithm is proposed as well as an outline for a fresh approach; (5) Industry reactions are then given to recommendations for network changes as a result of this investigation.

Keywords: TCP Congestion Control, Performance and QoS, Wired and Wireless Networks.

1 Introduction

1.1 Motivation

Wireless access to the internet has become increasingly pervasive and is now embedded in many devices – mobile phones, PDAs, media centres, laptops and MP3

J. Lillis and L. Guan
Department of Computer Science, Loughborough University, LE11 3TU, UK

X.G. Wang
School of Computing, Communication and Electronics, University of Plymouth, Plymouth, UK

A. Grigg
Systems Engineering Innovation Centre (SEIC), Loughborough University, LE11 3TU, UK

W. Dargie
Department of Computer Networks, Technical University of Dresden, Germany

L. Guan
Corresponding author: L.Guan@lboro.ac.uk

R. Lee, G. Hu, H. Miao (Eds.): Computer and Information Science 2009, SCI 208, pp. 285–294.
springerlink.com   © Springer-Verlag Berlin Heidelberg 2009

players are just some common examples. Despite the overwhelming trend of internet users migrating to some form of wireless connection, standard TCP implementations have not been reviewed.

The author's long-term involvement in a rural community broadband project, FramBroadband, provided the impetus to investigate how performance on the wireless network could be improved. FramBroadband was founded in 2003, aiming to supply broadband via wireless technologies, due to extremely slow deployment of ADSL in the small town of Framlingham and the surrounding rural villages located in Suffolk, UK. The network is formed of a number of wireless base stations, interconnected via wireless point-to-point links. The gateway to the internet is provided from the Technology Centre in Framlingham: an office building with fibre internet backhaul (c.f. Fig. 1). Although ADSL is now available in Framlingham and the surrounding villages, many users experience severe performance issues in rural areas and, in numerous cases, businesses and residents are located too far from the telephone exchange to receive a reliable service at adequate speed, or indeed any level of service at all. The initial aim was to supply an alternative to ADSL but it is now hoped to provide services superior to ADSL. The wireless infrastructure at time of writing has capacity to deliver up to 20mbps, pending an upgrade of fibre connections to the internet. At the present time FramBroadband supplies up to 12mbps.

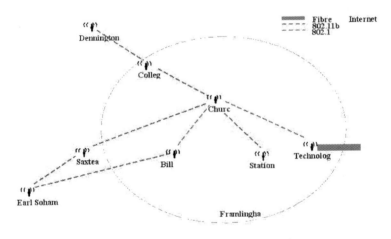

Fig. 1 FramBroadband Network Topology of Base Stations

Although wireless internet service providers are sparse in the UK at the moment, there is evidence to suggest this will change in the future. The major issue facing non-fibre connected areas is 'the last mile' of an internet connection – the telephone wire from the exchange to user premises. Factors such as line loss and distance from the exchange limit availability, performance and reliability. For services above 2mpbs, a distance within around 3.5km is required [1]. This is telephone line distance; therefore premises in urban areas that connect via an indirect cable route to the exchange are affected. ADSL Max/2+ [2] attempted to solve this problem by increasing the range and giving users their maximum possible line speed up to 8mbps. However, on average, an 'up to 8mbps' broadband connection in the UK was found to deliver only 2.7mbps [3]. Although primary infrastructure owner BT so far has failed to provide a solution to the limitations of ADSL, they have recognised the challenge and have been conducting research into WiMax as an alternative [4]. This provides encouraging evidence that wireless technologies will play a crucial role in coming years, and supports the wider value of this project.

1.2 Supporting Research with Frambroadband

Congestion control [5] is a serious issue for FramBroadband, especially over bottlenecks created by point-to-point links connecting network segments. Congestion can also be created by wireless link performance, which is the result of intermittent factors such as rain, wind and even humidity.

A large operational challenge for FramBroadband is web browsing performance over the long distance point-to-point links. Therefore, the focus was to identify potential for TCP performance improvements for existing users and provide recommendations for how FramBroadband may address TCP congestion control on its network. The project would further aim to link the context of results to the future role of wireless providing the 'last mile' of broadband delivery. This project outcomes would also be directly applicable to all users of wired broadband who connect through a home wireless access point.

A comparison of TCP on wired and wireless networks was required. Many papers investigated alternative TCP versions on wired networks, but there was little research into a comprehensive comparison of their performance in a wired and wireless environment. TCP was originally developed as a congestion control protocol for wired networks and makes the assumption that all packet loss is due to congestion. Therefore this project aimed to show that this assumption was not transferable to a wireless environment, and results in an inefficient utilization of available bandwidth.

A commonly used TCP version, NewReno, performs congestion control on the basis of packet loss. An alternative version, Vegas [6], uses round trip times instead [7]. Veno [8], which was specifically developed with wireless networks in mind was subsequently added to the investigation project. It was also agreed with FramBroadband, that results from a simulation replicating a portion of their network would be most valuable.

The paper continues with a review of network simulators, congestion control and wireless medium access control. The simulation was developed by writing several NS2 scenarios in a wired environment, before moving on to a wireless domain. Results from both wired and wireless scenarios are then critically evaluated. Final recommendations are then given for FramBroadband, along with the implications of these findings and suggestions for future research.

2 Simulation Investigation Results

2.1 Wired Scenario Network Setup

Figure 2 shows the topology used for wired network scenarios. Two UDP sources with a Pareto traffic distribution were used to create congestion over the bottleneck between nodes 3 and 4 during the FTP transfer. The Pareto distribution provides a more suitable traffic model as it incorporates traffic burstiness. This allowed comparison of TCP Vegas and NewReno with a greater degree of realism.

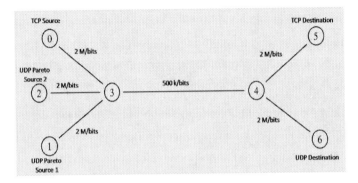

Fig. 2 Wired Scenario Network Topology

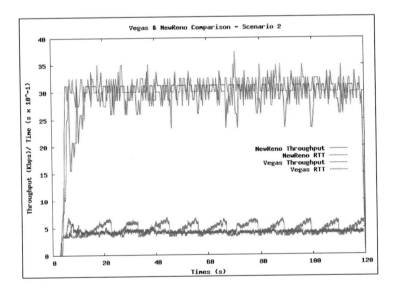

Fig. 3 Performance in Wired Scenario

In a wired network environment, with a number of randomly generated traffic sources, Vegas was shown to have enhanced performance when compared to NewReno, in particular, maintaining a higher level of throughput but with a lower and more constant round trip time.

Figure 3 and Table 1 show that Vegas achieves higher throughput than NewReno, with the differentiating factor being the avoidance of packet loss. The Pareto traffic sources showed that Vegas was also able to react to changing congestion levels, rather than just converging to a static window size when faced with a constant rate of background traffic.

Table 1 Performance in Wired Scenario

	Reno	Vegas
TX Packets	3317	3491
RX Packets	3294	3491
Dropped Packets	23	0
Drop Ratio	0.693	0
Throughput (Bytes)	3294000	3491000
Average Delay (ms)	337	262
Average RTT (ms)	488	413

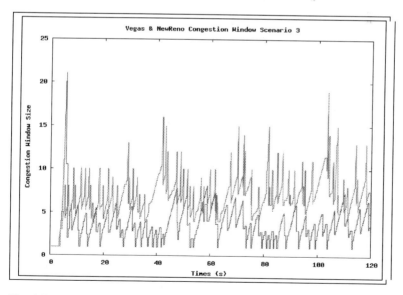

Fig. 4 Congestion Window Comparison

Table 2 Congestion Window Performance

	NewReno	Vegas
TX Packets	1066	1740
RX Packets	922	1668
Dropped Packets	74	72
Drop Ratio	6.942	4.138
Throughput (Bytes)	992000	1668000
Average Delay (ms)	196	194
Average RTT (ms)	347	345

The wired scenario was then extended to include a loss model to evaluate how the two protocol versions performed in an environment with a high level of loss. Due to Vegas' fundamentally different approach to congestion control, based on round trip time instead of loss, the performance of the protocol when presented with random loss was investigated. It was shown that Vegas had potential for substantial performance increase over NewReno, with a nearly 70% higher throughput.

The loss model introduced a very high loss rate for a wired network of 5% on the last link of the TCP route. Table 2 shows that in terms of throughput Vegas achieved much higher performance; however the delay for the connection is approximately the same. NewReno is forced to reduce its sending rate due to periodic loss, rather than being allowed to fill up buffers. Figure 4 shows that the difference in throughput was clearly due to Vegas maintaining a consistently larger congestion window. This behaviour can be explained by the difference in Vegas' loss recovery scheme – because Vegas keeps a much more accurate record of RTT it is able to respond to the losses caused by the error model much quicker than NewReno. It is also possible that NewReno could experience multiple losses at a given congestion window size, especially due to its approach of expanding window size until loss occurs, combined with the error model introduced. In the event of multiple loss, again NewReno is disadvantaged by the potential of the congestion window being reduced multiple times for loss at a particular rate. The error model scenario gives strong evidence that Vegas could outperform NewReno in environments of high loss. This provided justification to continue the comparison in a wireless environment.

2.2 Wireless Simulation

Before results generation took place, research was conducted into NS2's ability to model signal loss. Signal loss has a high impact on throughput and random loss over a wireless network. A simulation was written to evaluate two propagation models in NS2 over a varying distance between source and receiver. The 'Shadowing' model was shown to produce seemingly realistic degradation. The commonly used TwoRayGround, did not use a realistic random probability function, and cut off at a threshold value rather than a gradual decline.

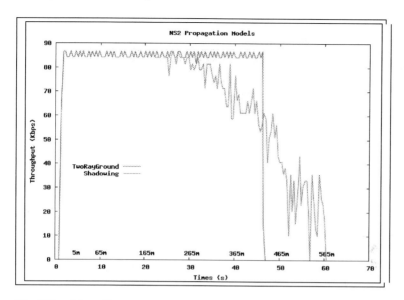

Fig. 5 Signal Loss Test

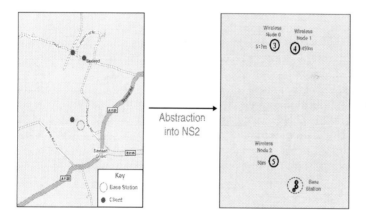

Fig. 6 Topology of Wireless Simulation

A portion of the FramBroadband network in Saxtead was then modelled in NS2. Attention was paid to ensure that the simulation replicated the real life setup of the network. This included examining how to specify the physical properties of equipment used, such as antenna gain. Positions of a number of current users were approximated using mapping, then replicated in the topology area.

The 802.11b standard was configured in NS2 and then validated against a live network test and found to produce throughput to within 10kbytes/s.

TCP Veno was then added to NS2, with the aim of confirming increased performance claims by the developers on wireless networks. However, testing of this protocol showed negligible performance increase, highlighting inaccurate modelling in NS2 of random packet loss in a wireless environment. Although NS2 appeared to model signal fading realistically, this was at the physical level and was based on the power generated by the reception of each wireless frame. This manifested as a drop in available bandwidth at the transport layer, without actually modelling random errors/loss (supported by wider research). Writing a random error model for NS2 was not in scope of the project, but it was key to show that standard implementations of TCP could be inefficient on wireless networks. This was achieved by migrating the effects of noisy wireless conditions on packet loss to a wired simulation, as well as the characteristics of the FramBroadband network such as bandwidth.

The loss model was then modified in the following ways:

(1). Link bandwidth was set to 5meg, approximately that of an optimum B link; delays were updated to be more realistic inline with FramBroadband network;
(2). Fu [8] claims throughput increase of up to 80%, on a typical wireless network with 1% random bit error and hence the error model was updated from 5% to 1%;
(3). TCP Sinks were changed to SACK1 to support Veno implementation.

In Table 3, Veno was shown to outperform NewReno in a variety of performance metrics, notably throughput. The results from the modified scenario were much more in line with the performance expectations of Veno. With a 1% random loss, Veno performs over 60% better than NewReno in terms of throughput – with a marginally lower average delay. Veno also outperforms Vegas, in terms of throughput. Perhaps the most important issue this scenario highlighted was the fact that NewReno, the most common implementation of TCP was beaten by both Vegas and Veno in all performance metrics. This scenario also appeared to validate the shortcoming in NS2 with regards to modelling random wireless bit errors. The error model essentially replicated random loss experienced in a wireless environment, so although this scenario is on a wired network, it is justifiable to assume these trends would be transferable.

Table 3 Performance Comparison in Wireless Test

	NewReno	Vegas	Veno
TX Packets	23432	34192	37696
RX Packets	23156	33833	37301
Lost Packets	276	359	395
Loss Ratio	1.178	1.050	1.048
Throughput (Bytes)	23156000	33833000	37301000
Average Delay (ms)	31	30	30
Average RTT (ms)	56	54	56

3 Live Testing

To optimize the results of this investigation, it was desirable to test alternative TCP versions on the network. Research was conducted into compiling swappable algorithms into the Linux operating system kernel, and the methodology provided to FramBroadband. This was also used to perform preliminary tests on the network that showed potential benefits of alternative TCP versions, as illustrated by Table 4.

Table 4 Live Network Testing

Test Number	NewReno (kbps)	Vegas (kbps)
1	436	460
2	322	280
3	484	560
4	419	487
5	504	593

4 Proposed Veno Modification

This paper has shown the need to review current implementations of TCP algorithms for wireless networks, as well as a possible review of the approach of congestion control. The following recommendations for a new TCP algorithm are shown below.

Proposed Veno Modifications

- Change slow start algorithm so that it expands congestion window exponentially until actual throughput drops below expected throughput.

- At the point where the actual throughput drops below expected, initiate Veno's congestion avoidance algorithm, but at a much slower rate of expansion, for example expand the window every four acknowledged packets.

- Modify recovery mechanism – if loss is encountered during uncongested state, reduce window by 1/5. If loss is encountered during congestion, initially reduce the window by ¼, then by another ¼ if congestion persists.

---

5 Industry Reaction

FramBroadband have reacted to recommendations for network changes as a result of this investigation, and are investigating migrating backhaul links to different frequency bands to commonly used 2.4GHz. This was suggested to reduce the random loss that would adversely affect standard TCP implementations, resulting in dropped web pages and poor FTP throughput. FramBroadband have also recognized their potential role as a test bed for new wireless network technologies and importantly as a proof of concept for the future direction of the UK broadband network.

6 Conclusions and Future Work

This paper has presented an argument for a completely different approach to TCP congestion control and provided substantial performance comparison based on simulation and live network testing results. The evidence provided has real impacts for industry. It shows the need to develop congestion control for wireless networks; enabling more effective and reliable communication that will lead to both increased and cheaper levels of service. This paper has also highlighted congestion control as a scalable challenge for provision of broadband over wireless. In future work, the outlined modification to TCP with advanced features in congestion control protocol fairness on wireless networks could be implemented.

Reference

[1] Thinkbroadband. Technical FAQ (2008),
    http://www.thinkbroadband.com/faq/sections/technical.html#166
[2] Riezenman, M.J.: Extending broadband's reach. IEEE Spectrum 40(3), 20–21 (2003)
[3] Times Online, Broadband speeds 'slower than advertised'. Times Newspapers Ltd. (2007a),
    http://ieeexplore.ieee.org/search/wrapper.jsp?arnumber=1184890
[4] Times Online, BT raises prospect of entry into Wi-Max technology, TimesNewspapers Ltd. (2007b),
    http://ieeexplore.ieee.org/search/wrapper.jsp?arnumber=1184890
[5] Welzl, M.: Network Congestion Control: Managing Internet Traffice. John Wiley & Sons, Chichester (2005)
[6] Brakmo, L.S., Peterson, L.L.: TCP Vegas: End to End Congestion Avoidance on a Global Internet. IEEE Journal on selected areas in communications 13(8), 1465–1480 (1995)
[7] Kurata, K., Hasegawa, G., Murata, M.: Fairness Comparisons Between TCP Reno and TCP Vegas for Future Deployment of TCP Vegas, Osaka University Japan (2007), http://www.isoc.org/inet2000/cdproceedings/2d/2d_2.htm
[8] Fu, C.P., Liew, S.C.: TCP Veno: TCP Enhancement for Transmission Over Wireless Access Networks. IEEE Journal on Selected Areas in Communications 24(2), 216–228 (2003)

Capacity of Memory and Error Correction Capability in Chaotic Neural Networks with Incremental Learning

Toshinori Deguchi, Keisuke Matsuno, Toshiki Kimura, and Naohiro Ishii

Abstract. Neural networks are able to learn more patterns with the incremental learning than with the correlative learning. The incremental learning is a method to compose an associative memory using a chaotic neural network. In the former work, it was found that the capacity of the network increases along with its size, with some threshold value and that it decreases over that size. The threshold value and the capacity varied by two different learning parameters. In this paper, the capacity of the networks was investigated by changing the learning parameter. Through the computer simulations, it turned out that the capacity also increases in proportion to the network size and that the capacity of the network with the incremental learning is above 11 times larger than the one with correlative learning. The error correction capability is also estimated in 100 neuron network.

1 Introduction

The incremental learning proposed by the authors is highly superior to the auto-correlative learning in the ability of pattern memorization[1, 2]. The idea of the incremental learning is from the automatic learning[3]. The neurons used in this learning are the chaotic neurons, and their network is called the chaotic neural network, which was developed by Aihara[4].

In the former work, we investigated the capacity of the networks[5]. Through the simulations, we found that the capacity of the network grows up along with its size, with some threshold value and that it falls off over the size. The threshold value and the capacity varied by two different learning parameters.

Toshinori Deguchi, Keisuke Matsuno, and Toshiki Kimura
Gifu National College of Technology,
e-mail: deguchi@gifu-nct.ac.jp

Naohiro Ishii
Aichi Institute of Technology,
e-mail: ishii@aitech.ac.jp

R. Lee, G. Hu, H. Miao (Eds.): Computer and Information Science 2009, SCI 208, pp. 295–302.
springerlink.com   © Springer-Verlag Berlin Heidelberg 2009

In this paper, first, we explain the chaotic neural networks and the incremental learning and refer to the former work on the capacities with two learning parameters[5], then examine the maximum capacity of the network with simulations changing the learning parameter and show that the capacity is also in proportion to the network size with appropriate parameters.

We also estimate the error correction capability in 100 neuron network.

2 Chaotic Neural Networks and Incremental Learning

The incremental learning was developed by using the chaotic neurons. The chaotic neurons and the chaotic neural networks were proposed by Aihara[4].

The incremental learning provides an associative memory. The network is an interconnected network, in which each neuron receives one external input, and is defined as follows[4]:

$$x_i(t+1) = f(\xi_i(t+1) + \eta_i(t+1) + \zeta_i(t+1)) \tag{1}$$

$$\xi_i(t+1) = k_s \xi_i(t) + \upsilon A_i(t) \tag{2}$$

$$\eta_i(t+1) = k_m \eta_i(t) + \sum_{j=1}^{n} w_{ij} x_j(t) \tag{3}$$

$$\zeta_i(t+1) = k_r \zeta_i(t) - \alpha x_i(t) - \theta_i(1 - k_r) \tag{4}$$

where $x_i(t+1)$ is the output of the i-th neuron at time $t+1$, f is the output sigmoid function described below in (5). k_s, k_m, k_r are the time decay constants. $A_i(t)$ is the input to the i-th neuron at time t, υ is the weight for external inputs, n is the size—the number of the neurons in the network, w_{ij} is the connection weight from the neuron j to the neuron i, and α is the parameter that specifies the relation between the neuron output and the refractoriness.

$$f(x) = \frac{2}{1 + \exp(\frac{-x}{\varepsilon})} - 1 \tag{5}$$

The parameters in the chaotic neurons are assinged as follows:

$$\upsilon = 2.0, k_s = 0.95, k_m = 0.1, k_r = 0.95, \alpha = 2.0, \theta_i = 0, \varepsilon = 0.015.$$

In the incremental learning, the network has each pattern inputted during fixed steps—it is 50 steps in this paper—before moving to the next one. After all the patterns are inputted, the first pattern comes repeatedly. In this paper, a set is defined as a period through all patterns inputted from the first pattern to the last pattern.

During the learning, a neuron which satisfies the condition of (6) changes the connection weights as in (7)[1].

$$\xi_i(t) \times (\eta_i(t) + \zeta_i(t)) < 0 \tag{6}$$

$$w_{ij} = \begin{cases} w_{ij} + \Delta w, & \xi_i(t) \times x_j(t) > 0 \\ w_{ij} - \Delta w, & \xi_i(t) \times x_j(t) \le 0 \end{cases} \quad (i \ne j) \tag{7}$$

where Δw is the learning parameter.

In this learning, the initial values of the connection weights can be 0, because some of the neurons' outputs are changed by their external inputs and this makes the condition establish in some neurons. Therefore, all initial values of the connection weights are set to be 0 in this paper. $\xi_i(0)$, $\eta_i(0)$, and $\zeta_i(0)$ are also set to be 0.

To confirm that the network has learned a pattern after the learning, the pattern is inputted to the usual Hopfield's type network which have the same connection weights as the chaotic neural network. That the Hopfield's type network with the connection weights has the pattern in its memory has the same meaning that the chaotic neural network recalls the pattern quickly when the pattern inputted. Therefore, it is a convenient way to use the Hopfield's type network to check the success of the learning.

3 Capacity

In this section, we retrace the simulations in the former work[5]. In the simulations, we settled the learning parameter Δw to be 0.05 which was used in the former works[1, 2]. The simulations investigated the number of success, which means the number of patterns that the network learned in it successfully, after 50 sets of learning in the networks composed of 50, 100, 200, 300, or 400 neurons.

In each network, the number of patterns to be learned moved from 10 to 300. These patterns are the random patterns generated with the method that all elements in a pattern are set to be -1 at first, then the half of the elements are chosen at random to turn to be 1.

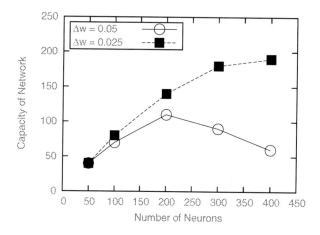

Fig. 1 Capacity of network

The results of the simulations are shown in Fig. 1.

In Fig. 1, the horizontal axis is the number of neurons in the network and the vertical axis is the "capacity of network" which is the maximum number of patterns at which the network can learn all of them.

In the case of $\Delta w = 0.05$, the capacity moved to larger value along with the network size until 200 neurons. Although it is a natural thinking that the capacity grows up as the size increases, it falls down from 200 to 400 neurons.

In the case of $\Delta w = 0.025$, the capacity increases as the number of neurons grows.

4   Capacity in Appropriate Parameters

In the preceding section, the capacity of network varies with the learning parameter Δw.

In this section, the simulations investigate the capacity after 100 sets of learning along with Δw in the networks composed of 50, 100, 200, 300, or 400 neurons. In the simulations, we change Δw from 0.001 to 0.1 in increments of 0.001.

The results of these simulations with the network composed of 100 neurons are shown in Fig. 2.

The horizontal axis is Δw and the vertical axis is the number of success which is how many patterns the network learned. The key "80 patterns" means that the network received 80 patterns for input and the line shows how many patterns the network learned when 80 patterns are inputted.

From Fig. 2, all the 80 input patterns were learned within the range of Δw from 0.004 to 0.036—"80 patterns" line reaches to 80—and so did the 89 patterns with the range from 0.009 to 0.012, but neither 90 nor 100 reached 90 or 100. In the case of "90 patterns", the line reached 89, but it doesn't reach 90. Thus, the maximum capacity was figured out to be 89 with Δw from 0.009 to 0.012.

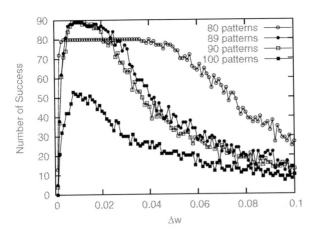

Fig. 2 Number of success with 100 neuron network

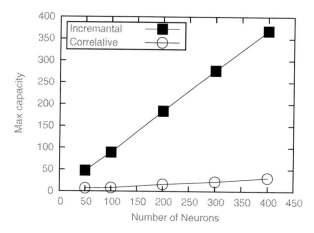

Fig. 3 Maximum capacity of network

In this way, we can find a maximum capacity at each size of network. Fig. 3 shows these maximum capacities with squares. For comparison, the capacities with the auto-correlative learning using the same patterns are also shown in Fig. 3 with circles. It should be restated that the capacity means the maximum number of stored patterns while the network can learn all the input patterns, in this paper.

Both of the capacities are seen to be proportional to the size of network, whereas the capacity of the incremental learning is above 11 times higher than that of the correlative learning.

In Fig. 4, Δw which gives the maximum capacity is shown.

In this results, the appropriate Δw is inverse proportional to the size of network.

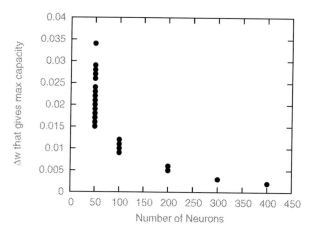

Fig. 4 Δw which gives the maximum capacity

5 Error Correction

Using the incremental learning, the capacity of the networks was almost directly proportional to the size and was above 11 times larger than the one with correlative learning.

It is a natural thinking that the capacity of the network conflicts with the error correction capability of the network. Therefore, to use the networks that learned many patterns with the incremental learning, it is important to estimate their error correction capability.

In this section, we investigate the capability with the following simulations.

Let $p^{(i)}$ be one of the patterns that the network learned, which means that $p^{(i)}$ is a n dimensional vector of 1 or -1, where n is the size of the network. To investigate the error correction capability, we examine how different patterns the network can converge to $p^{(i)}$ from. These differences are measured by hamming distance.

Our simulations are carried out by giving the patterns that differ from $p^{(i)}$ by d in hamming distance. But, to calculate with all the possible input patterns is difficult for combinatorial explosion.

Fortunately, the learned patterns are random patterns. Therefore, if you have to change d elements in a pattern, there is not much difference in choosing which elements to change. In these simulations, we select these elements from first. When a simulation is about hamming distance d, we use the following inputs $q_j^{(i)}(d)$.

$$q_j^{(i)}(d) = \begin{cases} -p_j^{(i)} & (k \leq d) \\ p_j^{(i)} & (k > d) \end{cases} \tag{8}$$

where $p_j^{(i)}$ is the j-th element of the i-th pattern that the network learned and $q_j^{(i)}(d)$ is the j-th element of the i-th input generated from $p_j^{(i)}$ by hamming distance d.

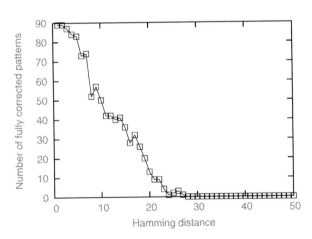

Fig. 5 Number of fully corrected patterns

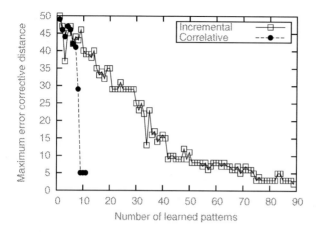

Fig. 6 Maximum error corrective distance

When $q^{(i)}(d)$ is inputted to the network, and the output of the network converges to $p^{(i)}$, it is fully corrected.

When the network is given $q^{(i)}(d)$ for all i, the number of fully corrected patterns becomes an estimate of the error correction capability.

In these simulations, the network with 100 neurons are used. Because the maximum capacity of this network is 89 patterns as in previous section, each simulation counts the number of fully corrected patterns with 89 patterns, varying hamming distance d.

The result is shown in Fig.5. The horizontal axis is the hamming distance between an input and the learned pattern that the input is made from. The vertical axis is the number of fully corrected patterns.

From Fig.5, all the inputs within hamming distance 2 are fully corrected. Above distance 2, fully corrected patterns are gradually decreasing. Let us take the maximum distance within which all the inputs are fully corrected as the indicator of the error correction capability, and call it the "maximum error corrective distance."

The next simulations are searching the maximum error corrective distance, changing the number of learned patterns with the network composed of 100 neurons.

The result is shown in Fig.6. The horizontal axis is the number of the patterns that the network learned, and the vertical axis is the maximum error corrective distance. For comparison, the maximum error corrective distance of the correlative learning is also investigated by simulation.

While these values are almost the same under 7 patterns, the network with the correlative learning loses its capability rapidly over 8 patterns, the network with the incremental learning loses most of the capability gradually until 40 patterns. For example, when 10% of the error in inputs is expected, the network can be used under 41 patterns to be learned by it.

6 Conclusion

The capacity of the networks was investigated by changing the learning parameters. It turned out that the capacity of the network with the incremental learning increases in proportion to the size with appropriate parameter and that it is above 11 times larger than the one with correlative learning. The appropriate learning parameter is in inverse proportion to the size.

The error correction capability was also estimated in 100 neuron network by the maximum error corrective distance, changing the number of learned patterns. The network with the incremental learning loses the capability gradually.

To investigate the error correction capability in different sized networks is remained as the future work.

References

1. Asakawa, S., Deguchi, T., Ishii, N.: On-Demand Learning in Neural Network. In: Proc. of the ACIS 2nd Intl. Conf. on Software Engineering, Artificial Intelligence, Networking & Parallel/Distributed Computing, pp. 84–89 (2001)
2. Deguchi, T., Ishii, N.: On Refractory Parameter of Chaotic Neurons in Incremental Learning. In: Negoita, M.G., Howlett, R.J., Jain, L.C. (eds.) KES 2004. LNCS, vol. 3214, pp. 103–109. Springer, Heidelberg (2004)
3. Watanabe, M., Aihara, K., Kondo, S.: Automatic learning in chaotic neural networks. In: Proc. of 1994 IEEE symposium on emerging technologies and factory automation, pp. 245–248 (1994)
4. Aihara, K., Tanabe, T., Toyoda, M.: Chaotic neural networks. Phys. Lett. A 144(6,7), 333–340 (1990)
5. Deguchi, T., Sakai, T., Ishii, N.: On storage capacity of chaotic neural networks with incremental learning. Memoirs of Gifu national college of technology (40), pp. 59–62 (in Japanese) (2005)

Author Index